Constants Useful in Quantum Mechanics

Planck's constant

$$h = 6.626 \times 10^{-34} \text{ J} \cdot \text{s} = 4.136 \times 10^{-15} \text{ eV} \cdot \text{s}$$
$$\hbar \equiv h/2\pi = 1.055 \times 10^{-34} = 6.58 \times 10^{-16} \text{ eV} \cdot \text{s}$$
$$hc = 1240 \text{ eV} \cdot \text{nm}$$
$$\hbar c = 197.3 \text{ eV} \cdot \text{nm}$$

Electrostatic constant $\quad ke^2 = 1.44 \text{ eV} \cdot \text{nm}$

Bohr radius $\quad a_0 \equiv \hbar^2/m_e ke^2 = 0.0529 \text{ nm}$

Hydrogen ground energy $\quad E_1 \equiv -ke^2/2a_0 = -13.61 \text{ eV}$

Proton/electron mass ratio $\quad m_p/m_e = 1836.2$

Special Units and Conversion Factors

1 eV = 1 electronvolt = potential energy change of an electron during a potential change of 1 V
$$= 1.602 \times 10^{-19} \text{ J}$$
1 u = atomic mass unit $\equiv \frac{1}{12}$(mass of $^{12}_{6}$C) $= 1.6604 \times 10^{-27} \text{ kg} = 931.48 \text{ MeV}/c^2$

Masses of Basic Quantons

Quanton	kg	MeV/c^2	u
Electron	9.109×10^{-31}	0.5110	0.0005486
Proton	1.6726×10^{-27}	938.27	1.007277
Neutron	1.6749×10^{-27}	939.57	1.008665
Hydrogen atom	1.6734×10^{-27}	938.77	1.007825
Helium atom	6.6459×10^{-27}	3728.3	4.002603

Complex Numbers

$c \equiv a + ib$, where a and b are real, $i = \sqrt{-1}$

$c^* \equiv$ complex conjugate of $c \equiv a - ib$

$|c|^2 \equiv$ absolute square of $c = c^*c = a^2 + b^2$

$e^{i\theta} \equiv \cos\theta + i\sin\theta \quad$ (note that $|e^{i\theta}|^2 = 1$)

The Inner (Dot) Product of Complex Vectors

If $|u\rangle = \begin{bmatrix} u_1 \\ u_2 \\ \vdots \end{bmatrix}$ and $|w\rangle = \begin{bmatrix} w_1 \\ w_2 \\ \vdots \end{bmatrix}$ then $\langle u \mid w \rangle \equiv u_1^* w_1 + u_2^* w_2 + \cdots$

Other Useful Information

Speed of sound (at normal temperature, pressure) = 343 m/s

Visible light wavelengths: 700 nm (red) to 400 nm (violet)

Electrons, protons, and neutrons have spin $s = \frac{1}{2}$, photons have spin $s = 1$

Six Ideas That Shaped Physics

Unit Q: Particles Behave Like Waves

Physics

Second Edition

Thomas A. Moore

McGraw Hill

Boston   Burr Ridge, IL   Dubuque, IA   Madison, WI   New York   San Francisco   St. Louis
Bangkok   Bogotá   Caracas   Kuala Lumpur   Lisbon   London   Madrid   Mexico City
Milan   Montreal   New Delhi   Santiago   Seoul   Singapore   Sydney   Taipei   Toronto

McGraw-Hill Higher Education

A Division of The McGraw-Hill Companies

SIX IDEAS THAT SHAPED PHYSICS, UNIT Q: PARTICLES BEHAVE LIKE WAVES
SECOND EDITION

5 6 7 8 9 0 QPD/QPD 0 9

ISBN–13: 978–0–07–239713–0
ISBN–10: 0–07–239713–6

Publisher: *Kent A. Peterson*
Sponsoring editor: *Daryl Bruflodt*
Developmental editor: *Spencer J. Cotkin, Ph.D.*
Marketing manager: *Debra B. Hash*
Senior project manager: *Susan J. Brusch*
Senior production supervisor: *Sandy Ludovissy*
Media project manager: *Sandra M. Schnee*
Designer: *David W. Hash*
Cover/interior designer: *Rokusek Design*

Cover image: *© IBM Corporation, Research Division, Almaden Research Center*
Senior photo research coordinator: *Lori Hancock*
Photo research: *Chris Hammond/PhotoFind LLC*
Supplement producer: *Brenda A. Ernzen*
Compositor: *Interactive Composition Corporation*
Typeface: *10/12 Palatino*
Printer: *Quebecor World Dubuque, IA*

Credit List: **Chapter 1** Q1.10: Courtesy Pasco Scientific; Q1.14a, b: Photo Courtesy of Richard Pelland, Pelland Pipe Organ Company; Q1.15: © James L. Amos/Corbis; Q1.16: Corbis/R-F Website. **Chapter 2** Q2.1: © Richard Megna/Fundamental Photographs; Q2.4b: © Erich Schrempp/Photo Researchers; Q2.7c, Q2.10b: © Richard Megna/Fundamental Photographs; Q2.12: © Tom Pantages; Q2.14: © Springer-Verlag. **Chapter 3** Q3.1c, Q3.3b: © McGraw-Hill Companies, Inc./photo by Chris Hammond. **Chapter 4** Q4.1: Lawrence Berkeley Radiation Laboratory; Q4.4: © Zeitschrift fur Physik, Springer-Verlag; Q4.5: From O. Carnal and J. Mlyneck, Physical Review Letters, 66, 2691, 1991. Copyright © 1991 by the American Physical Society. Q4.6: M.R. Andrews, C.G. Townsend, H.J. Miesner, D.S. Durfee, D.M. Kurn, and W. Ketterle: Observation of interference between two Bose condensates. Science 275, 637-641 (1997). **Chapter 5** Q5.2b: P.G. Merli, G. F. Missiroli and G. Pozzi, Journal of American Physics, Vol. 44, No. 3, March 1976, Fig. 1, pg. 306. **Chapter 8** Q8.5c: © BAse de donnees Solaire Sol 2000, Solar Survey Archive; Q8.6: Courtesy of NRAO/AUI. **Chapter 11** Q11.7b: Courtesy LBL. **Chapter 15** Q15.1: © Roger Ressmeyer/Corbis Images; Q15.4b Q15.5: © Bettmann/Corbis; Q15.7a: Courtesy of Princeton Plasma Physics Laboratory; Q15.7b: © PhotoDisc/ Vol. 6.

Library of Congress Cataloging-in-Publication Data

Moore, Thomas A. (Thomas Andrew)
   Six ideas that shaped physics. Unit Q: Particles behave like waves. / Thomas A. Moore. — 2nd ed.
      p.   cm.
   Contents: [1] Unit C: Conservation laws constrain interactions — [2] Unit N: The laws of physics are universal — [3] Unit R: The laws of physics are frame-independent — [4] Unit E: Electric and magnetic fields are unified — [5] Unit Q: Particles behave like waves — [6] Unit T: Some processes are irreversible.
   Includes bibliographical references and index.
   ISBN 0–07–239713–6 (acid-free paper)
   1. Physics—Study and teaching (Higher) 2. Physics—Problems, exercise, etc. I. Title: Particles behave like waves. II. Title.

QC32 .M647   2003
530—dc21

2002032552
CIP

www.mhhe.com

Dedication

For Terry, Maribeth, and Chris,
with whom it is fun to ponder the world's strangeness.

Table of Contents for
Six Ideas That Shaped Physics

Contents: Unit Q
Particles Behave Like Waves

About the Author

Thomas A. Moore graduated from Carleton College (*magna cum laude* with Distinction in Physics) in 1976. He won a Danforth Fellowship that year that supported his graduate education at Yale University, where he earned a Ph.D. in 1981. He taught at Carleton College and Luther College before taking his current position at Pomona College in 1987, where he won a Wig Award for Distinguished Teaching in 1991. He served as an active member of the steering committee for the national Introductory University Physics Project (IUPP) from 1987 through 1995. This textbook grew out of a model curriculum that he developed for that project in 1989, which was one of only four selected for further development and testing by IUPP.

He has published a number of articles about astrophysical sources of gravitational waves, detection of gravitational waves, and new approaches to teaching physics, as well as a book on special relativity entitled *A Traveler's Guide to Spacetime* (McGraw-Hill, 1995). He has also served as a reviewer and an associate editor for *American Journal of Physics*. He currently lives in Claremont, California, with his wife, Joyce, and two college-age daughters, Brittany and Allison. When he is not teaching, doing research in relativistic astrophysics, or writing, he enjoys reading, hiking, scuba diving, teaching adult church-school classes on the Hebrew Bible, calling contradances, and playing traditional Irish fiddle music.

Preface

Introduction

This volume is one of six that together comprise the text materials for *Six Ideas That Shaped Physics*, a fundamentally new approach to the two- or three-semester calculus-based introductory physics course. *Six Ideas That Shaped Physics* was created in response to a call for innovative curricula offered by the Introductory University Physics Project (IUPP), which subsequently supported its early development. In its present form, the course represents the culmination of more than a decade of development, testing, and evaluation at a number of colleges and universities nationwide.

Opening comments about *Six Ideas That Shaped Physics*

This course is based on the premise that innovative approaches to the presentation of topics and to classroom activities can help students learn more effectively. I have completely rethought from the ground up the presentation of every topic, taking advantage of research into physics education wherever possible, and have done nothing just because "that is the way it has always been done." Recognizing that physics education research has consistently underlined the importance of active learning, I have provided tools supporting multiple opportunities for active learning both inside and outside the classroom. This text also strongly emphasizes the process of building and critiquing physical models and using them in realistic settings. Finally, I have sought to emphasize contemporary physics and view even classical topics from a thoroughly contemporary perspective.

I have not sought to "dumb down" the course to make it more accessible. Rather, my goal has been to help students become *smarter*. I intentionally set higher-than-usual standards for sophistication in physical thinking, and I have then used a range of innovative approaches and classroom structures to help even average students reach this standard. I don't believe that the mathematical level required by these books is significantly different from that in most university physics texts, but I do ask students to step beyond rote thinking patterns to develop flexible, powerful conceptual reasoning and model-building skills. My experience and that of other users are that normal students in a wide range of institutional settings can, with appropriate support and practice, meet these standards.

The six volumes that comprise the complete *Six Ideas* course are

The six volumes of the *Six Ideas* text

Unit C (**Conservation Laws**):	Conservation Laws Constrain Interactions
Unit N (**Newtonian Mechanics**):	The Laws of Physics Are Universal
Unit R (**Relativity**):	The Laws of Physics Are Frame-Independent
Unit E (**Electricity and Magnetism**):	Electric and Magnetic Fields Are Unified
Unit Q (**Quantum Physics**):	Particles Behave Like Waves
Unit T (**Thermal Physics**):	Some Processes Are Irreversible

I have listed these units in the order that I recommend that they be taught, though other orderings are possible. At Pomona, we teach the first three units during the first semester and the last three during the second semester of a year-long course, but one can easily teach the six units in three quarters or even over three semesters if one wants a slower pace. The chapters of all these texts have been designed to correspond to what one might realistically discuss in a single 50-minute class session at the *highest possible pace:* while one might design a syllabus that covers chapters at a slower rate, one should *not* try to discuss more than one chapter in a 50-minute class.

For more information than I can include in this short preface about the goals of the *Six Ideas* course, its organizational structure (and the rationale behind that structure), the evidence for its success, and information about how to cut and/or rearrange material, as well as many other resources for both teachers and students, please visit the *Six Ideas* website (see the next section).

Important Resources

Instructions about how to use this text

I have summarized important information about how to read and use this text in an Introduction for Students immediately preceding the first chapter. Please look this over, particularly if you have not seen other volumes of this text.

The *Six Ideas* website

The *Six Ideas* website contains a wealth of up-to-date information about the course that I think both instructors and students will find very useful. The URL is

www.physics.pomona.edu/sixideas/

Essential computer programs

One of the most important resources available at this site are a number of computer applets that illustrate important concepts and aid in difficult calculations. In several places, this unit draws on some of these programs, and past experience indicates that students learn the ideas much more effectively when these programs are used both in the classroom and for homework. These applets are freeware and are available for both the Mac (Classic) and Windows operating systems.

Some Notes Specifically About Unit Q

The goal of this unit

This unit presents a basic introduction to the concepts of quantum physics and its application, most particularly to nuclear physics. This unit is structured to give the instructor a lot of flexibility in adjusting its length.

Chapters Q1 through Q8 are the irreducible core of the unit, discussing the basic behavior of waves, interference and diffraction experiments involving light, experiments that display the quantum nature of reality, the rules of quantum mechanics, and the basic concept of the wavefunction, the particle-in-a-box and Bohr models and energy quantization, and finally how spectra are connected with these models (but *not* the Schrödinger equation). This part of the unit depends on previous units as follows: in addition to basic mechanics, students need to know a few things about waves (discussed near the end of unit E), a bit about how electric fields are related to potential differences, and Coulomb's law. There is a part of chapter Q4 that refers to relativistic energy, but one can skip over this if necessary.

New in this edition is a discussion of complex numbers and the formal rules of quantum mechanics expressed in terms of vectors, using spin as an example. I recommend that instructors visit the website for more information about my goals for chapters Q5 and Q6; but in summary, I found by experience that tiptoeing around these issues (as I did in the last edition) was, I think, ultimately more confusing and less satisfying than confronting them head on.

Chapter Q9 discusses some general principles of atomic structure in a fairly superficial manner. I spend only a small amount of time on these topics because atomic physics, though important, is genuinely difficult, and I have not found treatments of atomic physics at this level to be either theoretically satisfactory or illuminating to students. I have tried to extract a few useful insights to give students some qualitative understanding of how atoms work without going too far into the details. This chapter can be omitted without loss of continuity.

By contrast, students can understand a *lot* of interesting things about nuclear structure (and do some decent physics) without knowing much more about quantum mechanics than the Pauli exclusion principle and the particle-in-a-box model. Nuclear physics also provides an excellent opportunity to apply ideas from other units in the course. Finally, nuclear physics has so many important social and historical implications that well-educated students ought to know something about it. Chapters Q12 through Q15 therefore provide a fairly detailed exploration of nuclear structure, nuclear stability, radioactivity, and nuclear technology. In addition to basic mechanics, this part of the unit draws on the Pauli exclusion principle from chapter Q8; the concepts of energy quantization and energy levels from chapters Q7 and Q8; the concept of relativistic energy and how it is related to mass from unit R; and electrostatic potential energy, potential energy diagrams, and concepts concerning bonds from unit C. This part of the unit, while I think it is fascinating and socially important, can be completely omitted if necessary.

Chapters Q10 and Q11 also constitute an independent set of chapters that discuss the one-dimensional time-independent Schrödinger equation and its solutions. These chapters deemphasize finding mathematical solutions to the Schrödinger equation and instead focus more on helping students see how this equation is a generalization of the de Broglie relation and on helping them develop an *intuitive* understanding of its solutions and their implications. This part culminates in a discussion of the covalent bond, which students can understand by applying carefully developed but qualitative wavefunction-sketching skills. This part draws on the core material presented in chapters Q1 through Q8 and basic mechanics (especially potential energy diagrams). It may be omitted, as no other material depends on it. It also can be discussed anytime after chapter Q7 (and doing it before chapter Q9 could make the material about the radial wavefunctions in that chapter more plausible).

There is a computer program (called *SchroSolver*) that helps make the ideas in this part of the course clearer. This program solves the Schrödinger equation for a variety of potential energy functions and is discussed explicitly in chapter Q10. Since it can generate a solution for any arbitrary energy, it vividly illustrates why energy must be quantized by showing how unrealistic solutions are for energies other than the quantized values. You can download this program from the *Six Ideas* website.

Here is a table summarizing how one might adjust the length of this unit:

Class days	Chapters	Comments
8	Q1–Q8	Just the basics
9	Q1–Q9	The basics + atomic physics
10	Q1–Q8, Q10–Q11	The basics + the Schrödinger equation
11	Q1–Q11	The above + atomic physics
12	Q1–Q8, Q12–Q15	The basics + nuclear physics
13	Q1–Q9, Q12–Q15	The basics + atomic and nuclear physics
14	Q1–Q8, Q10–Q15	Everything but atomic physics
15	Q1–Q15	Everything

One should also budget a day to talk about waves if unit E does not precede this unit.

Please see the Instructor's Manual for more detailed comments about this unit and suggestions about how to teach it effectively.

Thanks!

Appreciation

A project of this magnitude cannot be accomplished alone. I would first like to thank the others who served on the IUPP development team for this project: Edwin Taylor, Dan Schroeder, Randy Knight, John Mallinckrodt, Alma Zook, Bob Hilborn, and Don Holcomb. I'd like to thank John Rigden and other members of the IUPP steering committee for their support of the project in its early stages, which came ultimately from an NSF grant and the special efforts of Duncan McBride. Users of the texts, especially Bill Titus, Richard Noer, Woods Halley, Paul Ellis, Doreen Weinberger, Nalini Easwar, Brian Watson, Jon Eggert, Catherine Mader, Paul De Young, Alma Zook, Dan Schroeder, David Tanenbaum, Alfred Kwok, and Dave Dobson, have offered invaluable feedback and encouragement. I'd also like to thank Alan Macdonald, Roseanne Di Stefano, Ruth Chabay, Bruce Sherwood, and Tony French for ideas, support, and useful suggestions. Thanks also to Robs Muir for helping with several of the indexes. My editors Jim Smith, Denise Schanck, Jack Shira, Karen Allanson, Lloyd Black, J. P. Lenney, and Daryl Bruflodt as well as Spencer Cotkin, Donata Dettbarn, David Dietz, Larry Goldberg, Sheila Frank, Jonathan Alpert, Zanae Roderigo, Mary Haas, Janice Hancock, Lisa Gottschalk, Debra Hash, David Hash, Patti Scott, Chris Hammond, Rick Hecker, and Susan Brusch have all worked very hard to make this text happen, and I deeply appreciate their efforts. I'd like to thank all the reviewers, including Edwin Carlson, David Dobson, Irene Nunes, Miles Dressler, O. Romulo Ochoa, Qichang Su, Brian Watson, and Laurent Hodges, for taking the time to do a careful reading of various units and offering valuable suggestions.

I also wish to thank the following panel of reviewers for providing careful and insightful comments on the second edition of this unit:

Amy Bug, *Swarthmore College*

Shane Burns, *Colorado College*

Zhigang Chen, *San Francisco State University*

Richard Gelderman, *Western Kentucky University*

Keith Honey, *West Virginia Institute of Technology*

Michael Jackson, *University of Wisconsin—La Crosse*

Renee James, *San Houston State University*

Elizabeth McCormack, *Bryn Mawr College*

Ann Schmiedekamp, *Pennsylvania State University—Abingdon*

Brian Raichle, *Appalachian State University*

Bill Titus, *Carleton College*

John Walkup, *California State University—San Luis Obispo*

Brock Weiss, *Sonoma State University*

Thanks to Connie Wilson, Hilda Dinolfo, Connie Inman, and special student assistants Michael Wanke, Paul Feng, Mara Harrell, Jennifer Lauer, Tony Galuhn, Eric Pan, and all the Physics 51 mentors for supporting (in various ways) the development and teaching of this course at Pomona College. Thanks also to my Physics 51 students, and especially Win Yin, Peter Leth, Eddie Abarca, Boyer Naito, Arvin Tseng, Rebecca Washenfelder, Mary Donovan, Austin Ferris, Laura Siegfried, and Miriam Krause, who have offered many suggestions and have together found many hundreds of typos and other errors. Eric and Brian Daub, Nate Smith, and Ryan McLaughlin were indispensable in helping me put this edition together. Finally, very special thanks to my wife, Joyce, and to my daughters, Brittany and Allison, who contributed with their support and patience during this long and demanding project. Heartfelt thanks to all!

Thomas A. Moore
Claremont, California

Introduction for Students
How to Read and Use This Text Effectively

Introduction

Welcome to *Six Ideas That Shaped Physics!* This text has been designed using insights from recent research into physics learning to help you learn physics as effectively as possible. It thus has many features that may be different from those in science texts you have probably encountered. This section discusses these features and how to use them effectively.

Why Is This Text Different?

Research consistently shows that people learn physics most effectively if they participate in *activities* that help them *practice* applying physical reasoning in realistic situations. This is so because physics is not a collection of facts to absorb, but rather is a set of *thinking skills* requiring practice to master. You cannot learn such skills by going to factual lectures any more than you can learn to play the piano by going to concerts!

This text is designed, therefore, to support *active learning* both inside and outside the classroom by providing (1) resources for various kinds of learning activities, (2) features that encourage active reading, and (3) features that make it easier for the text (as opposed to lectures) to serve as the primary source of information, so that more class time is available for active learning.

The Text as Primary Source

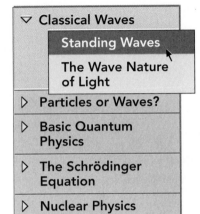

Features that help the text serve as the primary source of information

To serve the last goal, I have adopted a conversational style that I hope will be easy to read, and I tried to be concise without being so terse that you need a lecture to fill in the gaps. There are also many text features designed to help you keep track of the big picture. The unit's **central idea** is summarized on the front cover where you can see it daily. Each chapter is designed to correspond to one 50-minute class session, so that each session is a logically complete unit. The two-page **chapter overview** beginning each chapter provides a compact summary of that chapter's contents to consider before you are submerged by the details (it also provides a useful summary when you review for exams). An accompanying **chapter location diagram** uses a computer menu metaphor to display how the current chapter fits into the unit (see the example at the upper right). Major unit subdivisions appear as gray boxes, with the current subdivision highlighted in color. Chapters in the current subdivision appear in a submenu with the current chapter highlighted in black and indicated by an arrow.

All technical terms are highlighted using a **bold** type when they first appear, and a **Glossary** at the end of the text summarizes their definitions. Please also note the tables of useful information, including definitions of common symbols, that appear inside the front cover.

A physics *formula* is both a mathematical equation and a *context* that gives the equation meaning. Every important formula in this text appears in a **formula box.** Each contains the equation, a **purpose** (describing the formula's meaning and utility), a definition of the **symbols** used in the equation, a description of any **limitations** on the formula's applicability, and possibly some other useful **notes.** Treat everything in such a box as an *indivisible unit* to be remembered and used together.

Active Reading

What it means to be an active reader

Like passively listening to a lecture, passively scanning a text does not really help you learn. *Active* reading is a crucial study skill for effectively learning from this text (and other types of technical literature as well). An active reader stops frequently to pose internal questions such as these: *Does this make sense? Is this consistent with my experience? Am I following the logic here? Do I see how I might use this idea in realistic situations?* This text provides two important tools to make this easier.

Tools to help you become an active reader

Use the **wide margins** to (1) record *questions* that occur to you as you read (so that you can remember to get them answered), (2) record *answers* when you receive them, (3) flag important passages, (4) fill in missing mathematics steps, and (5) record insights. Doing these things helps keep you actively engaged as you read, and your marginal comments are also generally helpful as you review. Note that I have provided some marginal notes in the form of *sidebars* that summarize the points of crucial paragraphs and help you find things quickly.

The single most important thing you can do

The **in-text exercises** help you develop the habits of (1) filling in missing mathematics steps and (2) posing questions that help you *practice* using the chapter's ideas. Also, although this text has many examples of worked problems similar to homework or exam problems, *some* of these appear in the form of in-text exercises (as you are more likely to *learn* from an example if you work on it some yourself instead of just scanning someone else's solution). Answers to *all* exercises appear at the end of each chapter, so you can get immediate feedback on how you are doing. Doing at least some of the exercises as you read is probably the *single most important thing you can do* to become an active reader.

Active reading does take effort. *Scanning* the 5200 words of a typical chapter might take 45 minutes, but active reading could take twice as long. I personally tend to "blow a fuse" in my head after about 20 minutes of active reading, so I take short breaks to do something else to keep alert. Pausing to fill in missing mathematics also helps me to stay focused longer.

Class Activities and Homework

End-of-chapter problems support active learning

The problems at the end of each chapter are organized into categories that reflect somewhat different active-learning purposes. **Two-minute problems** are short, concept-oriented, multiple-choice problems that are primarily meant to be used *in* class as a way of practicing the ideas and/or exposing conceptual problems for further discussion. (The letters on the back cover make it possible to display responses to your instructor.) The other types of problems are primarily meant for use as homework *outside* class. **Basic** problems are simple drill-type problems that help you practice in straightforward applications of a single formula or technique. **Synthetic** problems are more challenging and realistic questions that require you to bring together multiple formulas and/or techniques (maybe from different chapters) and to think carefully about physical principles. These problems define the level of sophistication that you should strive to achieve. **Rich-context** problems are yet more challenging problems that are often written in a narrative format and ask you to answer a practical, real-life question rather than explicitly asking for a numerical result. Like situations you will encounter in real life, many provide too little information and/or too much information, requiring you to make estimates and/or discard irrelevant data (this is true of a some *synthetic* problems as well). Rich-context problems are generally too difficult for most students to solve alone; they are designed for *group* problem-solving sessions. **Advanced** problems are very sophisticated problems that provide supplemental discussion of subtle or advanced issues related to the material discussed in the chapter. These problems are for instructors and truly exceptional students.

Read the Text *Before* Class!

You will be able to participate in the kinds of activities that promote real learning *only* if you come to each class having already read and thought about the assigned chapter. This is likely to be *much* more important in a class using this text than in science courses you may have taken before! Class time can also (*if* you are prepared) provide a great opportunity to get your *particular* questions about the material answered.

Class time works best if you are prepared

Q1 Standing Waves

Chapter Overview

Section Q1.1:   Introduction to the Unit

This unit is focused on *quantum mechanics*, the revolutionary theory of microscopic systems that lies at the foundation of most of 20th-century physics. This theory grew out of the observation that in certain circumstances, *matter behaves like waves*. Each of the unit subdivisions shown in the menu to the left explores a crucial aspect of this great idea. See the section for a more detailed description of each of the five subunits.

Section Q1.2:   Tension and Sound Waves

In this chapter, we will focus primarily on one-dimensional waves that we can describe with a disturbance function $f(x, t)$ of position and time alone. **Tension waves** on a stretched string and **sound waves** in a tube are common and accessible examples of one-dimensional classical waves. A sound wave involves disturbances of the pressure and density of a gas away from the ambient atmospheric pressure and density.

Section Q1.3:   The Superposition Principle

The **superposition principle** for waves states that

> If two traveling waves are moving through a given medium, the disturbance function $f(x, t)$ for the combined wave at any time t and any position x is simply the algebraic sum of the functions $f_1(x, t)$ and $f_2(x, t)$ that describe the individual waves: $f(x, t) = f_1(x, t) + f_2(x, t)$.

This is not always strictly true, but for almost all small-amplitude mechanical waves, it is an excellent approximation.

Section Q1.4:   Reflection

When a medium's characteristics change significantly and suddenly at a certain **boundary,** waves will at least be partially reflected by that boundary. Waves are *completely* reflected at boundaries where their disturbance values are either *fixed* (for example, a string whose end is attached to a fixed point) or *free* (for example, a string whose end is allowed to move freely up and down). The wave reflected from a fixed boundary is *inverted,* but the wave reflected from a free end is *upright.* For sound waves in a tube, an opening in the tube acts as a fixed end on a string (because the air pressure at the opening is constrained to be the same as atmospheric pressure), while a closed end acts as the free end of a string does.

Section Q1.5:   Standing Waves

Sinusoidal waves reflected from a boundary will interfere with incoming waves in such a way as to create a standing wave described by the disturbance function

$$f(x, t) = 2A \sin kx \cos \omega t \qquad (Q1.9)$$

Such a wave does not move, but amounts to a fixed sinusoidal disturbance $\sin kx$ whose overall amplitude oscillates with time. The disturbance is always zero at points where $\sin kx = 0$; we call such points **nodes** of the standing wave. The disturbance oscillates maximally at positions where $\sin kx = \pm 1$; we call such points **antinodes** of the standing wave.

When a standing wave is trapped between two fixed boundaries a distance L apart, only waves having certain frequencies will match the specified boundary conditions. For example, standing waves on a string with two fixed ends must go to zero at both boundaries, so L must correspond to an integer number of half wavelengths, and the constraint on the wavelength constrains the standing wave's frequency as well. In general, the frequencies are such that

$$f = \frac{v}{2L} n \qquad \text{where } n = 1, 2, 3, \ldots \qquad \left(\begin{array}{c} \text{when the boundaries are analogous} \\ \text{to either two free or two fixed ends} \end{array} \right)$$

$$\text{(Q1.12}a\text{)}$$

$$f = \frac{v}{4L} n \qquad \text{where } n = 1, 3, 5, \ldots \qquad \left(\begin{array}{c} \text{when the boundaries are analogous} \\ \text{to one fixed and one free end} \end{array} \right)$$

$$\text{(Q1.12}b\text{)}$$

Purpose: These equations describe the frequencies f of the fundamental modes of standing waves in a medium between two reflecting boundaries.

Symbols: L is the distance between the boundaries, v is the speed of traveling waves in the medium, and n is a positive integer (a positive *odd* integer in the second case).

Limitations: This equation applies only when the medium is uniform and the waves are essentially one-dimensional. The boundaries must be perfectly reflecting if the wave is to sustain itself.

We call the allowed standing waves in such a system the system's **normal modes** (of oscillation). We call the $n = 1$ normal mode the system's **fundamental mode** and its frequency the system's **fundamental frequency.** We call modes where $n > 1$ the **harmonics** of the fundamental mode.

Section Q1.6: The Fourier Theorem

The Fourier theorem implies that we can think of *any* disturbance as being a sum of sinusoidal disturbances, and in particular, a disturbance in a medium between reflecting boundaries can be written as a superposition of waves corresponding to the system's normal modes. The section illustrates this in the case of a square wave (whose disturbance value is $+A$ for one-half of the distance between the boundaries and $-A$ for the other half).

Section Q1.7: Resonance

A disturbance of a medium between two boundaries most effectively transfers energy to that system if the frequency of the disturbance most closely matches one of the system's normal-mode frequencies. For example, a violin bow sliding on a string disturbs the string in complicated random ways, which the Fourier theorem teaches us we can think of as being a superposition of sinusoidal waves with many frequencies. Those frequencies that match the string's normal-mode frequencies will transfer energy to the string, causing the string to vibrate in a mixture of normal-mode frequencies that we hear as a fundamental tone "colored" by higher harmonics.

Q1.1   Introduction to the Unit

Quantum mechanics is based on the idea that matter behaves like waves

Perhaps the most important revolution in 20th-century physics resulted from the discovery that newtonian mechanics was unable to adequately describe the behavior of systems of particles roughly the size of atoms and smaller. The behavior of such systems is better described by the theory of *quantum mechanics*, which first began to take its modern shape in the late 1920s after decades of work by many physicists. Quantum mechanics is the foundation on which virtually all modern physics and chemistry rest.

Quantum mechanics is based on the idea that under certain conditions, objects (even those considered to be point particles, such as electrons) exhibit *wavelike* behavior: this strange thought is the "great idea" of this unit. The wavelike aspect of matter has a variety of surprising implications. For example, one implication is that the energy of a bound system of particles (such as an atom) must be *quantized* (that is, its energy can have only certain discrete and distinct values). This provides the key for understanding atomic spectra and certain details of molecular, atomic, and nuclear structure that have no newtonian explanation.

The world seen through the eyes of quantum mechanics is a very strange one, where a "particle" seems to follow more than one path in getting from point *A* to point *B* and interferes with itself, where the best predictions we can make are statistical, and where the simple act of observing a system irrevocably changes its physical state. Indeed, quantum mechanics is so strange that almost no one thinks that physicists completely understand it yet. Even so, it has proved to be indispensable in modern physics.

The goals of this unit are to provide a limited introduction to the theory of quantum mechanics, specifically tracing how the wavelike aspect of matter is linked to the phenomenon of energy quantization and its consequences for the structure of the atomic nucleus. The unit has five major subdivisions, as shown in Figure Q1.1:

1. *Classical waves* (chapters Q1 and Q2). The theory of quantum mechanics is based on making an analogy between the wavelike aspects of matter and "classical" waves that we can more directly view in nature (such as water and sound waves). This subdivision provides foundations for understanding the analogy by discussing the behavior of classical waves and the similar behavior of electromagnetic (light) waves.

2. *Particles or waves* (chapters Q3 and Q4). Once we have come to understand how classical waves behave, we can explore more fully how both light and matter behave in ways that are both wavelike and particlelike. This subdivision lays the foundation for a *quantum* theory that embraces the wavelike nature of matter.

3. *Basic quantum physics* (chapters Q5 through Q9). In this subdivision, we will explore the basic structure of quantum mechanics, focusing on why the wavelike behavior of matter implies that the energy of a bound system must be quantized and exploring the implications of this idea for atomic and molecular spectra.

These subdivisions comprise the unit's core. The remaining subdivisions are independent of each other and explore extensions and applications of this core.

4. *The Schrödinger equation* (chapters Q10 and Q11) extends the ideas presented in the core chapters to create a more sophisticated model of the quantum behavior of bound systems. This subunit closes with a discussion of the phenomena of *tunneling* and the *covalent bond*.

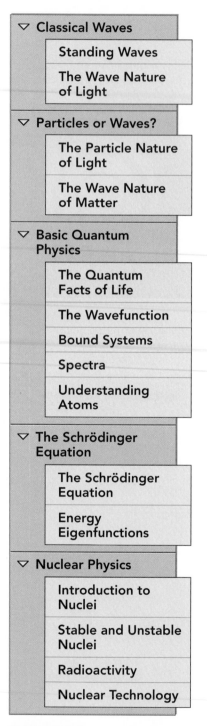

Figure Q1.1

The structure of unit Q.

5. *Nuclear physics* (chapters Q12 through Q15). This subunit focuses on how
    the ideas presented in the core chapters illuminate the structure and be-
    havior of atomic nuclei. Studying nuclei will also give us a chance to re-
    view and apply some results from units R and E and an opportunity to
    discuss radioactivity and nuclear power, which have important techno-
    logical applications and societal implications.

Q1.2   Tension and Sound Waves

A **classical wave,** you will recall from unit E, is fundamentally a *disturbance*   Review of basic wave concepts
that moves through a medium that remains basically at rest. The most common
examples of classical waves are *water waves* that disturb the surface of a body
of water, *tension waves* that move along a stretched string or spring, and *sound
waves* that move through fluids and/or solid objects.

In this chapter, we will focus our attention on waves that move in one dimension (which we take to be the x axis) and that we can accurately describe
by some function $f(x, t)$ that expresses the disturbance caused by the wave
as a function of position x along the x axis and time t (note that this function
essentially ignores the shape of the wave in the y and z directions).

A **tension wave** on a stretched string is an excellent example of such a
one-dimensional wave. In this case, the function $f(x, t)$ describes the transverse displacement from its equilibrium position (which we assume to lie
along the x axis) at position x and time t (see figure Q1.2a). Such a wave can
clearly travel only along the string (and thus the $\pm x$ direction), so a simple
function of x and t completely describes such a wave.

A **sound wave** in the air inside a narrow cylindrical tube is another   Sound waves
example of a one-dimensional wave. A sound wave is a fluctuation in the
density of the medium, so in this case, the function $f(x, t)$ describes the density of the air above or below its equilibrium value (see figure Q1.2b).[†] Since

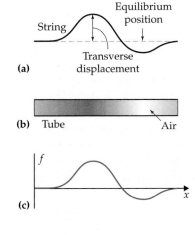

(a)

(b) Tube   Air

(c)

Figure Q1.2
(a) A transverse wave on a string. (b) A sound
wave in a tube. (c) The mathematical
function describing either of the two waves.

[†]Alternatively, one can describe a sound wave in terms of the extent to which air molecules are
displaced from their rest position by the passing wave. While this displacement representation
has a few advantages, I think that the density/pressure representation is more intuitive in general. Rather than confuse the issues by discussing both representations, I will use the density/
pressure representation exclusively.

the *pressure* of a gas depends strongly on its density, we can also think of a sound wave in a tube as causing a variation in the air pressure in the tube. If the tube is fairly narrow, the variation in the density across the tube's width is negligible, and the wave is adequately described by a function of x and t alone.

Q1.3   The Superposition Principle

The thing that most sharply distinguishes how waves behave from how particles behave (and thus the aspect of wave behavior that is most crucial in this unit) is how passing waves *interfere* with each other as they overlap. Our fundamental model for understanding this interference is the **superposition principle** for waves, which states that

> If two traveling waves are moving through a given medium, the function $f(x, t)$ that describes the combined wave at any time t and any position x is simply the algebraic sum of the functions $f_1(x, t)$ and $f_2(x, t)$ that describe the individual waves: $f(x, t) = f_1(x, t) + f_2(x, t)$.

The superposition principle for waves is a theoretical *assertion* about how interfering waves behave. We find experimentally, however, that *most* of the waves encountered in nature obey this principle. Waves that do obey this principle are called **linear** waves, and those that don't are called **nonlinear** waves. In a given medium, waves that represent *small* disturbances typically obey the superposition principle. For example, the gentle sound waves generated by conversation (or even loud music) obey the superposition principle to a high degree of accuracy, but shock waves (produced by an explosion or a jet moving at supersonic speeds) noticeably do not. In this text, we will study only linear waves.

The superposition of two traveling waves is illustrated in figure Q1.3. In this diagram, a traveling pulse wave moving in the $+x$ direction meets a weaker pulse wave traveling in the opposite direction. The figure shows the pulses' motions using a series of "snapshots" arranged vertically, like a spacetime diagram, with time increasing *upward*. Each graph's vertical axis represents the degree to which the medium is disturbed from its normal value. In the case of a tension wave on a string, this axis would correspond to the string's transverse displacement; in the case of a sound wave, it would correspond to the air pressure in the wave compared to normal air pressure, and so on. Figure Q1.3a shows the case where the disturbance in both cases is positive. Figure Q1.3b shows the case where one of the disturbances is negative (that is, in the wave, the quantity represented by the vertical axis is less than the normal value of that quantity in the medium). In each case and at all times, we find the function representing the combined wave by simply adding the functions representing the two pulses.

An important implication of the superposition principle, as illustrated in figure Q1.3, is that linear traveling waves can pass through each other without modification. For example, two pebbles dropped in a pond produce rings of ripples that pass through each other without affecting each other. Similarly, two widely separated people calling to each other can hear and interpret the details of each other's calls even though the two sets of sound waves may cross in transit.

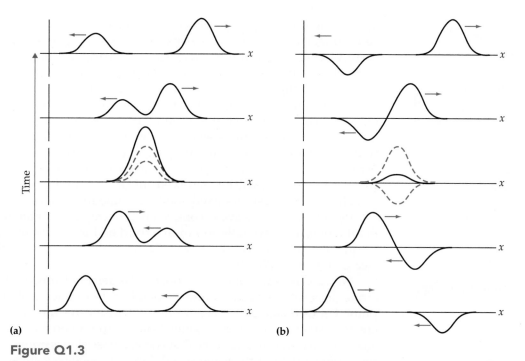

(a)   (b)

Figure Q1.3

Illustration of the superposition principle for traveling pulse waves. Read this diagram from the bottom up. Note that in (b), where the waves have opposite signs, the disturbance when the waves cross is smaller than the absolute magnitude of either wave.

Exercise Q1X.1

Two transverse wave pulses, A and B on the diagram below, are moving with the same constant speed of 10 cm/s but in opposite directions along a stretched string. The graph shows the shape of the function describing the two waves at $t = 0$. On the same graph, sketch the function describing the combined wave at $t = 2$ s and $t = 3$ s.

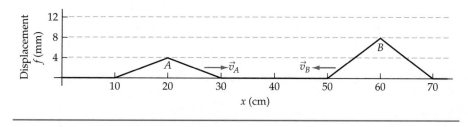

Q1.4   Reflection

When a traveling wave encounters a *boundary,* part of or all the wave will be reflected at that boundary. A **boundary** in this context is somewhere where the medium's characteristics suddenly change. A wave on a string encounters a boundary if the mass density of the string changes at a point. A sound wave moving through a tube encounters a boundary if the tube suddenly widens or narrows. A light wave moving through transparent glass is reflected at both the front and back surfaces of the glass, because glass has different characteristics than air as a medium for light waves.

Extreme cases of boundaries between media

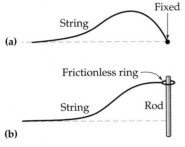

(a)

(b)

Figure Q1.4

A pulse wave encountering (a) a fixed end and (b) a free end. (In the latter case, the ring and rod ensure that the string's displacement at the free end is entirely transverse.)

A model for understanding reflection

We can understand this phenomenon better if we consider two extreme cases. Imagine a traveling wave on a stretched string, and let one end of the string be *fixed,* so that it cannot displace at all in response to the wave. (You can think of this situation as being the extreme case of one string being connected to a second string of infinite mass density.) The wave cannot move beyond the fixed point, but its energy must go somewhere. The only way that the wave's energy can be conserved is if the boundary entirely reflects the wave.

The opposite extreme is seen when one end of the string is completely *free* (this is the extreme case of one string being connected to a second string with zero mass density). Since again the wave cannot move at all beyond this boundary, its energy will be conserved only if the boundary entirely reflects it.

Figure Q1.4 illustrates fixed and free boundaries for a string.

Even though the wave is completely reflected at both the fixed and free boundaries, the wave reflected from a fixed end is *opposite* in sign to that of the initial wave, whereas the wave reflected from a free end has the *same* sign (see figure Q1.5). Why?

We can use the following model to understand this. Pretend that the string does *not* end at the boundary point x_B, but rather continues past that point. Let us imagine that as our original pulse wave moves toward x_B, we create another pulse wave the same distance from x_B on the other side whose shape is the mirror image of the first wave, whose sign is opposite, and which moves toward x_B at the same speed. Since these waves always have equal magnitudes but opposite signs at x_B, they will exactly cancel each other there, keeping the string at x_B fixed at all times as they move past each other.

Now the string to the *left* of x_B cannot tell whether the string is motionless at x_B because it is fixed there or because another upside-down mirror-image wave happens to be coming in from the right. Therefore, it must behave in the same way in either case. Since in our imaginary case the upside-down mirror image wave will continue to move toward the left, this must be what happens in the fixed-end situation also. Therefore, we see that the fixed end must reflect an upside-down, mirror-image version of the wave that hits it (see figure Q1.5a).

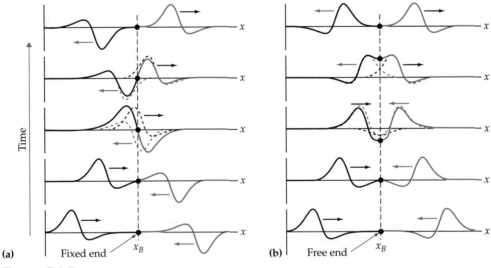

(a)   Fixed end x_B **(b)**   Free end x_B

Figure Q1.5

A model for understanding reflection of pulse waves on a string. We *imagine* the string to continue beyond the boundary (the imaginary part is in color here) and that the pulse is met at the boundary by a mirror-image pulse. Dotted lines show the individual pulse waves; solid lines show their sum. Read the diagram from the bottom up.

We can use the same model to help us understand reflection from a free end. Here, the trick is to understand that the string's *slope* as we approach the free end must go to zero. Why? Let the tension on the string be F_T and its mass per unit length be some constant μ. Consider a tiny hunk of string of length dx at the end of the string: the mass of this hunk will be $m = \mu\,dx$. The transverse (vertical) component of the force exerted by the rest of the string on this part of the string will be $-F_T \sin\theta$, which is approximately $-F_T \tan\theta = F_T \cdot$ (slope of the string) if θ is small (see figure Q1.6). As $dx \to 0$, $m \to 0$, so this transverse force must also go to zero so that the hunk's acceleration does not grow to infinity in this limit. Thus, the *slope* at the end of the string must go to zero as dx gets small.

Now pretend that the string continues past the boundary, and imagine that as our original pulse moves toward the boundary point, a mirror-image *upright* pulse also moves toward the boundary point from the other side. Since the slopes of the original pulse and its mirror-image are always equal in magnitude and opposite in sign at x_B, the combined wave's slope will always add to zero at x_B, just as if the end were free. Since the left part of the string cannot tell whether the end is really free or just *behaving* as if it were free because of a pulse coming in from the right, it will behave the same way in either case. Thus the free end must reflect an *upright* mirror image of any wave that encounters it (figure Q1.5b).

If the boundary is intermediate between these extremes (a connection between a thin and thick string, for example), a wave will be partially transmitted and partially reflected (see figure Q1.7). The reflected part of the wave is upside down if the medium beyond the boundary is more like a fixed point (a denser string, for example), and right side up if it is more like a free end (a less dense string, for example).

The *open* end of a cylindrical tube is to a sound wave what a fixed end is to a wave on a string (contrary to what one might intuitively expect!). This is so because while the air pressure can vary dramatically when the air is trapped in the tube (and thus cannot escape regions of high pressure easily), air under even a small amount of pressure near an open end can just expand into the surrounding atmosphere, dissipating that pressure. Therefore the air pressure is essentially *fixed* at the value of atmospheric pressure at the tube's open end.

On the other hand, the air's density and pressure can vary freely at a closed end of the tube (indeed, the pressure can become quite high as air is crammed against the closed end by a wave). A *closed* end to a tube is thus to a sound wave inside the tube what a *free* end is to a wave on a string. This is illustrated in figure Q1.8. (Make sure that you remember this analogy!)

Exercise Q1X.2

A ripple on the surface of water in a tub will reflect off the solid wall of the tub. Will the reflected wave be an upright or inverted version of the original wave?

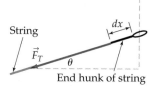

Figure Q1.6
Close-up of the free end of a string.

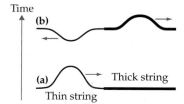

Figure Q1.7
When a wave on a string passes a boundary where the string's thickness changes, part of the wave is reflected and part is transmitted. (Read this diagram from the bottom *up*.)

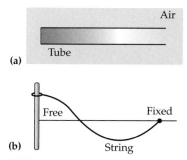

Figure Q1.8
The closed and open ends of a tube are to a sound wave inside like the free and fixed ends of a string are to a wave on the string.

Q1.5   Standing Waves

Up to this point, we have been considering mostly pulselike traveling waves (because the ideas of superposition and reflection are easier to think about and illustrate in the case of pulse waves). However, many naturally occurring waves are similar to *sinusoidal* waves.

A review of the characteristics of sinusoidal waves

In unit E, we saw that we can describe a one-dimensional sinusoidal traveling wave by the mathematical function

$$f(x, t) = A \sin(kx - \omega t) \tag{Q1.1}$$

where A is the amplitude of the wave and k and ω (which we call the wave's *wavenumber* and *angular frequency*, respectively) are constants related to the wave's wavelength λ and period T by the expressions

$$k = \frac{2\pi}{\lambda} \qquad \omega = \frac{2\pi}{T} \tag{Q1.2}$$

A wave described by equation Q1.1 is a *traveling* sinusoidal wave. We can see this by focusing on a given crest of the wave, say, the one where the phase value in the parentheses is $\pi/2$. This crest's location at all times is given by

$$\frac{\pi}{2} = kx_{\text{crest}} - \omega t \quad \Rightarrow \quad x_{\text{crest}} = \frac{\pi}{2k} + \frac{\omega}{k}t \tag{Q1.3}$$

Taking the time derivative of this equation, we find that the crest's x-velocity is

$$v_x = \frac{dx_{\text{crest}}}{dt} = 0 + \frac{\omega}{k} = +\frac{\omega}{k} \tag{Q1.4}$$

So we see that the wave moves in the $+x$ direction with speed ω/k.

Using the same kind of approach, you should be able to show that

$$f(x, t) = A \sin(kx + \omega t) \tag{Q1.5}$$

describes a sinusoidal wave that moves in the $-x$ direction with speed ω/k.

Exercise Q1X.3

Verify this last claim.

Exercise Q1X.4

Show that the speed of the wave described by either equation Q1.1 or equation Q1.5 can also be written (where $f \equiv 1/T$ is the wave's frequency).

$$v = \frac{\lambda}{T} = \lambda f \tag{Q1.6}$$

A standing wave

Now imagine that we have a string whose left end (at $x = 0$) is fixed. Imagine then that we start a sinusoidal wave moving leftward toward this fixed point. When the wave reaches this fixed point, it will reflect off the fixed end, creating an *inverted* sinusoidal wave moving to the right. In the region where the left- and right-going waves overlap, the total wave function (according to the superposition principle) is

$$f(x, t) = A \sin(kx + \omega t) + A \sin(kx - \omega t) \tag{Q1.7}$$

The first term on the right represents the original left-going wave, while the second term represents the inverted reflected wave. The latter does represent an *inverted* wave, even though its sign in the formula is positive: note that at the boundary point $x = 0$, the second term becomes $A \sin(-\omega t) = -A \sin(+\omega t)$, so it exactly cancels the first term at all times, keeping $f(0, t) = 0$ (consistent with the fact that this point is fixed).

Now there is a trigonometric identity

$$\sin(A \pm B) = \sin A \cos B \pm \cos A \sin B \qquad \text{(Q1.8)}$$

Using this identity, you can show that we can rewrite equation Q1.7 as follows:

$$f(x, t) = 2A \sin kx \cos \omega t \qquad \text{(Q1.9)}$$

Exercise Q1X.5

Verify equation Q1.9.

Note that this is *not* the equation of a traveling wave: rather this is the equation of a *stationary* sinusoidal wave $f(x) = \sin kx$ multiplied by a factor $2A \cos \omega t$ that oscillates with time (see figure Q1.9). We call such a wave a **standing wave.**

Note that the string's displacement function $f(x, t)$ is zero at *all* times wherever $\sin kx = 0$. We call these positions of zero movement **nodes.** On the other hand, points where $\sin kx = \pm 1$ oscillate up and down with a larger amplitude than any other points on the string: we call these positions **antinodes.**

If both ends of the string are fixed a distance L apart, sinusoidal waves can bounce back and forth between those ends, creating a self-sustaining standing wave (figure Q1.10 displays such a standing wave on an actual vibrating string). The requirement that the string be fixed at *both* ends, however, implies that the standing waves can have only certain special wavelengths: since $\sin kx$ has to be zero at $x = L$ as well as $x = 0$, we must have

$$kL = n\pi \quad \Rightarrow \quad k = \frac{2\pi}{\lambda} = n\frac{\pi}{L} \quad \Rightarrow \quad L = n\frac{\lambda}{2} \qquad \text{(Q1.10)}$$

where n is some (nonzero) integer. As illustrated in figure Q1.11, this essentially means a sinusoidal standing wave on this string has to fit exactly n half-wavelengths between the two fixed endpoints. The frequency of oscillation f of such a standing wave is given by

$$f = \frac{\omega}{2\pi} = \left(\frac{\omega}{k}\right)\frac{k}{2\pi} = \frac{v}{2L}n \qquad \text{(Q1.11a)}$$

Standing waves on a string with boundaries at both ends

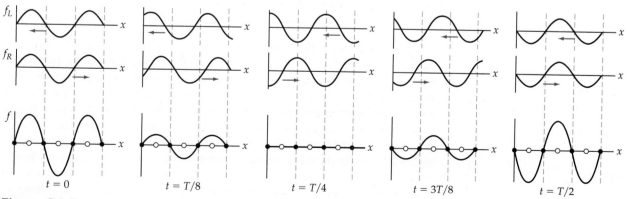

$t = 0 \qquad t = T/8 \qquad t = T/4 \qquad t = 3T/8 \qquad t = T/2$

Figure Q1.9
How left-going and right-going sinusoidal waves combine to form a standing wave. Black dots correspond to nodes, white dots to antinodes.

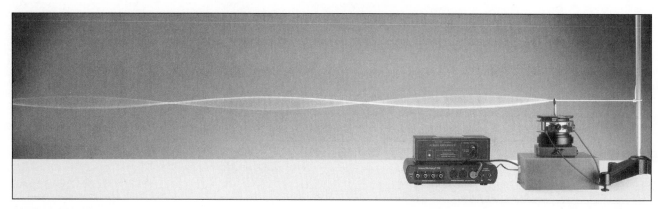

Figure Q1.10
A photograph of standing waves on a vibrating string.

Figure Q1.11
Normal modes of oscillation of a string fixed at both ends. Note that n corresponds to the number of half-wavelengths of the wave that fits between the fixed points.

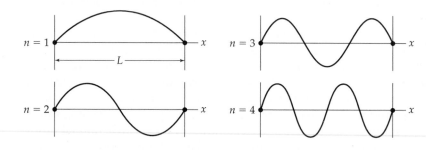

where v is the velocity of waves on the string. Note that the frequency f of any viable standing wave on this string will be an integer multiple of the **fundamental frequency** $v/2L$, which happens to be the number of times a wave can move down the string and back (a distance of $2L$) per second.

We call the standing wave corresponding to each value of n here a **normal mode** of the string's oscillation. The wave corresponding to $n = 1$ we call the string's **fundamental mode,** and its other modes are the **harmonics** of the fundamental mode.

Exercise Q1X.6

Use equations Q1.2, Q1.4, and Q1.10 to fill in the missing steps leading to the last equality in equation Q1.11*a*.

If the left end of the string is fixed while the other end is *free*, we can still set up standing waves on this string, but the characteristics of the standing wave are somewhat different. Equations Q1.7 and Q1.9 still apply, but the condition that has to be satisfied at $x = L$ is not that $\sin kx = 0$ but rather that the *slope* of $\sin kx$ must be equal to zero. The slope of $\sin kx$ is zero at the maxima and minima of the sine function, which occur where $kx = \pi/2$, $3\pi/2, 5\pi/2$, and so on. Therefore for a string with one fixed end and one free end (or for air in a tube with one open end and one closed end), we have

$$kL = n\frac{\pi}{2} \quad \Rightarrow \quad L = n\frac{\lambda}{4} \quad \text{and} \quad f = \frac{v}{4L}n \quad n \text{ is an odd integer}$$

$$(Q1.11b)$$

We see that the frequency of the normal-mode oscillations in this case is still an integer multiple of the fundamental frequency $v/4L$, but the fundamental frequency is *one-half* that of a string fixed at both ends, and only odd multiples of this frequency correspond to viable normal modes. Figure Q1.12 illustrates some of these modes.

So, in summary,

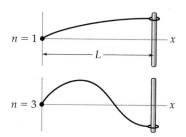

$$f = \frac{v}{2L}n \qquad \text{where } n = 1, 2, 3, \ldots$$

$$\left(\begin{array}{l}\text{when the boundaries are analogous} \\ \text{to either two free or two fixed ends}\end{array}\right) \quad (Q1.12a)$$

$$f = \frac{v}{4L}n \qquad \text{where } n = 1, 3, 5, \ldots$$

$$\left(\begin{array}{l}\text{when the boundaries are analogous} \\ \text{to one fixed and one free end}\end{array}\right) \quad (Q1.12b)$$

Purpose: These equations describe the frequencies f of the fundamental modes of standing waves in a medium between two reflecting boundaries.

Symbols: L is the distance between the boundaries, v is the speed of traveling waves in the medium, and n is a positive integer (a positive *odd* integer in the second case).

Limitations: This equation applies only when the medium is uniform and the waves are essentially one-dimensional. The boundaries must be perfectly reflecting if the wave is to sustain itself.

Figure Q1.12

The first two standing wave modes for a string with one end fixed and one end free. Note that in this case, an odd number of quarter-wavelengths must fit between the boundaries.

Q1.6   The Fourier Theorem

In section Q1.5, we saw that traveling *sinusoidal* waves on a string can set up standing waves between the string's ends, *if* they have the right wavelength and frequency. What if the shape of the initial disturbance on the string is *not* sinusoidal?

During the decade of the 1820s, Jean Baptiste Joseph Fourier showed that one can construct *any* periodic and reasonably continuous wave function $f(x, t)$ by superposing a sufficiently large number of sinusoidal waves with appropriately chosen amplitudes and frequencies. In particular, he showed that one can consider *any* waveform of a vibrating string to be a superposition of appropriately weighted sinusoidal waves corresponding to the *normal modes* of the string.

For example, imagine that we have a string fixed at both ends whose shape at time $t = 0$ is what one might call a **square wave**

$$f(x) = \begin{cases} +A & \text{if } 0 \leq x < \frac{1}{2}L \\ -A & \text{if } \frac{1}{2} \leq x \leq L \end{cases} \qquad (Q1.13)$$

It turns out that this wave is equivalent to the following infinite sum of sinusoidal waves (see figure Q1.13):

$$f(x) = A\frac{4}{\pi}\left(\sin kx + \frac{1}{3}\sin 3kx + \frac{1}{5}\sin 5kx + \cdots\right) \qquad (Q1.14)$$

where $k = \pi/L$ is the wavenumber of the fundamental mode. Note that this sum involves only sinusoidal waves whose wavenumbers correspond to normal modes of a string fixed at both ends (see equation Q1.9). Each of these

Statement of the Fourier theorem

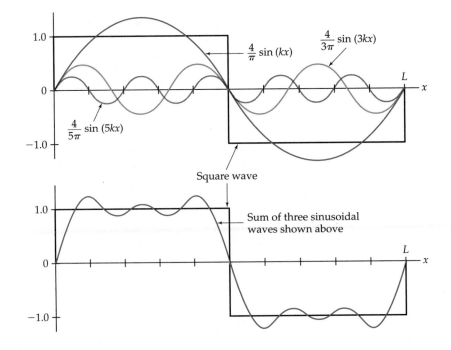

Figure Q1.13
This figure illustrates how we can construct a square wave from sinusoidal waves. The top graph shows the square wave and the sinusoidal waves corresponding to the first three terms of equation Q1.14. The bottom graph shows that the sum of just these three terms is a recognizable approximation to the square wave. The approximation gets better as more terms are added.

normal-mode standing waves, in turn, can be considered to be a sum of sinusoidal traveling waves (see equations Q1.6 and Q1.8).

One implication of this theorem is that if we deeply understand the behavior of sinusoidal waves, we effectively understand the behavior of any kind of wave, since we can consider any wave to be a sum of sine waves. *This theorem is perhaps the single most important principle of wave behavior.* While its mathematical proof is beyond our means at present, you may well see it proved several times in different ways if you take higher-level physics or math classes (an indication of its importance). In our present context, this theorem implies that no matter what the shape of the initial disturbance of the string might be, we can think of its subsequent oscillation as being a combination of normal-mode oscillations.

Understanding sinusoidal waves is the key to understanding all waves

Q1.7   Resonance

The system responds most strongly to disturbances at normal-mode frequencies

Imagine that we disturb a string by wiggling it at a frequency different from one of its normal-mode frequencies. We will find that sometimes a given wiggle happens to be timed correctly to give the string some energy, but just as often the string is moving in such a way as to push back as we attempt to wiggle it, transferring energy back to us. The result is that on the average, our wiggling transfers very little energy to the string, even if we wiggle it violently. On the other hand, if we wiggle the string at one of its normal-mode frequencies, then we can time it so that when we tug on the string, the string is moving in just the right way to accept energy from the push. As a result, the string picks up more and more energy from our efforts, increasing its oscillation amplitude (until drag and other effects dissipate energy at the same rate as we supply it).

By way of analogy, imagine that you give a child on a swing a series of pushes at a rate of, say, two pushes per second. Pushes this frequent are not going to effectively get the child moving. Sometimes the child is moving forward as you push forward, and so you transfer energy to the child. Just as often, though, the child is moving backward, so your push actually *extracts* energy from the child's motion. Except for frustrating the child, your efforts do very little on the average.

On the other hand, if you synchronize your pushes with the natural motion of the swing (delivering them with a frequency of more like once every 2 s), you are always pushing on the swing when it moves forward, and therefore you always transfer energy *to* the swing instead of the reverse. Each successive oscillation, therefore, the swing will gain more energy from your push, and (to the child's delight) the amplitude of oscillation increases.

This tendency to react strongly to disturbances at normal-mode frequencies but ignore disturbances at other frequencies is a general feature of oscillating systems: we call this phenomenon **resonance.**

The concepts of resonance and the Fourier theorem help us understand how many kinds of musical instruments work. For example, a bow sliding across a violin string disturbs the string in a complicated, random fashion. The Fourier theorem teaches us that we can think of this random disturbance as being a superposition of many *sinusoidal* disturbances having a wide range of frequencies and amplitudes. *Some* of these frequencies closely match the normal-mode frequencies of the violin string: the string thus extracts energy from the bow at these frequencies and begins to vibrate. The resulting wave on the string is a complex superposition of normal-mode oscillations at the string's fundamental frequency and integer multiples of that frequency. Our ears and brains process the complicated sound produced by the vibrating string as a musical tone at the fundamental frequency "colored" by the harmonics, which give it the distinctive sound that makes us think "violin."

Many musical instruments use columns of air in tubes instead of strings as the resonating system that creates the sound. When you blow across the top of a bottle, for example, you are randomly disturbing the air trapped in the bottle, and some of the energy in the hiss of the air goes to exciting normal modes of the air in the bottle, creating a tone.

Organ pipes work in a similar manner: a stream of air blowing against a sharp edge creates vortices that disturb the air in the pipe in a complex way. An organ pipe that is open at both ends (see figure Q1.14a) is analogous to a string fixed at both ends, and so the sound generated by the pipe consists of a tone at the fundamental frequency of the air in the pipe plus contributions from all higher harmonics. This gives the pipe a rich sound that is much like that of a violin.

Other kinds of organ pipes are closed at one end (see figure Q1.14b), which is analogous to a string that is free at one end (see figure Q1.8). Not only does this give the pipe a fundamental frequency that is *one-half* that of a doubly open pipe of equivalent length, but also the air in a closed pipe can only vibrate at the odd harmonics of its fundamental tone (see equation Q1.12b). This absence of even harmonics gives the sound produced by such pipes a very distinctive "flutey" tone color.

Finally, the phenomenon of resonance helps explain why light, small wooden buildings generally do very well in an earthquake, while nearby buildings that are six stories tall (and certain multiples of six stories) can be severely damaged. A six-story building constructed using ordinary techniques happens to have a fundamental natural frequency of oscillation that is close to that of a certain type of earthquake wave, so these buildings effectively absorb energy and oscillate wildly in response to an earthquake. The same waves can effectively transfer energy to higher-frequency normal modes in a building whose height is an appropriate multiple of six stories. Small, wooden-frame buildings, on the other hand, have a much higher

Musical instruments

Resonances during earthquakes

Figure Q1.14
(a) A photograph showing organ pipes that are open at both ends.
(b) A photograph showing organ pipes that are closed at one end.
(The handle at the end of each allows a tuner to adjust the length of the air column in the pipe.)

fundamental frequency of oscillation and so do not effectively absorb energy from an earthquake.

We will see in chapters Q7 and Q8 that resonance helps us understand the peculiar behavior of bound quantum systems.

Example Q1.1

Problem   The air column in a certain organ pipe open at both ends has a fundamental frequency of 260 Hz (this corresponds to middle C). How long must such a pipe be?

Model   The air column in an organ pipe open at both ends is analogous to a string fixed at both ends. According to equation Q1.10, the fundamental frequency of such a system is given by

$$f = \frac{v}{2L} \tag{Q1.15}$$

where in this case L is the length of the air column (which is the length of the pipe) and v is the speed of a wave moving in this column (which is the speed of sound, or 343 m/s at normal temperature and pressure).

Solution   Therefore, the required length of the pipe is

$$L = \frac{v}{2f} = 0.66 \text{ m} \approx 2.2 \text{ ft} \qquad\qquad (Q1.16)$$

(The pipe will actually be a bit shorter than this: the effective "fixed end" of the air column is actually a bit outside of the open end.)

TWO-MINUTE PROBLEMS

Q1T.1   A linear traveling wave can be partially reflected when it encounters another linear traveling wave, true (T) or false (F)?

Q1T.2   A sound wave traveling in air hits the surface of a body of water. Is the reflected wave (A) inverted or (B) upright? (Make a guess). The reflection will be total, T or F?

Q1T.3   Imagine that we create a traveling compression wave in a spring that has one end fixed. When the wave reflects from the fixed end, it will be inverted, T or F?

Q1T.4   Imagine that you are near one end of a 150-m-long cylindrical tunnel open to the air on both ends. If you give a shout, you might hear an echo, T or F?

Q1T.5   If you face a cliff or a large concrete wall and give a shout, you will hear an echo. Are the sound waves of the echo inverted or upright compared to the waves of your original shout?
A.   Inverted
B.   Upright

Q1T.6   The frequencies of the normal modes of a string that is free at both ends are the same as those of a string that is fixed at both ends, T or F?

Q1T.7   A sinusoidal standing sound wave inside a tube that is open at both ends must fit between the tube's ends
A.   An integer number of wavelengths
B.   An integer number of half-wavelengths
C.   An odd integer number of quarter-wavelengths

Q1T.8   When the frequency of a standing wave on a string is doubled, its wavelength is multiplied by a factor of
A. $\frac{1}{4}$
B. $\frac{1}{2}$
C. $1/\sqrt{2}$
D. $\sqrt{2}$
E.   2
F.   4
T. λ is unchanged

Q1T.9   The period T of the fundamental mode of the air in a pipe open at one end and closed at the other is equal to what multiple or fraction of the time Δt required for a sound wave to travel from one end of the tube to the other?
A. $T = \frac{1}{4}\Delta t$
B. $T = \frac{1}{2}\Delta t$
C. $T = \Delta t$
D. $T = 2\,\Delta t$
E. $T = 4\,\Delta t$
F.   Other (specify)

Q1T.10   A certain organ pipe is open at both ends. Another pipe of the same length is open at one end and closed at the other. Which will have the lower pitch?
A.   The pipe open at both ends.
B.   The pipe closed at one end.
C.   Both will have the same pitch.
D.   It depends on the pipes' diameters.

Q1T.11   A certain stretched string has a fundamental frequency of 165 Hz (E below middle C). If someone sings a note at a frequency of 495 Hz (B above middle C), the string will *not* respond significantly to the disturbing influence of the sound waves, T or F?

HOMEWORK PROBLEMS

Basic Skills

Q1B.1   Consider the triangle-shaped waves shown in the drawing below. Each wave moves with a speed of 5 cm/s in the direction indicated. Draw separate graphs showing what the superposition principle

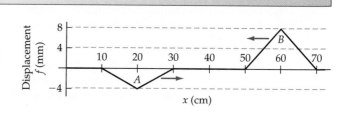

implies that the combined wave should look like at $t = 2$ s, 3 s, 4 s, and 6 s.

Q1B.2 Consider the sawtooth-shaped waves shown in the drawing below. Each wave moves with a speed of 5 cm/s in the direction indicated. Draw separate graphs showing what the superposition principle implies that the combined wave should look like at $t = 2$ s, 3 s, 4 s, and 6 s.

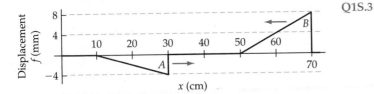

Q1B.3 Imagine that we have a string 1.2 m long and fixed at both ends. We adjust the tension on the string until the speed of waves on the string is 24 m/s. What is the frequency of the string's fundamental mode of oscillation?

Q1B.4 An organ pipe open at both ends is 2.2 m long. What is the frequency of the fundamental mode of the air in the pipe?

Q1B.5 An organ pipe open at both ends has a fundamental frequency of 440 Hz (concert A). What is the length of this pipe? What are the frequencies of its first three harmonics?

Q1B.6 An organ pipe closed at one end has a fundamental frequency of 220 Hz (A below middle C). What is the length of this pipe? What are the frequencies of its first three harmonics?

Q1B.7 Imagine that we have a string 1.5 m long that is fixed at both ends. We adjust the string's tension so that the string's fundamental frequency is 100 Hz. What is the frequency of the normal mode of the string's oscillation that has three antinodes?

Q1B.8 Imagine that we have an organ pipe closed at one end. The length of the pipe is such that the fundamental frequency of the air in the pipe is 230 Hz. What is the frequency of the normal mode of the air having two internal antinodes (not counting the antinode at the closed end)?

Synthetic

Q1S.1 When you tune a woodwind instrument, you pull apart or push together two sections of the instrument.
(a) Why does this change the instrument's pitch?
(b) How is this related to the purpose of the slide on a trombone?

Q1S.2 Imagine that a string on an acoustic guitar is 25 in. long between its fixed ends. According to example E15.2, the speed of waves on a stretched string

is $v = \sqrt{F_T/\mu}$. The highest E string on such a guitar has a pitch of about 329 Hz. Assume that the particular string used has a mass per unit length of $\mu = 0.2$ g/m.
(a) What tension force must be applied to this string?
(b) By what fraction would we have to increase the tension to tune the string up to G (392 Hz)?

Q1S.3 Here is a way to demonstrate the Fourier theorem. Find a piano and open it so that you can clearly hear the strings. Press the damper (often the right-most) pedal: this allows the strings to vibrate freely. Now sing "uuuuu" loudly but at a definite pitch for a brief time. You should be able to hear the strings play the same note back to you. If you touch various strings with your finger, you may be able to convince yourself that only the one set of two or three strings is significantly vibrating, the set closest in natural frequency to the pitch you sang. The other strings essentially did not respond to your note, since they did not have the right natural frequency to resonate with it.

　　Now clap your hands loudly, or slam a book on the floor, or otherwise make a sudden loud sound with no definite pitch. What do the piano strings do? How does what you hear imply that we can think of the single complicated pulse of your clap (or whatever), even though it has no discernible pitch, as the sum of sine waves that *do* have definite frequencies?

Q1S.4 A concert flute (see figure Q1.15) is about 2 ft long, and its lowest pitch is middle C (about 261 Hz). Should we consider a flute to be a pipe that is open at both ends or at just one end? (One end of the flute seems to be clearly closed, so if you choose the former, you should try to explain where the other open end is.)

Figure Q1.15
A photograph of a standard concert flute. Does this represent a tube open at one end or both ends? (See problem Q1S.4.)

Q1S.5 You may know that if you inhale helium, your voice sounds strange and high-pitched if you talk as you exhale the helium. Why is this? (*Hints:* Your sinuses are resonating chambers of air that emphasize certain of the pitches produced by your vocal cords. The speed of sound is almost 3 times higher in helium than in normal air.) *Note:* Inhaling helium can be dangerous because while the helium is in your lungs, your body is not getting the oxygen it needs to survive.

Q1S.6 Imagine that you have a string that has one end that is fixed and the other end is perfectly absorbing, so that traveling waves moving past that end are not reflected at all. Imagine that we wiggle the string sinusoidally near the absorbing end (this will send traveling sinusoidal waves in both directions along the string). Can we set up standing waves on this string? If so, at what frequencies (or is there *any* limit on the frequency)?

Q1S.7 Consider an organ pipe 34.3 cm long that has one end open and one end closed. What is the fundamental pitch of this pipe? Where are the nodes (relative to the closed end) of the normal mode of the air in this pipe whose frequency is 1250 Hz?

Q1S.8 Consider an organ pipe 1.72 m long that has one end open and one end closed. What is the fundamental pitch of this pipe? Where are the nodes (relative to the closed end) of the normal mode of the air in this pipe whose frequency is 150 Hz?

Q1S.9 In the equal-temperament tuning system (the most common system today for tuning musical instruments), each half-step on the musical scale has a frequency that is $2^{1/12}$ higher than the previous note.
(a) Make a list of the frequencies of the 12 half-steps (C#, D, D#, E, F, F#, G, G#, A, A#, B, C) above middle C, given that A is defined to be 440 Hz.
(b) Argue that any note 12 half-steps above another note will have exactly twice the frequency of the lower note. (We describe such pitches as being an *octave* apart.)
(c) Certain combinations of notes sound "harmonic" because their frequencies are very nearly simple integer ratios of each other. As an example, consider a C major chord (C, E, G). What are the simplest ratios that are close to the actual ratios of the frequencies of E to C and G to C? (In the equal-temperament tuning system, these ratios are not quite exact. Other tuning systems make these ratios more pure in certain keys, but the equal-temperament system, because of its symmetry, has the advantage that no key is favored.)
(d) Sets of notes with simple frequency ratios also correspond to the harmonic frequencies of a single note. If C, E, and G are adjacent harmonics of some fundamental tone, what is the frequency of that tone?

Q1S.10 The speed of the wave on a flexible string can, if you think about it, depend on only two quantities: the tension force F_T on the string (which tells you how strongly the string is pulled back toward equilibrium when it is disturbed) and the mass per unit length μ of the string (which tells how quickly or slowly the string *responds* to that restoring push). Let us guess that the speed depends on some power of F_T multiplied by some power of μ. Show that if this is so, the speed v of a wave on the string *must* depend on these quantities as follows

$$v = C\sqrt{\frac{F_T}{\mu}} \qquad\qquad (Q1.17)$$

(where C is an unknown constant with no units), since this is the only such way to combine these quantities that yields the correct units. (In chapter E15, we did a much more difficult formal derivation of this wave speed and found that $C = 1$.)

Q1S.11 The speed of a sound wave in air plausibly depends on the ambient pressure p_0 of the air (which expresses how strongly a bit of compressed air wants to return to equilibrium) and the ambient density ρ_0 of the air (which expresses how slowly or quickly the air responds to pressure changes). Assuming that the sound velocity is a product of powers of these quantities, find the only possible such product that has the correct units. Since standard air pressure is 1.0×10^5 N/m^2, the density of air at this pressure and 20°C is 1.2 kg/m^3, and the speed of sound has a measured value of 343 m/s under such circumstances, determine the value of any unitless constant that might be in your equation. (*Hint:* Use dimensional analysis.)

Rich-Context

Q1R.1 You and a companion are trying to escape from some bad guys one dark night. With the sounds of pursuit close behind, you come upon an open pipe that appears to cut through a hillside. If the pipe is open at the other end, you may be able to escape this way. If it is closed, you will be trapped. Your companion says, "I know how to find out," and sticks his head in, yells something, and then listens. He then pulls his head out and says, "It must be open at the other end. When I yelled 'hello,' I heard the reflection come back 'olleh.' Thus the reflection was inverted, and since the open end of a pipe is like the fixed end of a string, it will reflect the sound inverted." With a thrill of fear, you realize that your companion is *lying* to you and thus is possibly in cahoots with the bad guys. How do you know this?

Q1R.2 A bugle (see figure Q1.16) is simply a coiled length of pipe, without slides or valves. One plays different notes on a bugle by buzzing one's lips at different frequencies.

Figure Q1.16
A photograph of a bugle (see problem Q1R.2).

(a) How does this change the pitch (frequency) of the sound the bugle makes?
(b) The bugle can play only certain pitches and not others (for example, think of the piece "Taps," which is entirely constructed of only four different pitches). What are these allowed pitches, and why are other pitches impossible?
(c) Imagine that certain bugle plays in C, so that the pitches in "Taps" are low G (196 Hz), middle C (261 Hz), E (329 Hz), and G (392 Hz). How long would this bugle be if uncoiled? (*Hint:* Is the bugle effectively a tube with two open ends or a tube with one open end and one closed end? Think about the implications of either model. How could you get the pitches listed if the bugle is open at one end and closed at the other?)

Advanced

Q1A.1 Here is yet another way to derive equation Q1.17 in problem Q1S.10: Consider a pulse traveling left down a string at a constant velocity v. Imagine that we look at this situation from a reference frame where the pulse is at rest, and the string is moving to the right with speed v. This frame will be inertial, so we can apply Newton's laws. Imagine that we focus our attention on a little bit of string of length dL passing the crest of the pulse. We can find a circle with some radius R that approximates the curvature of this bit of string (see figure Q1.17). Argue that the net downward force on this piece of string is roughly $F_{net} = F_T(dL/R)$ (ignoring gravity). Since this force is what constrains the bit of string to follow a circular path of radius R as it moves over the crest with speed v, this net force must be equal to mv^2/R by Newton's second law, where m is the mass of the bit of string. Show that combining these equations implies equation Q1.17, with $C = 1$. (*Hint:* Note that if dL is small, then θ is small, implying that $\sin\theta \approx \theta$.)

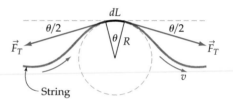

Figure Q1.17
A drawing illustrating how we can treat a small part of a string mass undergoing circular motion when the crest of the pulse wave passes.

ANSWERS TO EXERCISES

Q1X.1 The total wave function looks as shown below at time $t = 2$ s (gray lines) and $t = 3$ s (colored lines):

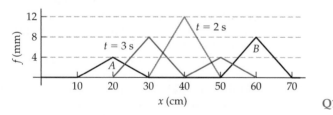

Q1X.2 The edge of the tub will be analogous to a free end of a string or the closed end of a tube: the water's amplitude is not limited, but the wave cannot continue beyond the boundary. Therefore, the reflected wave will be upright.

Q1X.3 Consider again the crest that corresponds to the argument of the sine function being $\pi/2$:

$$kx_{crest} + \omega t = \frac{\pi}{2} \quad \Rightarrow \quad x_{crest} = \frac{\pi}{2k} - \frac{\omega}{k}t \tag{Q1.18}$$

Taking the time derivative of both sides of this equation, we find that $v_x \equiv dx_{crest}/dt = -\omega/k$.

Q1X.4 Since $\omega = 2\pi/T$ and $k = 2\pi/L$, we have

$$v = |v_x| = +\frac{\omega}{k} = \frac{2\pi/T}{2\pi/\lambda} = \frac{\lambda}{T} \tag{Q1.19}$$

Since $1/T = f$, this also implies that $v = \lambda f$.

Q1X.5 According to equation Q1.8,

$$\sin(kx - \omega t) = \sin kx \cos \omega t - \cos kx \sin \omega t$$
$$(Q1.20a)$$

$$\sin(kx + \omega t) = \sin kx \cos \omega t + \cos kx \sin \omega t$$
$$(Q1.20b)$$

When we add these equations, the rightmost terms cancel:

$$\sin(kx - \omega t) + \sin(kx + \omega t) = 2 \sin kx \cos \omega t$$
$$(Q1.21)$$

Equation Q1.9 is simply A times this.

Q1X.6 According to the beginning of equation Q1.11a,

$$f = \left(\frac{\omega}{k}\right)\frac{k}{2\pi} = v\frac{k}{2\pi}$$
$$(Q1.22)$$

where the last step follows from equation Q1.4. Equation Q1.2, on the other hand, tells us that $k = 2\pi/\lambda$, and equation Q1.10 tells us that

$$L = n\frac{\lambda}{2} \quad \Rightarrow \quad \lambda = \frac{2L}{n}$$
$$(Q1.23)$$

Putting this all together, we find that

$$f = v\frac{2\pi/\lambda}{2\pi} = \frac{v}{\lambda} = v\frac{n}{2L}$$
$$(Q1.24)$$

which is the desired result.

Q2

The Wave Nature of Light

Chapter Overview

Introduction

In this chapter we examine interference and diffraction of waves moving in two and three dimensions and discuss how the observation of interference and diffraction effects in light offers strong evidence for the wavelike nature of light.

Section Q2.1: Two-Slit Interference

Two-dimensional waves going through a small opening in a barrier (a **slit**) fan out into **circular waves** (whose crests are circles centered on the slit), as if the slit were a point source for the waves. This phenomenon is called **diffraction.** The superposition principle implies that circular waves emerging from two closely spaced slits will interfere with each other in the region beyond the slits where the waves overlap, **constructively interfering** at some points and **destructively interfering** at others. A straightforward geometric argument implies that if wave crests emerge from both slits simultaneously, we can locate points where the waves interfere constructively by using the following equation:

$$d \sin \theta_{nc} = n\lambda \quad \Rightarrow \quad \theta_{nc} = \sin^{-1} \frac{n\lambda}{d} \qquad (Q2.1)$$

Purpose: The set of all points at which waves emerging from two slits in a barrier *constructively* interfere forms lines that radiate from the point halfway between the slits. This equation specifies the angles θ_{nc} that those lines make with the direction perpendicular to the barrier.

Symbols: λ is the wavelength of the waves, d is the center-to-center distance between the slits, and n is an integer.

Limitations: This equation is accurate only at points very distant from the slits compared to d. The slits must be arranged so that a given crest of the line wave moves through both slits simultaneously. The slits need to be comparable in width to λ to generate circular waves that are strong at all physically reasonable angles (see section Q2.3).

This equation also applies to three-dimensional waves (such as sound waves) if we measure the waves on a plane that is perpendicular to the barrier containing the slits but contains the line connecting the centers of the slits.

Section Q2.2: Two-Slit Interference of Light

In the first decade of the 1800s, Thomas Young was able to demonstrate that light with a well-defined color on the rainbow (**monochromatic** light) emerging from two slits in an opaque barrier created an interference pattern that was accurately described by equation Q2.1. This provided the first strong evidence that light might consist of *waves* (not particles, as Newton had presumed).

Section Q2.3: Diffraction

Huygens's principle states that each point in the crest of a wave acts as if it were a point source of circular **wavelets,** and that the wave's crest at a later time is tangent to the wavelet crests emitted by all points on the crest at the earlier time. Using this principle, we can argue that waves moving through a single slit will create a pattern involving a central range of angles where the waves are strong flanked by **fringes** of weaker waves, all separated by specific angles where the waves suffer complete destructive interference:

$$\theta_{nd} = \sin^{-1} \frac{n\lambda}{a} \qquad (Q2.10)$$

Purpose: The points at which waves emerging from a single rectangular slit *completely cancel out* lie along lines that radiate from the slit's center. This equation specifies the angles θ_{nd} between those lines and the direction of motion of the original waves moving through the slit.
  Symbols: λ is the wavelength of the waves, a is the width of the slit, and n is a *nonzero* integer.
  Limitations: This equation applies to a rectangular slit whose height is much greater than its width a. It is accurate only at points very distant from the slit compared to a. The slit must be oriented so that a given wave crest moves through all parts of the slit simultaneously.

The great majority of the diffracted wave's energy is contained in the central range $-\theta_{1d} < \theta < \theta_{1d}$ between the innermost angles of destructive interference: the wave amplitude in any of the fringes is much smaller than that in the central region.

Section Q2.4: Optical Resolution

Light passing through a *circular* opening is diffracted into a bright central circle flanked by dimmer circular fringes, separated by dark rings where completely destructive interference occurs. The angle of the innermost ring of destructive interference in this case turns out to be

$$\sin \theta_{1d} = 1.22 \frac{\lambda}{a} \qquad (Q2.13)$$

This means that light from a point source will spread out into a diffraction pattern after going through an optical instrument's aperture, blurring the image somewhat (even if the instrument is perfectly focused). An instrument can resolve two point sources only if the angle θ between those sources is such that

$$\theta > \theta_{1d} \approx \sin^{-1} \frac{1.22\lambda}{a} \qquad (Q2.14)$$

Purpose: This equation specifies the minimum angular separation θ that sources can have if they are to be resolved by an optical instrument whose aperture has diameter a.
  Symbols: λ is the average wavelength of the light from the source, and θ_{1d} is the innermost angle of destructive interference.
  Limitations: This equation is an approximation that becomes better if the instrument's sensor screen is far from the aperture compared to the aperture's diameter.
  Note: Equation Q2.14 expresses what is called the **Rayleigh criterion.**

Instruments that can resolve sources at this limit are said to be **diffraction-limited.** Human eyes are very nearly diffraction-limited when operating in bright light.

Q2.1   Two-Slit Interference

In chapter Q1, we extensively explored the behavior of one-dimensional waves such as waves on a string or sound waves in a tube. In this chapter, we will consider the behavior of two-dimensional waves, such as waves on the surface of a body of water. Three-dimensional waves (such as sound waves) evaluated on a specified surface in space will behave in the same way.

Imagine a sinusoidal wave in water whose successive crests are parallel straight lines: we call such a two-dimensional wave a **line wave.** Imagine that such a line wave approaches a small gap (usually called a **slit**) in some kind of barrier. We empirically observe that waves traveling through a sufficiently small slit spread out to form essentially **circular waves** (i.e., waves whose successive crests are concentric circles) centered on the slit, as if the slit were a point source of waves (see figure Q2.1). This phenomenon is called **diffraction.**

One can intuitively see why this has to happen with water waves at least. As the wave moves through the slit, its sides are sheared off by the slit's sides. After the wave emerges from the slit, it will spread out to smooth out the sharp vertical edges where its sides used to be, creating the circular pattern. We will discuss diffraction in greater detail in section Q2.3: for our purposes now, it is enough to know that line waves going through a sufficiently small slit become circular waves.

With this in mind, imagine that we send line waves through *two* narrow slits a short distance apart, and assume that a given wave crest arrives at each slit simultaneously. The circular waves emerging from the slits will overlap each other, as shown in figure Q2.2. The superposition principle implies that the total wave at any given point will be the algebraic sum of the waves from each slit. At some points in the overlap region, wave crests from each slit will arrive at the same time, and the sum will be a wave with twice the amplitude of the waves from each slit: we say that the waves **constructively interfere** with each other at such a point. At certain other points, a crest from one slit will arrive at the same time as a trough from the other, and the sum of the waves will be zero: we say that the wave **destructively interfere** with each other at such a point. Constructive interference and destructive interference are illustrated in figure Q2.3.

The phenomenon of diffraction

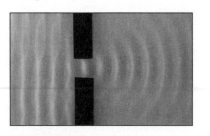

Figure Q2.1
The diffraction of water waves in a ripple tank. Line waves moving from the left disperse in a circular pattern after going through a small slit in a barrier.

A qualitative introduction to two-slit interference

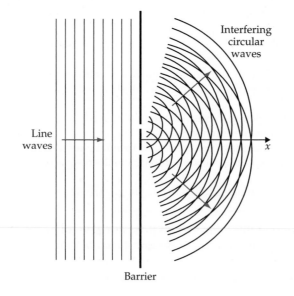

Interfering circular waves

Line waves

x

Barrier

Figure Q2.2
Circular waves emerging from two closely spaced slits will interfere with each other in the region where the waves overlap.

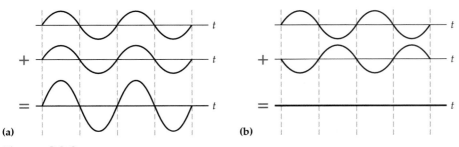

Figure Q2.3

(a) If at a given point, wave crests from the two slits arrive at the same time, the resulting wave at that point has twice the amplitude of the original waves: this is *constructive* interference. (b) If a wave crest from one slit arrives at the same time as the trough from the other, the resulting wave has zero amplitude: this is *destructive* interference.

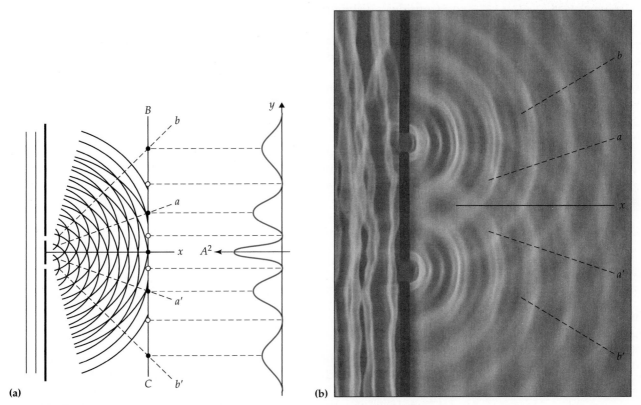

Figure Q2.4

(a) If we measure the total wave amplitude A along the line BC and plot A^2 (a nonnegative measure of the total wave's strength) as a function of y, we get the graph shown (sideways) at the right edge. (b) A photograph of a two-slit interference pattern for real water waves showing the lines x, a, a', b, and b'.

An examination of figure Q2.2 shows that the points on the water's surface where the waves interfere constructively lie on lines that appear to radiate from a point midway between the two slits (these lines are labeled x, a, a', b, b' in figure Q2.4). The points where the waves interfere destructively lie on similar lines. If we were to graph the wave amplitude along the line marked BC on the diagram, we would find that points of constructive interference alternate with points of destructive interference, as shown in figure Q2.4a (the graph has been plotted sideways to make its connection with the diagram clear).

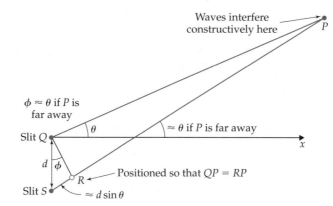

Figure Q2.5
Assume that P is a point where the waves interfere constructively. The *extra* distance that a wave from S has to go to get to point P compared to a wave from Q is the distance from S to R. This distance is roughly equal to $d \sin \theta$ (where d is the distance between slits) if P is very far away compared to d.

Determining the angles of maximal constructive interference

We can use a simple geometric argument to determine the angles θ that the lines x, a, a', b, b', etc. make with the horizontal direction (x axis). Consider a distant point P where the waves from each slit interfere constructively, and imagine lines connecting the point P to slits Q and S, as shown in figure Q2.5. Draw a line from slit Q to a point R on the line from S to P such that the distance between R and P is the same as the distance between Q and P (that is, $\triangle QRP$ is an *isosceles* triangle). The extra distance that a wave from slit S has to travel to get to P is then the distance between points S and R.

Now, if point P is very far from the slits, lines QP and SP make almost the same angle θ with the x axis, the angle ϕ in the small triangle $\triangle QRS$ is approximately equal to θ, and the largest angle in that triangle is approximately a right angle. This means that if the distance between the two slits is d, the *extra* distance that the wave from slit S has to cover to get to point P is approximately equal to $d \sin \theta$.

Now, if P is to be a point where the waves from S and Q interfere constructively, then the distance $d \sin \theta$ between R and S must be equal to an integer number of the wave's wavelengths, so that crests from Q and S arrive simultaneously at P. So the condition for constructive interference at point P is simply

The equation for the angles of maximal constructive interference in a two-slit interference experiment

$$d \sin \theta_{nc} = n\lambda \quad \Rightarrow \quad \theta_{nc} = \sin^{-1} \frac{n\lambda}{d} \qquad \text{(Q2.1)}$$

Purpose: The points at which waves emerging from two slits in a barrier constructively interfere most strongly form lines that radiate from the point halfway between the slits. This equation specifies the angles θ_{nc} that those lines make with the direction perpendicular to the barrier.

Symbols: λ is the wavelength of the waves, d is the center-to-center distance between the slits, and n is an integer.

Limitations: This equation is accurate only at points very distant from the slits compared to d. The slits must be arranged so that a given crest of the line wave moves through both slits simultaneously. The slits need to be comparable in width to λ to generate circular waves that are strong at all physically reasonable angles (see section Q2.3).

The value $n = 0$ yields an angle $\theta_{0c} = 0$, which corresponds to the line of constructive interference along the x axis in figure Q2.4. The values $n = \pm 1$ yield the angles θ_{1c} and θ_{-1c} of lines a and a', the values $n = \pm 2$ yield the angles θ_{-2c} and θ_{2c} of the lines b and b', and so on.

Exercise Q2X.1

Argue that in the limit that the large angles in the isosceles triangle $\triangle PQR$ are roughly 90° (that is, the angle $\angle QPR \approx 0$), then $\theta \approx \phi$.

Exercise Q2X.2

If the distance between point P and the slits is about 3 m and the distance d between slits is 4 cm, then the large angles in triangle $\triangle PQR$ are actually about 89.6°. What is the difference between θ and ϕ in this case?

Example Q2.1   Interference of Water Waves

Problem   Imagine that line water waves with a wavelength of 1.2 cm moving in the $+x$ direction go through two slits in a barrier parallel to the y axis. The slits are separated by 4.0 cm (center to center). Imagine that we measure the wave amplitude along a line parallel to the barrier but 1.0 m from it. At what points along this line will the waves constructively interfere most strongly?

Translation   The diagram below shows the situation and defines some useful symbols. Note that I have defined the midpoint between the slits to have coordinate $y = 0$.

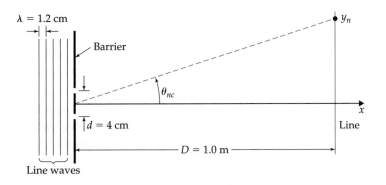

Model   Since $D = 1.0$ m $\gg 4.0$ cm $= d$, the condition on equation Q2.1 is satisfied here. We can thus use that equation to find the angles $\theta_{nc} = \sin^{-1}(n\lambda/d)$ of lines along which maximal constructive interference occurs. Basic trigonometry implies that these lines intersect the line along which we measure the waves at positions $y_n = D \tan \theta_{nc}$. We know D, λ, and d, so we know enough to solve for θ_{nc} and y_n for various values of n.

Solution   The specific angles are

$n = 0$: $\theta_{0c} = 0$

$n = \pm 1$: $\theta_{\pm 1c} = \sin^{-1}\left(\dfrac{\pm\lambda}{d}\right) = \sin^{-1}\left(\dfrac{\pm 1.2\text{ cm}}{4.0\text{ cm}}\right) = \sin^{-1} 0.30 = \pm 17.5°$

$n = \pm 2$: $\theta_{\pm 2c} = \sin^{-1}\left(\dfrac{\pm 2\lambda}{d}\right) = \sin^{-1}(\pm 0.60) = \pm 37°$ (Q2.2)

$n = \pm 3$: $\theta_{\pm 3c} = \sin^{-1}\left(\dfrac{\pm 3\lambda}{d}\right) = \sin^{-1}(\pm 0.90) = \pm 64°$

Values of n such that $|n\lambda/d| > 1.0$ yield no meaningful solution for θ_{nc}, so the series stops here. The corresponding positions along the measurement line where the waves are strongest are

$$
\begin{aligned}
n = 0: &\qquad y_0 = D\tan\theta_{0c} = 0 \\
n = \pm 1: &\qquad y_{\pm 1} = D\tan\theta_{\pm 1c} = (1.0\text{ m})\tan(\pm 17.5°) = \pm 0.32\text{ m} \\
n = \pm 2: &\qquad y_{\pm 2} = D\tan\theta_{\pm 2c} = (1.0\text{ m})\tan(\pm 37°) = \pm 0.75\text{ m} \\
n = \pm 3: &\qquad y_{\pm 3} = D\tan\theta_{\pm 3c} = (1.0\text{ m})\tan(\pm 64°) = \pm 2.05\text{ m}
\end{aligned}
\qquad\text{(Q2.3)}
$$

Evaluation   These results have the right units and are plausible.

While so far we have considered only water waves, the argument leading to equation Q2.1 applies to waves of all kinds, including three-dimensional waves (such as sound waves) as long as we evaluate the waves in a plane that contains the waves' direction of motion as well as both slits or sources, and as long as wave crests emerge from the slits or sources simultaneously.

Exercise Q2X.3

Imagine that you place your stereo speakers outside (to provide music for a party). They are placed distance of 1.5 m apart along a line we will call the y axis. Imagine that both speakers are reproducing the sound of a female singer holding a solo high A (880 Hz) for several seconds. Consider a line parallel to the y axis but 8.0 m from it. At what points along this second line would the sound be loudest? At what points would it be weakest?

Q2.2   Two-Slit Interference of Light

Studying light has motivated revolutions in physics

Few physical phenomena are so common and important to our daily experience as light. Yet light is so subtle that understanding its nature has challenged the scientific community for thousands of years. Newton himself wrote a book on the subject (*Opticks*, 1704), but this book is not as famous as his work on mechanics because his particle model of light was seemingly contradicted by research done in the early 1800s by Young, Fresnel, Fitzeau, Lloyd, and Kirchhoff, which demonstrated in what seemed to be a fairly conclusive fashion that light was a wave. Maxwell's greatest triumph was his claim (later supported by experiment) that light was in fact an *electromagnetic* wave. This triumph, in combination with work in the late 1800s that established the frame independence of the speed of light, prompted Einstein to develop the theory of relativity. At roughly the same time, research into how atoms absorbed and emitted light was beginning to lay the foundations for quantum mechanics. The theory of quantum electrodynamics that eventually resulted served in turn as the template for those who developed quantum field theories for the weak and strong nuclear interactions in the 1970s. The study of light (and the challenges it raised) thus provided either the impetus or the template for essentially all the major revolutions in physics since the 1840s!

It is appropriate, therefore, that we start our investigations of quantum mechanics by studying light. My goal in this chapter is to bring your understanding of light up to where most physicists stood in the late 1800s, the eve of the quantum revolution. At this time, physicists were firmly convinced that light was a wave. What was the evidence supporting such a conclusion?

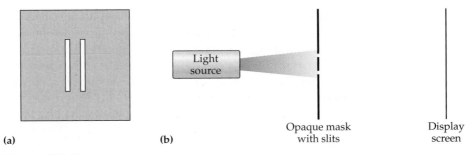

(a)   **(b)**

Figure Q2.6
(a) An opaque mask with two slits. (b) A top view of Young's two-slit interference experiment.

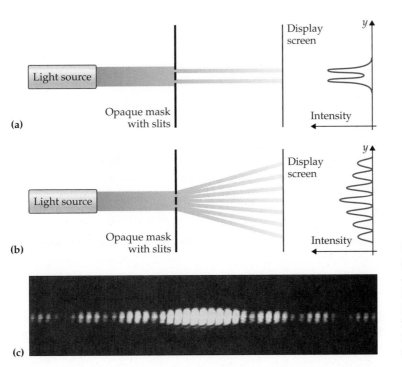

Figure Q2.7
(a) The outcome of the Young double-slit experiment predicted by Newton's particle model of light. (b) The actual outcome of the Young double-slit experiment. (c) A photograph of an actual two-slit interference pattern.

A Victorian physicist might have pointed to the two-slit interference experiments performed by Thomas Young during the first decade of the 1800s. Young directed a beam of light having essentially a single color (**monochromatic** light) on an opaque mask in which two narrow slits had been cut, as illustrated in figure Q2.6a. Light that passes through the slits falls on a screen placed some distance behind the mask, as illustrated in figure Q2.6b.

If light consisted of particles (as Newton thought), particles passing through the slits would travel in straight lines to the second screen, implying that we would see two isolated bright lines on the display screen, one for each slit, as shown in figure Q2.7a. If we take into account the fact that particles of light can follow slightly different straight paths from the source to the screen through each slit, we might expect the images of the slits be somewhat blurred; but there should be two and only two slit images, and a plot of the intensity of light as a function of distance y along the display screen would look something that shown at the far right of figure Q2.7a.

If we do this experiment, though, we see not two, but many bright spots, as shown in figure Q2.7b. The spot spacing does depend on the slit separation,

Young's two-slit interference experiment

What a particle model of light predicts

The actual results

but reducing the slit separation counterintuitively *increases* the separation of the bright spots! We also find that the spot separation depends on the color of the light used, decreasing as one goes from red to violet.

These results are clearly incompatible with a simple particle model of light, but are easily explained by a wave model. If monochromatic light consists of waves having a well-defined wavelength, Young's two-slit light interference experiment is essentially the same as the two-slit water wave interference experiments considered in section Q2.1. We can interpret the bright and dark spots on the screen as positions where light waves interfere constructively or destructively, respectively. The interference pattern displayed in figure Q2.7b is qualitatively quite similar to the one for water waves shown in figure Q2.4, and quantitative measurements show that equation Q2.1 accurately describes how the positions of these peaks change as we vary the distance between slits. This experiment therefore offers compelling evidence that *light is a wave.*

Example Q2.2   What Is the Wavelength?

Problem   Imagine that we have a laser that produces monochromatic red light of an unknown wavelength. We allow the laser's light to fall on a pair of slits whose center-to-center distance is $d = 0.050$ mm, and we display the resulting interference pattern on a screen a distance $D = 3.0$ m from the slits. If the distance between adjacent spots on the screen has a very nearly constant value of $s = 3.8$ cm for spots near the central dot, what is the laser light's approximate wavelength?

Translation   The drawing below illustrates the experiment.

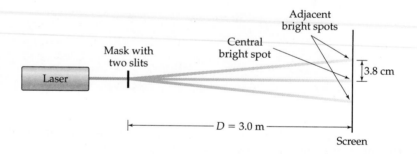

Model   In equation Q2.1, the central maximum will correspond to $n = 0$ (since $n = 0$ implies $\theta = 0$). The angle between the line connecting the slits with the central maximum and that connecting the slits and a bright spot at position y_n is

$$\theta_{nc} = \tan^{-1} \frac{y_n}{D} \approx \frac{y_n}{D} \qquad (Q2.4)$$

where I have used the fact that $\tan\theta \approx \theta$ for small angles measured in radians. Note that for the $n = 1$ spot, the angle comes out to be 0.038 m/3.0 m = 0.0127 rad = 0.7°, so the small-angle approximation will be justified in this case for spots with reasonably small n. For small angles we also have $\sin\theta \approx \theta$, so equation Q2.1 implies that

$$\frac{n\lambda}{d} = \sin\theta_{nc} \approx \theta_{nc} \approx \frac{y_n}{D} \qquad (Q2.5)$$

Note that in this approximation $y_n \propto n$, so the spots are evenly spaced. If we define $s \equiv y_{n+1} - y_n$, then equation Q2.5 implies that in the small-angle limit,

$$\frac{(n+1)\lambda}{d} - \frac{n\lambda}{d} \approx \frac{y_{n+1}}{D} - \frac{y_n}{D} \quad \Rightarrow \quad \frac{\lambda}{d} \approx \frac{s}{D} \qquad \text{(Q2.6)}$$

Since we know d, s, and D, we can solve for λ.

Solution Doing this, we find that

$$\lambda \approx \frac{sd}{D} = \frac{(0.038 \text{ m})(0.05 \times 10^{-3} \text{ m})}{3.0 \text{ m}} = 6.3 \times 10^{-7} \text{ m} = 630 \text{ nm} \quad \text{(Q2.7)}$$

Evaluation Note that the units are correct. The small value of this wavelength helps explain why wave aspects of light are not immediately obvious.

Note how the small-angle approximation helps simplify the mathematics in this example. Many practical interference experiments with light involve small angles, so equation Q2.6 is a variation of equation Q2.1 worth remembering.

Interference experiments provide a practical means of measuring the wavelength of light because they link λ to quantities such as d and D that are big enough to measure with a ruler. Such experiments show that the wavelengths of visible light range from about 700 nm (deep red) to 400 nm (deep violet).

Such experiments allow us to measure the wavelength of light

Exercise Q2X.4

Imagine that we replace the slits described in example Q2.2 with slits whose center-to-center spacing is 0.030 mm. Find the spacing between adjacent bright spots now. (*Hint:* You should find that the spacing gets *larger*.)

Exercise Q2X.5

Explain in *words* why the separation between the displayed bright spots *increases* when the separation between the slits decreases. Do *not* appeal to equations: rather, base your explanation on figure Q2.5.

Exercise Q2X.6

Imagine that we use a different laser in the experiment described in exercise Q2X.4, a laser that produces green light with a wavelength of 510 nm. What is the spacing between bright spots now?

Q2.3   Diffraction

You may have noticed that not all the dots in the interference pattern shown in figure Q2.7c are equally bright. To explain why this is so, we need to better understand what happens to waves going through a *single* slit.

We can understand this most easily with the help of a simplified model of wave propagation that we call **Huygens's principle** (after the Dutch

Huygens's principle

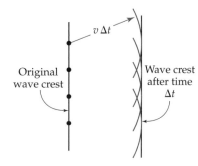

Figure Q2.8
Propagation of a line wave according to Huygens's principle.

physicist Christian Huygens who first proposed this model in 1678). This principle states

> We can model each point on a given wave crest to be a point source of circular (or spherical) **wavelets.** At a time Δt later, the new position of the wave's crest will be a curve (or surface) tangent to the wavelet crests.

This principle essentially expresses the idea that when each point in a medium is disturbed by the wave, the effects of that disturbance move radially away from that point as time passes. For example, if a bit of water is lifted above the surface of a pond by the crest of a wave, the water seeks to sink back downward; as it does, it pushes water on all sides outward, creating a circular ripple. The sum of all these circular ripples is what forms the crest as it moves forward.

Figure Q2.8 illustrates how this works for a line wave. We imagine each point on the line wave to be a point source for a circular wave (only a few points are shown for the sake of clarity). After time Δt, the wave from each point has expanded to a radius of $r = v\,\Delta t$, where v is the speed at which waves travel in the medium in question. The tangent to these circular waves forms another straight line.

Exercise Q2X.7

Show, using the same approach, that the crest of an initially circular wave will spread out to form a larger circle as time passes.

An analysis of what happens to a wave passing through a single slit

Now consider the situation shown in figure Q2.9a. Imagine that line waves with a well-defined wavelength go through a slit of width a, and imagine that we measure the wave amplitude at various points P along the y axis, which is (at its closest) a distance $D \gg a$ from the slit. When D is very large, the lines that waves must follow to get to P from various points within the slit are approximately parallel lines, as shown in figure Q2.9b.

Now, according to Huygens's principle, we can imagine that every point along the crest of a wave passing through the slit emits a circular wave. To keep things relatively simple, we consider in figure Q2.9b only 12 points equally spaced along the slit.

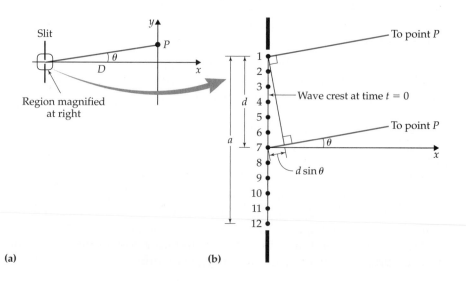

Figure Q2.9
(a) A top view of a single-slit diffraction experiment. (b) A magnified image of the slit region in this experiment. Here we treat the wave crest in the slit as if it were 12 points that emitted circular waves. Each circular wave contributes to the wave at point P.

(a)   (b)

When $\theta = 0$ (P is directly in the $+x$ direction relative to the slit), the distance from each point to the final destination at point P is equal, so the wavelets from *each* point interfere constructively when they reach P. But as we move our measurement point P up the y axis, the angle θ increases, and the distance that the wavelets have to move from each point to get to point P is no longer the same. In particular, if the distance between a given pair of points in the slit is d, then the *extra* distance s that a wavelet from the lower point has to go to get to P beyond the distance to P from the upper point is

$$s = d \sin\theta \qquad (Q2.8)$$

Thus as θ increases, the wavelets from each point get more and more out of phase with each other, and thus less and less strongly reinforce each other. The total wave amplitude will thus *decrease* as we move point P up the y axis.

When we reach the angle where $a \sin\theta = \lambda$, something interesting happens. Note that the distance between point 1 and point 7 is $\frac{1}{2}a$, as is the distance between points 2 and 8, points 3 and 9, and so on. This means that the wavelet from point 7 will have to travel a distance $s = \frac{1}{2}a \sin\theta = \frac{1}{2}\lambda$ farther to get to P than the wavelet from point 1 does, and similarly for points 8 and 2, 9 and 3, and so on. Since one-half of a wavelength from a wave crest is a trough, the crest from point 1 will be canceled by a trough from point 7, the crest from point 2 by a trough from point 8, and so on. In short, the wavelet from each point in the upper half of the slit is exactly canceled by that from a point on the bottom half! We will therefore detect *nothing* at the angle θ_{1d} such that

$$a \sin\theta_{1d} = \lambda \qquad \Rightarrow \qquad \theta_{1d} = \sin^{-1}\frac{\lambda}{a} \qquad (Q2.9)$$

(The subscript tells us that this is the first angle of complete *destructive* interference.)

As we move point P still farther up the y axis, we begin to see waves again at point P, but they are *much* weaker. For example, consider the angle such that $a \sin\theta = (3/2)\lambda$. In this case, the wavelet from point 1 cancels the wavelet from point 5, since the distance between points 1 and 5 is $\frac{1}{3}a$, so $s = \frac{1}{3}a \sin\theta = \frac{1}{2}\lambda$ again (the condition for destructive interference). Similarly, the wavelet from point 2 cancels that from point 6, the wavelet from point 3 cancels that from point 7, and the wavelet from point 4 cancels that from point 8. This leaves points 9 through 12 contributing to the net wave at point P at this angle. But even wavelets from these points do not completely reinforce each other: wavelets from point 9 are almost completely out of phase with those from point 12, so the net wave at point P at this angle comes mainly from points 10 and 11. Thus it has about $2/12 = 1/6$ of the amplitude of the wave at $\theta = 0$, where all 12 points contribute strongly. (A more exact calculation yields a ratio of $1/4.7$.) The point is that the wave is generally *much* smaller in amplitude for angles larger than θ_{1d} than it was near $\theta = 0$.

Exercise Q2X.8

Argue that the waves again cancel completely at the angle θ_{2d} such that $\sin\theta_{2d} = 2\lambda/a$. (*Hint:* The wavelets from each point again cancel in pairs. Which points cancel which points in this case?)

Indeed, you can argue in a similar way that we will have completely destructive interference at any angle satisfying

The equation describing the angles of completely destructive interference in single-slit diffraction

$$\theta_{nd} = \sin^{-1}\frac{n\lambda}{a} \qquad\qquad (Q2.10)$$

Purpose: The points at which waves emerging from a single rectangular slit *completely cancel out* lie along lines that radiate from the slit's center. This equation specifies the angles θ_{nd} between those lines and the direction of motion of the original waves moving through the slit.

Symbols: λ is the wavelength of the waves, a is the width of the slit, and n is a *nonzero* integer.

Limitations: This equation applies to a rectangular slit whose height is much greater than its width a. It is accurate only at points very distant from the slit compared to a. The slit must be oriented so that a given wave crest moves through all parts of the slit simultaneously.

This equation looks very much like equation Q2.1 for two-slit interference. Note, however, that equation Q2.1 refers to *two-slit interference*, while equation Q2.10 describes *single-slit diffraction*. Even more important, equation Q2.1 specifies angles along which the waves *constructively* interfere, but equation Q2.10 describes angles along which the waves *destructively* interfere (as the subscripts on the angles indicate)! Be aware of these differences!

If the wave model of light is correct, then light waves moving through a narrow rectangular slit must behave as we have argued. The brightness of any point on the display screen will depend on the energy per unit time per unit area (i.e., the *intensity*) of the light at that spot, which is proportional to the square of the wave amplitude. Figure Q2.10a shows a graph of the predicted intensity of light versus $\sin\theta$ for light moving through a narrow slit.

Figure Q2.10b shows a photograph of light emerging from a single slit. We can see that this pattern agrees completely with the prediction of the wave model. Experiments concerning the diffraction of light done in the early 1800s provided some of the crucial evidence that finally convinced the physics community that Young was correct about light being a wave.

Note that virtually all the light energy is contained within the angular range of $-\theta_{1d}$ to $+\theta_{1d}$ between the first angles of completely destructive interference. The bright regions flanking this central region (which are called **fringes** of the diffraction pattern) contain comparatively small amounts of energy.

The wave model leads to results consistent with experiment

Figure Q2.10

(a) A plot of wave intensity versus $\sin\theta$ for the waves of wavelength λ emerging from a single slit of width a. (Note that if θ is small, $\sin\theta \approx \theta$: in such a case, this will essentially be a plot of intensity versus θ.) (b) The actual diffraction pattern produced by light going through a narrow slit. [The spacing between fringes in this photograph is about half of that shown in the graph in part (a).]

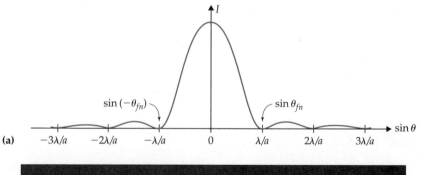

(a)

(b)

Example Q2.3   Diffraction of Light

Problem   Light with a wavelength $\lambda = 633$ nm goes through a narrow slit of width a and is then projected on a screen $D = 3.0$ m from the slit. The central bright region displayed on the screen is $w = 2.0$ cm wide. What is a?

Model   The total angle $\Delta\theta$ spanned by the central maximum goes from $-\theta_{1d}$ to $+\theta_{1d}$, where $\theta_{1d} = \sin(\lambda/a)$ (see equation Q2.8). Since angles are going to be small here, we can again use the approximation $\sin\theta \approx \tan\theta \approx \theta$, so we get

$$\frac{w}{D} = 2\frac{\frac{1}{2}w}{D} = 2\tan\theta_{1d} \approx 2\theta_{1d} = 2\sin^{-1}\frac{\lambda}{a} \approx \frac{2\lambda}{a} \qquad \text{(Q2.11)}$$

Since we know w, D, and λ, we can solve for a.

Solution: Doing this, we get

$$a \approx \frac{2D\lambda}{w} = \frac{2(3.0\ \text{m})(633 \times 10^{-9}\ \text{m})}{0.020\ \text{m}} = 1.9 \times 10^{-4}\ \text{m} = 0.19\ \text{mm} \qquad \text{(Q2.12)}$$

Evaluation: The units are right, and the result seems reasonable for a narrow slit.

Now we can understand the variation in the brightness of spots in the two-slit interference pattern. Light emerging from each slit fans out according to the *single*-slit diffraction pattern. If the two slits are the same width and are close together compared to the distance to the screen on which the patterns are to be displayed, their single-slit interference patterns will almost exactly overlap on the screen. If there were no interference *between* the waves emerging from two slits, the displayed intensity pattern would be essentially that of a single slit, as displayed in figure Q2.11a. Since the brightness of a bright spot in the *two*-slit interference pattern will depend on how much light is there in the first place to constructively interfere, the brightness of such a spot will be modulated by how bright the light in the *single*-slit pattern is at that point, as illustrated in figure Q2.11b.

The phenomenon of diffraction also explains why you can hear someone talking around a corner even when you can't see the person. Equation Q2.10 also suggests that longer wavelengths will diffract more widely than shorter

Explaining the brightness variation in the two-slit pattern

Why you can hear someone around a corner

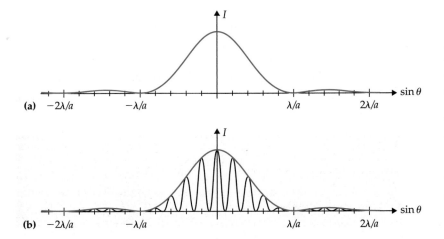

(a) $-2\lambda/a$ $-\lambda/a$ λ/a $2\lambda/a$

(b) $-2\lambda/a$ $-\lambda/a$ λ/a $2\lambda/a$

Figure Q2.11
(a) A graph of wave intensity versus position for light emerging from a single slit of width a. (b) A graph of the wave intensity versus position for light emerging from two slits of width a separated by distance $d = 5a$. Since the light required to generate the two-slit interference pattern has to come from the light provided by each single slit, the single-slit pattern provides the maximum possible intensity for the two-slit pattern.

wavelengths. Sound waves have wavelengths on the order of magnitude of 1 m, so when sound waves go through an open doorway (which is also about a meter wide), they diffract pretty well in all directions, but light (whose wavelength is much shorter) is hardly diffracted at all. This effect also explains why voices around a corner may sound "muffled": the higher-frequency components of sound that add sharpness and clarity to a voice may not diffract as well to your position as lower-frequency components do.

Q2.4   Optical Resolution

Diffraction by a circular aperture

Figure Q2.10 displays the diffraction pattern of light going through a rectangular slit that is much longer than its width a. It is more difficult to calculate what happens to light when it goes through a circular aperture, but figure Q2.12 shows the result: we see a bright circular central region flanked by weaker circular fringes. It turns out that the angle of the innermost dark ring where complete cancellation occurs is given by

$$\sin \theta_{1d} = 1.22 \frac{\lambda}{a} \qquad (Q2.13)$$

where a in this case is the aperture's diameter. This angle is only a bit larger than predicted by equation Q2.9.

Figure Q2.12
The diffraction pattern created by light after it has gone through a small circular aperture.

Exercise Q2X.9

A laser on the International Space Station emits green light of wavelength 510 nm from a hole 5.0 mm in diameter. When this beam reaches the ground 180 km below, what is the approximate diameter of the laser beam's central region?

Implications for resolution of optical instruments

The fact that light is diffracted by a small opening has important implications for the ability of an optical instrument (such as a telescope or an eye) to resolve two distant objects whose angular separation is small. For example, imagine looking at two point sources of light (perhaps two stars) separated by a small angle θ, as shown in figure Q2.13. The light from each source is diffracted somewhat as it goes through your pupil. This means that even if your lens focuses the light as well as possible, the light from each source creates a small but spread-out diffraction pattern on your retina.

Figure Q2.14 shows what the resulting diffraction patterns might look like on your retina for different angular separations between two sources. One can *just begin* to see that the sources are separate objects in the case shown in figure Q2.14b, where the central maximum of one diffraction pattern overlaps the first first minimum of the other. Therefore, we can see the sources as separate only when their angular separation θ is such that

Rayleigh's criterion for resolving point sources

$$\theta > \theta_{1d} \approx \sin^{-1} \frac{1.22\lambda}{a} \qquad (Q2.14)$$

Purpose: This equation specifies the minimum separation angle θ that sources can have if they are to be resolved by an optical instrument whose aperture has diameter a.

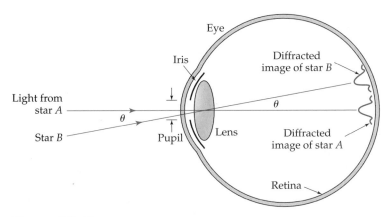

Figure Q2.13
If two stars are separated by a sufficiently large angle, then the diffraction patterns produced on the retina when the light from the stars goes through the pupil do not overlap very much, so the retina will register these images as being separate. (The "bumps" on the retina in this drawing are meant to indicate graphs of the light intensity versus position on the retina.) If, however, the angle becomes much smaller, the diffraction patterns will begin to overlap, causing the two stars to look like a single blob.

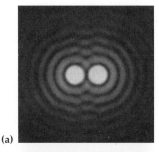

(a)

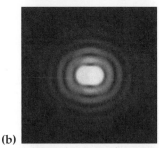

(b)

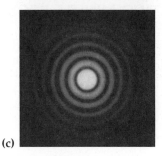

(c)

Figure Q2.14
What the diffraction patterns for two point sources look like as their angular separation decreases. In (b), the angular separation of the sources is such that the central maximum of one pattern overlaps the first minimum of the other pattern. In this case, the sources can be barely distinguished as being separate.

> **Symbols:** λ is the average wavelength of the light from the source, and θ_{1d} is the innermost angle of destructive interference.
> **Limitations:**   This equation is an approximation that becomes better if the sensor screen is far from the aperture compared to the aperture's diameter.
> **Note:**   Equation Q2.14 expresses what is called the **Rayleigh criterion.**

We call an instrument good enough to resolve sources separated by this angle **diffraction-limited.** The resolving power of a real instrument may be worse because of other factors, but it cannot be better than this. Since our ancestors depended so much on their eyes for staying alive, evolution has given us eyes that operate pretty close to the diffraction limit in daylight. The Hubble telescope, by virtue of its position above the atmosphere, also operates at close to the diffraction limit of its aperture. (The resolution of ground-based telescopes is limited more by unavoidable turbulence in the earth's atmosphere.)

Example Q2.4   The Resolving Ability of the Human Eye

Problem   At night, the pupil of a typical person's eye opens up to as wide as 8 mm. What would be the smallest possible angular separation between two stars in the sky that the human eye might be able to resolve?

Model   In this case, the aperture that the light goes through is the pupil, so $a = 0.008$ m. The eye is most sensitive to light whose wavelength is in the center of the visual range, so let's estimate that $\lambda \approx 550$ nm. If we assume that the eye is diffraction-limited, this gives us enough information to apply the Rayleigh criterion. Since the angle will be pretty small, we can also use the small-angle approximation $\sin \theta \approx \theta$.

Solution   Equation Q2.13 implies that

$$\theta_{min} \approx \sin^{-1} \frac{1.2\lambda}{a} \approx \frac{1.2\lambda}{a} \approx \frac{1.2(550 \times 10^{-9} \text{ m})}{0.008 \text{ m}} = 8 \times 10^{-5} \text{ (rad)}$$

$$= (8 \times 10^{-5} \text{ rad}) \left(\frac{360^{\circ}}{2\pi \text{ rad}}\right) \left(\frac{3600 \text{ arc seconds}}{1^{\circ}}\right) \approx 20 \text{ arc seconds}$$

(Q2.15)

Evaluation   I have expressed the answer to only one significant digit because the approximations we have made are not more accurate than that. For comparison, earth-based telescopes are capable of resolving stars on the order of 1 arc second apart, while the Hubble telescope can in principle resolve stars on the order of 0.1 arc second apart.[†]

TWO-MINUTE PROBLEMS

Q2T.1   Waves from two slits S and Q will *destructively* interfere and cancel at a point P if the distance between P and S is larger than the distance between Q and S by

A. λ
B. $n\lambda$
C. $\dfrac{n\lambda}{2}$ (where n is an integer)
D. $(n + \frac{1}{2})\lambda$ (where n is an integer)
E. $\dfrac{\lambda}{4}$
F. Other (specify)

Q2T.2   Consider a two-slit interference experiment like the one shown in figure Q2.6b. The distance between adjacent bright spots in the interference pattern on the screen (A) increases, (B) decreases, or (C) remains the same if
(a) The wavelength of the light increases.
(b) The spacing between the slits increases.
(c) The intensity of the light increases.
(d) The width of the slits increases.
(e) The distance between the slits and the screen increases.
(f) The value of n increases (careful!).

Q2T.3   We have seen that Huygens's principle implies that a circular wave front will remain circular and a line wave front will remain linear as time passes. Imagine that at an instant of time we set up a wave front that is shaped like a square moving away from its center. Huygens's principle implies that this wave front will also maintain its square shape as time passes, true (T) or false (F)?

Q2T.4   Imagine that sound waves with a certain definite wavelength flow through a partially opened sliding door. If the door is closed somewhat further (but not shut entirely), the angle through which the sound waves are diffracted
A. Decreases
B. Increases
C. Remains the same
D. Depends on quantities not specified (explain)

Q2T.5   Imagine that sound waves with a certain definite wavelength flow through a partially opened sliding door. If the wavelength of the waves increases, the angle through which the sound waves are diffracted
A. Decreases
B. Increases
C. Remains the same
D. Depends on quantities not specified (explain)

Q2T.6   Line waves with wavelength λ going through a slit with width a will be diffracted into circular waves with approximately equal amplitude in all forward directions
A. Always
B. Never
C. Only if $a \gg \lambda$
D. Only if $a \ll \lambda$

[†]The measured visual acuity of the eye at night is actually more like 200 arc s. This is so because the dark-adapted eye averages the response of many retinal cells in order to be able to respond to very dim light. This averaging reduces visual acuity. In bright light, the human eye does perform at close to the diffraction limit.

Q2T.7 If the two slits in a two-slit interference experiment were so far apart that their diffraction patterns did not overlap, the pattern displayed would be consistent with a particle model of light, T or F?

Q2T.8 Consider an experiment where we send monochromatic light to a distant screen through a single narrow slit. The distance between adjacent dark fringes in the diffraction pattern displayed on the screen (A) increases, (B) decreases, or (C) remains the same if
(a) The wavelength of the light increases.
(b) The intensity of the light increases.
(c) The width of the slit increases.
(d) The distance between the slit and the screen increases.
(e) We look at fringes farther from the central maximum.

Q2T.9 In the region where their beams overlap, two car headlights will create a clear interference pattern on a distant screen, T or F?

Q2T.10 If we shine white light through two slits onto a distant screen, we will see a clear interference pattern on the screen, T or F?

Q2T.11 Evolution has given eagles and other predatory birds very sharp eyesight. A friend claims to have read that an eagle's eye has 10 times the resolution of a human eye in broad daylight. This is physically impossible, T or F?

Q2T.12 With an optically perfect 200-power telescope with a 1.5-in.-diameter tube, you can resolve objects that are roughly how many times closer together than you could with your naked eye? (Choose the closest response, and ignore air turbulence.)
A.   500 times
B.   200 times
C.   100 times
D.   20 times
E.   5 times
F.   No better at all

HOMEWORK PROBLEMS

Basic Skills

Q2B.1 Water waves with an amplitude of 0.80 cm and a wavelength of 2.5 cm go through two openings in a barrier. Each opening is 1.2 cm wide, and the openings are separated by 12.0 cm (center to center). Find the angles of the lines along which the waves constructively interfere.

Q2B.2 Sound waves with a frequency of 320 Hz are emitted by two speakers 44 cm wide and 3.5 m apart. Find the angles of the lines along which the waves from the speakers constructively interfere (assuming that wave crests are emitted by the speakers simultaneously).

Q2B.3 Imagine that the distance between two slits in a given experiment is $d = 0.050$ mm and that the distance between the slit mask and the display screen is $D = 1.5$ m. If the distance between adjacent interference bright spots (for low n) is about 2.0 cm on the screen, what is the wavelength of the light involved?

Q2B.4 Imagine that the distance between two slits in a given experiment is $d = 0.040$ mm and that the distance between the slit mask and the display screen is $D = 2.5$ m. If the distance between adjacent interference bright spots (for low n) is about 3.0 cm on the screen, what is the wavelength of the light involved?

Q2B.5 Imagine that the light source for a two-slit interference experiment provides red light with a wavelength of 633 nm. If this light is sent through a pair of slits separated by a distance $d = 0.040$ mm, find the spacing between adjacent bright spots of the interference pattern when it is displayed on a screen 3.2 m from the slits.

Q2B.6 Imagine that the light source for a two-slit interference experiment provides yellow light with a wavelength of 570 nm. If this light is sent through a pair of slits separated by a distance $d = 0.030$ mm, show that the angle between the bright spot corresponding to $n = 0$ and the bright spot corresponding to $n = 1$ is about 1.1°. If the display screen is a distance $D = 2.4$ m from the slits, show that the distance between these bright spots on the screen is about 4.6 cm.

Q2B.7 Explain in terms of wave concepts why sounds emitted from a person's mouth can be heard almost equally well in all directions.

Q2B.8 As you are just about to round a corner, you hear two people talking some distance beyond the corner. One has a very deep voice, and the other has a high voice. Which voice more easily carries around the corner? Explain.

Q2B.9 A stereo speaker 30 cm wide sounds a pure 1250-Hz note. Within what angle from the forward direction will you be able to hear this note (in a perfectly absorbing room)?

Q2B.10 Sound waves from a stereo inside a house go through a partially open sliding patio door to the yard outside. If the door opening were 12 cm wide, what would be the lowest frequency of sounds that would *not* be diffracted in essentially all directions through that opening?

Q2B.11 Ocean waves with an amplitude of 2.0 m and a wavelength of 15 m go through a 55-m opening in a breakwater that is shaped like a line and protects a body of water shaped like a square 800 m on a side between the breakwater, the beach, and two rocky ridges on either side. Draw a sketch of this situation that accurately and quantitatively illustrates the angle through which the waves are diffracted. (Also show your work in computing that angle.) Indicate some points on your sketch where boats within the breakwater will feel little wave motion.

Q2B.12 Light with a wavelength of 633 nm goes through a narrow slit. The angle between the first minimum on one side of the central maximum and the first minimum on the other side is 1.2°. What is the width of the slit?

Q2B.13 Light of wavelength 441 nm goes through a narrow slit. On a screen 2.0 m away, the width of the central maximum of the diffraction pattern is 1.5 cm. What is the width of the slit?

Q2B.14 The beam of light emitted from a certain laser has a wavelength of 633 nm and an initial diameter of 1.0 mm. What is the diameter of the beam when it reaches the moon (384,000 km away)?

Synthetic

Q2S.1 You are setting up a pair of PA speakers on a field in preparation for an outdoor event. Each speaker is 0.65 m wide, and the speakers are separated by 8.2 m. To test the speakers, your coworker plays a single tone through the speakers whose frequency is 440 Hz. You are standing 52 m directly in front of the speakers and facing them. Roughly how far would you have to walk to your left or right to hear the sound amplitude drop almost to zero? How much farther would you have to go to hear the amplitude go back to its original strength?

Q2S.2 Two radio antennas 60 m apart broadcast a synchronized signal with a frequency of 100 MHz. Imagine that we have a detector 5.0 km from the antennas. At this distance, what is the separation between adjacent "bright spots" in the interference pattern along a line parallel to the line between the antennas?

Q2S.3 If we hold two flashlights parallel, will they create a clear interference pattern in the region where the two beams overlap? Carefully explain at least two reasons why *not*.

Q2S.4 When you connect stereo speakers to an amplifier, it is important that the speakers be connected in

phase, so that if a signal from the amplifier pushes the cone of one speaker out at a given time, it pushes the cone of the other speaker out at the same time. Reversing the two wires connecting one of the speakers to the amplifier will make it so that the same signal pushes one speaker cone out but pulls the other one in. Explain why this could be a problem, or at least undesirable. Would you still be able to hear the music?

Q2S.5 Two sets of sinusoidal line waves approach each other from opposite directions. These waves have exactly the same amplitude and wavelength. When they overlap, will they *constructively* interfere, cancel each other out, or do something else? Describe as carefully as you can what will happen when these waves meet.

Q2S.6 How big a speaker would you need to create a directed beam of sound waves with a frequency of 440 Hz whose total width increases by only 5 m for every 100 m the beam goes forward?

Q2S.7 Television sets (particularly older models) produce a "whistle" sound at about 15,800 Hz (this is the frequency with which the electron beam sweeps across the face of the screen). This sound is audible to most youngsters and adults who haven't lost their high-frequency hearing. Imagine that a TV set is on (with the "mute" on) in your sister's bedroom as you walk by in the hallway. If your sister's door is ajar, leaving an opening 6.0 cm wide, and your ear is about 1.5 m from the door as you walk by, for about how many centimeters of your walk will you be able to hear the TV "whistle" (if you can at all)?

Q2S.8 The two headlights on a certain approaching car are 1.4 m apart. At about what distance could you resolve them as being separate? Make appropriate estimates as needed.

Q2S.9 Under ideal conditions and when Mars is closest, estimate the linear separation of two objects on Mars that can barely be resolved (a) by the naked eye and (b) the Hubble telescope (whose main mirror is 1 m in diameter).

Q2S.10 Let's guess that a person can distinguish between letters of the alphabet, if he or she can resolve features of the letter roughly one-fourth the size of the letter. Let's assume that in bright light, a person's pupil is ≈3 mm in diameter, and that the eye is most sensitive to light with a wavelength of 550 nm. What is the approximate maximum distance that one could read letters that are 3 mm high? The letters in headings (TWO-MINUTE PROBLEMS, HOMEWORK PROBLEMS; exercise numbers, etc.) in this text are about this height: check your calculation by direct observation and report the results.

Q2S.11 The paintings of Georges Seurat consist of closely spaced small dots ($\approx$2 mm wide) of pure pigment. The illusion of blended colors occurs at least partly because the pupils of the observer's eyes diffract light entering them. Estimate the appropriate distance from which to view such a painting, considering the fact that art museums are usually very well lit. (Adapted from Serway, *Physics*, 3d ed., Saunders, Philadelphia, 1990, p. 1096.)

Q2S.12 In J. R. R. Tolkien's *The Lord of the Rings* (volume 2, p. 32), Legolas the Elf claims to be able to accurately count horsemen and discern their hair color (yellow) 5 leagues away on a bright, sunny day. Make appropriate estimates and argue that Legolas must have very strange-looking eyes, have some means of nonvisual perception, or have made a lucky guess. (1 league $\approx$ 3.0 mi.)

Rich-Context

Q2R.1 You are at sea on a foggy night. You are trying to find out how far you are from the shore, but the fog is too thick to see anything. After a certain time of aimless sailing, you dimly hear two separate foghorns on your port (left) side, perpendicular to your direction of travel. Each foghorn emits one short blast of sound at a pitch of 120 Hz every 2.00 s exactly, and each sounds about as loud as the other. Looking at your map, you see two foghorn locations marked plausibly near what you guess your location to be, and on the map the foghorns are 1700 m apart, flanking the entrance to a harbor. At a certain time, you notice that you hear the blasts from the foghorns simultaneously. After you have sailed at a steady heading for 22 min at a speed of 8.8 km/h (as measured by your boat's speedometer), you hear the foghorns exactly out of phase (one honks, then the other, then the first, etc.). Roughly how far are you from the foghorns now? (*Hint:* Treat this as a two-slit interference problem.)

Q2R.2 You are the prosecuting attorney in a case. You are eliciting testimony from your star witness, who has said that he was sitting on his porch 600 ft away from the crime scene when he saw the defendant commit the crime in broad daylight. You ask, "How did you know that it was the defendant?" Reply: "I recognized the Angels baseball cap the defendant often wears, the same cap the defendant is wearing now." Question: "How did you know that it was this Angels cap and not an ordinary cap of the same color?" Reply: "I could see the big A very clearly." Question: "You are absolutely sure of this?" Reply: "Yes, absolutely." With sudden shock, you realize that your star witness is *lying*. How do you know this?

Q2R.3 The phenomenon of *refraction* arises because light travels at different speeds in different media. The speed of light in a vacuum is $c = 299{,}792{,}458$ m/s (by definition of the meter). Light moves about 0.03 percent slower than this in air, 25 percent slower in water, 34 percent to 40 percent slower in glass, and so on. Physicists typically express the speed of light in a certain medium in terms of the medium's *index of refraction*, which we define to be

$$n_m = \frac{c}{v_m} \qquad (Q2.16)$$

where c is the speed of light in a vacuum and v_m is its speed in the medium. The index of refraction for air is 1.0003, water is 1.333, glass is 1.5 to 1.6, and so on. (Note that n_m is not necessarily an integer and is always > 1.)

Figure Q2.15 shows plane waves of light crossing the boundary between two transparent media. Let us say that the speed of light in the upper medium is v_1 but $v_2 < v_1$ in the lower medium. The plane waves approach the boundary in such a way that their direction of travel makes an angle of θ_1 with the respect to a line perpendicular to the boundary between the surfaces. Let us focus our attention on a certain wave crest that at $t = 0$ is just hitting the boundary at point A. The wave crest at point B will hit the boundary at point C a certain time Δt later such that $v_1 \Delta t$ is equal to the distance between points B and C. In the meantime, the Huygens wavelet from point A will have moved outward a distance of $v_2 \Delta t$ in the lower medium. The wave front must therefore go through point C and be tangent to the wavelet from point A. You can see from the diagram that this wave front after time Δt has been twisted somewhat from its original direction: the wave front's direction of motion now makes a smaller angle θ_2 with respect to the direction perpendicular to the boundary. Refraction is this bending of the direction of the motion of the light wave as it moves across the boundary.

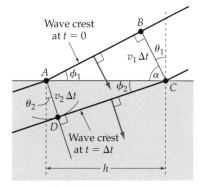

Figure Q2.15

A wave crest being refracted as it moves from a medium where its speed is v_1 to one where its speed is v_2. The arrows indicate the crest's direction of motion.

Use figure Q2.15 to determine the relationship between θ_1, θ_2, n_1, and n_2. (*Hint:* The relationship will involve the sines of the angles and should involve *only* θ_1, θ_2, n_1, and n_2.)

Advanced

Q2A.1 Consider a two-slit interference experiment where you measure the amplitude of the combined wave a certain constant distance D from the center of the slits. Using the superposition principle, show that the squared amplitude of the combined wave as a

function of θ is given by

$$A^2(\theta) = A^2 \cos^2 \psi \qquad \text{where } \psi = \pi \frac{d \sin\theta}{\lambda} \quad \text{(Q2.17)}$$

and A is some constant. (The wave intensity is proportional to its squared amplitude.) You may assume that the circular waves created by each slit have approximately constant amplitude at all angles. (*Hint:* An appropriate trigonometric identity might help.)

ANSWERS TO EXERCISES

Q2X.1 Let's call the angle between the x axis and the line QR angle α (see figure Q2.16). We are told that the angle $\angle PQR \approx 90°$. If this is so, then $\theta + \alpha \approx 90°$. But by construction, $\phi + \alpha = 90°$ as well. Subtracting these equations, we get $\theta - \phi \approx 0$, implying that $\theta \approx \phi$.

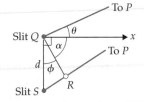

Figure Q2.16

See the answers to exercises Q2X.1 and Q2X.2.

Q2X.2 Consider figure Q2.16. We are given that $\theta + \alpha = 89.6°$, while $\phi + \alpha = 90°$ still. Subtracting these equations, we get $\phi - \theta = 0.4°$, so these angles are only 0.4° different.

Q2X.3 *Translation:* Figure Q2.17 shows a picture of this situation. Define $y = 0$ to be the point on the line that is directly opposite the speakers: to answer the

question, we essentially need to find positions y for the chair such that the sound should be loudest or almost disappear.
Model: We will assume that wave crests are emitted simultaneously from the speakers at a rate of 880 crests per second and diffract into essentially circular waves centered on the speakers. We can use equation Q1.6 to calculate the wavelength of the sound waves here:

$$\lambda = \frac{v}{f} = \frac{343 \text{ m/s}}{880 \text{ /s}} = 0.39 \text{ m} \quad \text{(Q2.18)}$$

According to equation Q2.1, the angles along which constructive interference will take place are given by $\theta_{nc} = \sin^{-1}(n\lambda/d)$, and the positions along the line at these angles will have values of y_n such that $\tan\theta_{nc} = y_n/D$. Since we know λ, d, and D, we can find θ_{nc} and y_n for any n.
Solution: According to equation Q2.1, the angles for maximally constructive interference are

$$n = 0: \qquad \theta_{0c} = 0$$

$$n = \pm1: \qquad \theta_{\pm1c} = \sin^{-1}\left(\frac{\pm\lambda}{d}\right)$$
$$= \sin^{-1}\left(\frac{\pm0.39 \text{ m}}{1.5 \text{ m}}\right)$$
$$= \sin^{-1} 0.26$$
$$= \pm15°$$

$$n = \pm2: \qquad \theta_{\pm2c} = \sin^{-1}\left(\frac{\pm2\lambda}{d}\right) \quad \text{(Q2.19)}$$
$$= \sin^{-1}(\pm0.52)$$
$$= \pm31°$$

$$n = \pm3: \qquad \theta_{\pm3c} = \sin^{-1}\left(\frac{\pm3\lambda}{d}\right)$$
$$= \sin^{-1}(\pm0.78)$$
$$= \pm51°$$

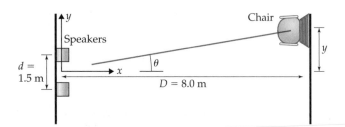

Figure Q2.17
Stereo speakers broadcasting a tone into open air.
(See the answer to exercise Q2X.3.)

The value of $4\lambda/d$ is greater than 1.0, so the series ends here. The positions along the line are then

$n = 0$: $y_0 = D\tan 0 = 0$

$n = \pm 1$: $y_{\pm 1} = D\tan(\pm 15°) = (8.0\text{ m})(\pm 0.27)$
$= \pm 2.1$ m

$n = \pm 2$: $y_{\pm 2} = D\tan(\pm 31°) = (8.0\text{ m})(\pm 0.60)$ (Q2.20)
$= \pm 4.8$ m

$n = \pm 3$: $y_{\pm 3} = D\tan(\pm 51°) = (8.0\text{ m})(\pm 1.23)$
$= \pm 9.9$ m

These are the positions where the sound would seem to be loudest (that is, where the waves constructively interfere maximally). The positions where the sound would appear to be softest (where the waves destructively interfere) would be angles where

$$d\sin\theta = n\lambda + \tfrac{1}{2}\lambda \qquad (Q2.21)$$

(so that the waves from each slit arrive a half-wavelength out of phase). The angles satisfying this equation are $\sin^{-1}(\pm\tfrac{1}{2}\lambda/d) = \sin^{-1}(\pm 0.13) = \pm 7.5°$, $\sin^{-1}[\pm(3/2)\lambda/d] = \sin^{-1}(\pm 0.39) = \pm 23°$, $\sin^{-1}[\pm(5/2)\lambda/d] = \sin^{-1}(\pm 0.65) = \pm 40.5°$, and $\sin^{-1}[\pm(7/2)\lambda/d] = \sin^{-1}(\pm 0.91) = \pm 65.5°$. The corresponding positions are ± 1.05 m, ± 3.4 m, ± 6.8 m, and ± 17.6 m.

Evaluation: Note that the two inner positions are almost exactly halfway between the points listed in equation Q2.20. The discrepancy gets larger as the angle gets larger because even though $n\lambda/d$ increases linearly with n, $y_n = D\tan[\sin^{-1}(n\lambda/d)]$ does not.

Q2X.4   *Model:* The small-angle approximation should still apply, so we can still use equation Q2.6. We have not changed the laser, so we should still have $\lambda = 630$ nm, and D has not changed either. Therefore we can solve equation Q2.6 for s.
Solution: Doing this, we find that

$$s \approx \frac{\lambda D}{d} = \frac{(630 \times 10^{-9}\text{ m})(3.0\text{ m})}{0.03 \times 10^{-3}\text{ m}}$$
$$= 6.3 \times 10^{-2}\text{ m} = 6.3\text{ cm} \qquad (Q2.22)$$

Q2X.5   As shown below, as the slits get closer together, the angle that yields one wavelength's worth of length

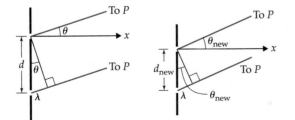

difference between the two paths to point P will increase. This increase in the angle implies an increase in the separation of bright spots on the screen.

Q2X.6   According to equation Q2.6, the spacing s will be

$$s = \frac{\lambda D}{d} = \frac{(510 \times 10^{-9}\text{ m})(3.0\text{ m})}{0.03 \times 10^{-3}\text{ m}}$$
$$= 5.1\text{ cm} \qquad (Q2.23)$$

Q2X.7   Figure Q2.18 shows the construction. We see that the tangent to the wavelets produced by the points on the original wave front is a larger ring centered on the same point: therefore a circular wave will remain circular according to Huygens's principle.

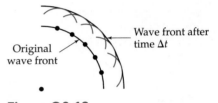

Figure Q2.18
Wavelets from a circular wave crest form a larger circular wave after some time has passed (see the answer to exercise Q2X.7).

Q2X.8   Note that the distance between point 1 and point 4 is $d = \tfrac{1}{4}a$. When $\sin\theta_{2d} = 2\lambda/a$, then the extra distance the wavelet from 4 has to go to get to P above the distance that the wavelet from 1 has to go to get to P is $s = d\sin\theta_{2d} = (\tfrac{1}{4}a)(2\lambda/a) = \tfrac{1}{2}\lambda$. Therefore the wavelets from these points will cancel when they get to point P. Similarly, the wavelet from 2 will cancel the wavelets from 5, those from 3 will cancel those from 6, and so on. Each point in the upper quarter of the slit is canceled by the corresponding point in the next quarter, and the same applies to the quarters below the middle. Thus the wavelets cancel in pairs, and we have completely destructive interference at this angle.

Q2X.9   Assuming the small-angle approximation is valid here, we can do pretty much the same thing that we did in equation Q2.11:

$$\frac{w}{D} = 2\frac{\tfrac{1}{2}w}{D} = 2\tan\theta_{1d} \approx 2\theta_{1d}$$
$$= 2\left(1.22\sin^{-1}\frac{\lambda}{a}\right) \approx \frac{2.44\lambda}{a}$$
$$\Rightarrow \quad w \approx \frac{2.44\lambda D}{a}$$
$$= \frac{2.44(510 \times 10^{-9}\text{ m})(180{,}000\text{ m})}{5 \times 10^{-3}\text{ m}} = 45\text{ m} \qquad (Q2.24)$$

Q3

The Particle Nature of Light

Chapter Overview

Introduction

In chapter Q2 we saw that light displays interference and diffraction effects, which provide compelling evidence that light is a wave. In this chapter we will see, however, that there is also very compelling evidence that light consists of particles.

Section Q3.1: The Photoelectric Effect

In the late 1800s, physicists discovered that ultraviolet light shining on a metal could dislodge electrons from the metal, an effect physicists called the **photoelectric effect.** This effect has a number of technological applications in video cameras, image intensifiers, and sensitive photodetectors.

Section Q3.2: Idealized Photoelectric Experiments

We might quantitatively explore the photoelectric effect by placing two metal plates in close proximity and illuminating one with light. If we connect an ammeter to the plates, we can measure the current that flows between them due to ejected electrons. If we connect a voltmeter across the plates, we can measure the potential difference between the plates. As electrons are ejected by the light from one plate (the **cathode**), some will make it to the other plate (the **anode**), making both plates charged. This charging process will continue until the electric field between the plates is so large that no electrons have enough energy to make it all the way to the other plate before being turned back by the field. If the voltmeter has essentially infinite resistance, the measured potential difference will be directly proportional to the maximum kinetic energy that an electron has when it leaves the plate. Indeed, if we express an electron's kinetic energy in electronvolts (abbreviated eV), then the maximum potential difference (in volts) has the same numerical value as the maximum electron kinetic energy (in electronvolts).

Experiments such as these can determine how the number and energy of electrons ejected depend on the intensity and wavelength of the light and/or characteristics of the metal.

Section Q3.3: Predictions of the Wave Model

If light really is a wave, then (as discussed in detail in the section) we can make the following predictions about the outcomes of such experiments:

1. At sufficiently high intensities, the rate at which electrons are ejected will be proportional to the intensity of the light.
2. At very low light intensities, there will be a significant delay between the illumination of the plate and the emission of electrons (maybe several seconds).
3. The rate of electron ejection might depend in a complicated and metal-specific way on the frequency of the light (due to resonance effects).
4. The maximum kinetic energy of the ejected electrons is generally likely to increase with intensity.

The first two predictions follow from very basic considerations involving conservation of energy.

Section Q3.4:   Confronting the Facts

When the experiments are actually performed, the following results emerge:

1. At sufficiently high intensities and at a fixed light frequency, the number of ejected electrons is proportional to the intensity (as predicted).
2. However, electrons are ejected essentially *instantly* (delays of less than 10^{-8} s), *no matter how low the intensity is* (!).
3. If the intensity is held constant, the number of electrons ejected *decreases* with increasing frequency in a similar way for all metals.
4. If the light frequency is below a certain value, *no* electrons are ejected, no matter how intense the light is! (Physicists call this metal-specific value the metal's **cutoff frequency.**)
5. If the light frequency is held constant, the electrons' maximum kinetic energy is *independent* of intensity!
6. Above the cutoff frequency, the electrons' maximum kinetic energy is directly *proportional* to the frequency of the light, with the same slope for *all* metals.

As discussed in the section, most of these observations directly contradict the wave model.

Section Q3.5:   The Photon Model of Light

In 1905, Einstein proposed a particle model of light to address this problem. In this model, Einstein claimed that light consists of particles (later called **photons**) whose energy was given by

$$E = \frac{hc}{\lambda} \qquad \text{(Q3.7)}$$

Purpose: This equation specifies the energy E carried by a single photon of monochromatic light whose collective wavelength is λ.
Symbols: h is Planck's constant, and c is the speed of light.
Limitations: This expression only applies to *photons*.
Note: The quantity hc has an easily remembered value in units of electronvolts times nanometers: $hc = 1240 \text{ eV} \cdot \text{nm}$.

Since $f = c/\lambda$ for light, equation Q3.7 also means that $E = hf$. Einstein proposed that each electron ejected from the metal received its energy by absorbing a single photon, and that the kinetic energy K of the ejected photon was simply equal to $E - W$, where W is the metal's work function (i.e., the energy required simply to remove the electron from the metal). This model provided a simple and straightforward explanation for all the observed features of the photoelectric effect.

In spite of its simplicity, physicists resisted this model for a long time, because it did not offer an explanation for the wavelike aspects of light discussed in chapter Q2. We will begin to deal with this issue in chapter Q5.

Section Q3.6:   Detecting Individual Photons

This section describes a device called a **photomultiplier** tube that uses the photoelectric effect to register individual photons, vividly illustrating the particle nature of light.

Q3.1   The Photoelectric Effect

In chapter Q2, we saw compelling evidence that light consists of waves. For nearly the entire 19th century, physicists considered this evidence conclusive, and the wave nature of light was never in serious doubt. But in the first decade of the 1900s, evidence began to emerge that in certain circumstances, it was more useful to treat light as a *particle*.

In 1901, for example, Max Planck showed that if one modeled a glowing hot object as emitting light energy in discrete bundles instead of continuously as waves, it would resolve a nasty problem physicists were having with matching the observed emissions from a glowing body with theoretical predictions. Planck and most other physicists of the time, though, considered this just a handy calculation trick that approximated some wave process not yet fully understood. But other evidence of the particle nature of light soon came from observations of what physicists called the *photoelectric effect*. This effect is somewhat easier to understand than the issue that Planck was worried about, so we will begin our study of the particle nature of light with an examination of this effect.

A basic description of the photoelectric effect

The photoelectric effect was uncovered in bits and pieces by several physicists in the latter years of the 19th century, and as a result, it is not easy to credit any single person with its discovery. In 1887, Heinrich Hertz (who, ironically, was working at the time on experiments that verified important aspects of Maxwell's theoretical model of electromagnetic waves) first observed that ultraviolet light shining on metallic electrodes made it easier for a spark to jump between them. Further work by J. J. Thomson, Philipp Lenard, and others established by about 1900 that many kinds of metal surfaces actually emit electrons when illuminated by light. The ability of light to dislodge electrons from a metallic surface is called the **photoelectric effect.**

Technological applications

Quite aside from its importance in physics (which we will discuss shortly), the photoelectric effect has a wide variety of technological applications. The electrons ejected by light incident on a surface can be electronically detected and the resulting signals amplified. The photoelectric effect is thus exploited in a wide variety of devices that electronically detect, measure, and record visual images. A standard analog video camera, for example, employs the photoelectric effect to create an electronic "image" of the scene presented to it. Image intensifiers (which are used in astronomical and military applications to enhance a very dim image) use the photoelectric effect to convert the image to an electronic equivalent, which is then amplified before being reconverted to a visual image. The most sensitive photodetectors available also employ the photoelectric effect (as we'll see in section Q3.6).

Q3.2   Idealized Photoelectric Experiments

As soon as the effect was qualitatively described, physicists began investigating its quantitative features. How many electrons are ejected in a given time? How does this number depend on the wavelength (i.e., the color) of the light and its intensity (i.e., power delivered to a unit area on the metal)? How energetically are the electrons ejected, and upon what does the ejection energy depend? Are electrons ejected essentially instantly after the surface was illuminated, or is there a measurable time delay?

Figure Q3.1 illustrates a pair of idealized experiments that can be performed to answer these questions. In both of these experiments, light falls on

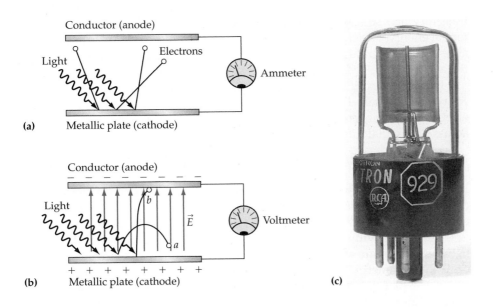

(a)

(b)

(c)

Figure Q3.1
(a) An idealized apparatus for studying the rate at which electrons are dislodged from a metallic plate by light. (b) An idealized apparatus for measuring the kinetic energy of the liberated electrons. (c) A practical photoelectric sensor contains a curved cathode and a central wirelike anode, mounted in a vacuum tube. This makes it easy for the light to get to the cathode.

a metallic plate, liberating electrons. (We call this plate a **cathode,** a word which essentially means "something that acts as a source of electrons.") Many of these electrons are collected by a nearby conducting plate (called an **anode,** which means "something that collects electrons").

In the first experiment, these plates are connected by an ammeter having essentially no resistance. Electrons collected by the anode thus freely flow back to the cathode through the ammeter, which measures the number of electrons flowing through it per second and thus reflects the number of electrons collected by the anode every second. While not every electron liberated from the cathode will end up at the anode, the number collected by the anode should at least be proportional to the number actually liberated from the cathode. This apparatus can thus be used to explore how the number of electrons liberated per unit time depends on the intensity or color of the light, and how long it takes for the electrons to be liberated (i.e., how long light has to shine on the cathode before a current begins to flow through the meter).

An experiment that investigates the rate of electron ejection

In the second experiment, the two plates are connected by a voltmeter that has essentially infinite resistance. This means that any electrons collected by the anode must remain there. As light liberates electrons from the cathode, the cathode becomes more and more positively charged, while the anode becomes negatively charged. This creates an electric field between the plates in the direction shown in figure Q3.1b.

An experiment that investigates the maximum electron ejection energy

This electric field makes it harder for liberated electrons to reach the anode. Whether a given electron is able to make it to the anode against this developing electric field depends on how much kinetic energy the electron has available to convert to electrical potential energy. If the potential difference between the plates is $\Delta\phi = 2.0$ V, for example, the difference between an electron's electrostatic potential energy at the anode and its potential energy at the cathode has a magnitude of

$$|q\,\Delta\phi| = (1.6 \times 10^{-19}\text{ C})(2.0\text{ V}) = 3.2 \times 10^{-19}\text{ J} \qquad (Q3.1)$$

remembering that $1\text{ V} \equiv 1\text{ J/C}$. An electron will therefore be able to make it to the anode across this potential difference only if it has an original kinetic energy that is greater than this value. (Even if it does, it might not make it to the anode if its trajectory is too flat: because of its horizontal motion, the electron marked a in figure Q3.1b has a large horizontal speed at the peak of its trajectory, and so was not able to convert all its original kinetic energy to potential energy. The electron marked b was emitted more vertically and so was able to convert enough of its kinetic energy to potential energy to make it to the anode.)

In this context it is natural to measure electron energies in electronvolts. As you may recall, 1.0 eV of energy is defined to be equal to the change in electrical potential energy that an electron experiences when it is transported across a potential difference of 1.0 V:

$$1 \text{ eV} \equiv (1.6 \times 10^{-19} \text{ C})(1.0 \text{ V}) = 1.6 \times 10^{-19} \text{ J} \qquad (Q3.2)$$

In the case at hand, we could say that if the potential difference between the cathode and anode is 2.0 V, then an electron must have an initial kinetic energy of at least 2.0 eV at the cathode to make it all the way to the anode.

The anode will stop collecting electrons when the potential difference between the anode and cathode becomes so great that *no* electron has sufficient kinetic energy to make it to the anode. The potential difference between the plates cannot increase beyond this point, since electrons can no longer travel from one plate to the other. The final potential difference between the plates is thus a direct measure of the maximum kinetic energy of the ejected electrons.

Exercise Q3X.1

In a certain experiment of the type shown in figure Q3.1b, the final potential difference between the plates is measured to be 3.32 V. What is the maximum kinetic energy of the ejected electrons (in electronvolts and joules)?

Exercise Q3X.2

Why does the potential difference developed between the plates in the experiment shown in figure Q3.1b reflect the *maximum* kinetic energy of the ejected electrons and not, say, the *average* kinetic energy?

Q3.3   Predictions of the Wave Model

What does the wave model of light predict about the results of such experiments? To answer this question, we have to imagine how an electromagnetic wave might eject an electron from the metal. A passing electromagnetic wave causes the electric field in the electron's vicinity to oscillate, which will cause the electron to oscillate in response. Now, the intensity I of an electromagnetic wave expresses the energy delivered per unit time to a unit area perpendicular to the beam. We saw in chapter E16 that the intensity of a sinusoidal electromagnetic wave is $I = cE_0^2/4\pi k$, where E_0 is the electric field amplitude. Thus, the more intense the light is, the stronger its electric field and thus the more violent the electron oscillations it will cause.

How the rate of electron ejection should depend on light intensity

We can even make quantitative predictions about this based on conservation of energy. A typical atom has a diameter $d \approx 0.1$ nm. Let us assume that all the energy falling on an atom gets channeled to one of its electrons. The approximate area on the surface corresponding to each atom is roughly d^2. If an atom has to absorb at least an energy E_{eject} from the light to be ejected, then the *maximum* possible rate at which electrons could be ejected will depend on the light intensity as follows:

$$\frac{\text{Electrons ejected}}{\text{Time}} = \frac{(\text{energy delivered per electron})/(\text{time})}{\text{energy needed per electron}}$$

$$= \frac{\dfrac{\text{energy delivered}}{\text{time} \cdot \text{area}} \cdot (\text{area per electron})}{\text{energy needed per electron}} \approx \frac{Id^2}{E_{\text{eject}}} \quad (Q3.3)$$

We see that the electron ejection rate will be proportional to the light intensity.

Time delay for electron ejection at low intensities

If the light intensity is extremely low, this analysis also leads us to expect that it may take a long time for an electron to accumulate enough energy to break out of the metal. If an atom has to absorb some energy E_{eject} to eject an electron, and energy Id^2 is delivered to that atom's area per unit time, then the time required for an atom to absorb enough energy to eject an electron will be roughly

$$\frac{\text{Time}}{\text{Electron}} \approx \frac{E_{\text{eject}}}{Id^2} \quad (Q3.4)$$

In a typical experimental situation, this time could be several seconds (see exercise Q3X.4), so in such a case we would not expect *any* electrons to be ejected until several seconds have passed.

Other predictions

These conclusions are based only on the principle of conservation of energy and the fact that the energy in a wave is spread essentially evenly throughout the wave. To make predictions about the average kinetic energy that an electron would have after being ejected or the dependence of the ejection rate on frequency, we need a more detailed model of exactly how the electrons are ejected. We saw in chapter Q1 that oscillating systems (such as a child on a swing) accumulate energy from an external disturbance most effectively if the frequency of the disturbance is close to a resonant frequency of the system. If this model applies to this situation, we might expect to see that the number and energy of ejected electrons depend in a complicated and metal-specific way on the frequency of light used.

It is plausible as a general trend, however, that the kinetic energy of the ejected electrons will increase as the intensity of the wave increases, because the stronger electric fields in intense light will generally cause a more violent response. As an analogy, we might consider debris and spray thrown around by an ocean wave crashing on some rocks: the kinetic energy of this ejecta clearly will get larger as the wave gets bigger.

Basic predictions of the wave model

In short, conservation of energy and the basic principle that energy is spread continuously throughout a wave imply that

1. At sufficiently high intensities, the rate at which electrons are ejected will be proportional to the intensity of the light.
2. At very low light intensities, there will be a significant delay between the illumination of the plate and the emission of electrons.

Predictions that may depend on details of the model

In addition, the following conclusions are probable, but may depend on the microscopic details of how the wave actually ejects the electrons:

3. The rate of electron ejection might depend in a complicated and metal-specific way on the frequency of the light (due to resonance effects).
4. The maximum kinetic energy of the ejected electrons is generally likely to increase with increasing intensity (and may also depend in a complicated way on the light's frequency).

Exercise Q3X.3

Which of the two experimental setups shown in figure Q3.1 would we use to check the first prediction?

Exercise Q3X.4

Imagine that a lamp radiates 40 W of power at an average wavelength of 590 nm. If such a light is placed about 1.0 m away from a zinc plate, the intensity of light falling on the plate from the lamp ≈ 3.2 W/m^2. If we assume that zinc atoms on the surface of the plate are about 0.1 nm apart, about how much energy (on the average) falls on a single zinc atom every second? Assume that *all* the energy falling on it eventually goes into ejecting one electron, and it takes a few electronvolts of energy to pry the electron out of the metal (as suggested by a variety of other considerations). About how long would it take for an atom to accumulate enough energy to eject one electron?

Q3.4   Confronting the Facts

The actual experimental results

When the experiments are actually performed, the following results emerge:

1. At sufficiently high intensities and at a fixed frequency, the number of ejected electrons is proportional to the intensity (as predicted).
2. However, electrons are ejected essentially *instantly* (delays of less than 10^{-8} s), *no matter how low the intensity is* (!).
3. If the intensity is held constant, the number of electrons ejected *decreases* with increasing frequency in a similar way for all metals.
4. If the frequency is below a certain value, *no* electrons are ejected, no matter how intense the light is! (Physicists call this metal-specific value the metal's **cutoff frequency.**)
5. If the frequency is held constant, the electrons' maximum kinetic energy is *independent* of intensity!
6. Above the cutoff frequency, the electrons' maximum kinetic energy is directly *proportional* to the frequency of the light, with the same slope for *all* metals (see figure Q3.2).

Instantaneous ejection is a problem for wave model

Only the first of these results is consistent with the wave model of light. The other results are troublesome at best and flatly contradict the wave model at worst. Observation 2 (that electrons start coming off within 10 ns) is particularly troubling, particularly in low-intensity experiments where the delay should be many seconds. If the energy of the wave is truly dispersed continuously throughout the wave (as Maxwell's wave theory would imply), there is *no way* that an electron could gather the energy required to become ejected in such a short time.

Exercise Q3X.5

Consider again the low-intensity experiment described exercise Q3X.4. How many atom's worth of light energy would one zinc atom have to grab if it is to be able to eject an electron within 10 ns (10^{-8} s) of the light reaching the zinc plate? Does this seem possible?

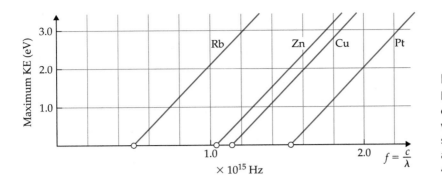

Figure Q3.2
Plot of maximum electron kinetic energy versus frequency for various types of metals. The slope of the line is the same for all metals and has a value $h = 4.15 \times 10^{-15}$ eV · s.

Also troubling are observations 5 and 6. No matter how we model the interaction of a light wave with electrons in the metal, we would not expect the electrons' maximum kinetic energy to be *independent* of intensity. That is like saying that a 50-ft tsunami crashing on the beach will do exactly the same damage that a 1-ft ankle splasher would. The strictly linear, substance-independent frequency dependence of the kinetic energy suggests the existence of a simple physical explanation (that would seem to have to do more with light itself than the metal), but the wave model does not offer anything promising. Physicists in the early 1900s spent quite a bit of effort looking, but were simply unable to come up with any modification of the wave model that could explain these basic observations.

Also hard-to-understand results on electron energy

Q3.5   The Photon Model of Light

Not all these facts were known in 1905: observations 2, 4, and 5 were the only things known and these only qualitatively. On the basis of this information and some other observations concerning the light emitted by heated objects, Albert Einstein in 1905 proposed a simple alternative to the wave model of light that explained these qualitative observations and essentially predicted the other observations listed above. These predictions were experimentally verified by Robert Millikan and described in a paper that he published in 1916. It is interesting to note that Einstein won the Nobel Prize in 1921 not for either special relativity or general relativity but rather "for his services to theoretical physics and especially for his discovery [sic] of the photoelectric effect."[†]

Einstein's model was simply this. Imagine that light consists of *particles* (which were later called **photons**), each of which carries a certain amount of energy. The energy of each photon in a beam of monochromatic light is

The energy carried by a photon

$$E = hf \qquad (Q3.5)$$

where f is the frequency of the light and h is a universal constant (equal to the slope of the graph in figure Q3.2). The value of h measured by Millikan in 1916 was 6.57×10^{-34} J · s. This constant is now called **Planck's constant,** and its currently accepted value is 6.63×10^{-34} J · s $= 4.15 \times 10^{-15}$ eV · s.

We commonly express different colors of visible light in terms of wavelengths instead of frequencies (partly because the frequencies of visible light waves are inconveniently high). Since the frequency f of a light wave is related to its speed c and wavelength λ by

$$f = \frac{c}{\lambda} \qquad (Q3.6)$$

[†]Quoted in A. Pais, *Subtle Is the Lord*, Oxford, 1982, p. 527.

we can rewrite equation Q3.5 in terms of the wavelength of light as follows:

$$E = \frac{hc}{\lambda} \qquad (Q3.7)$$

Purpose: This equation specifies the energy E carried by a single photon of monochromatic light whose collective wavelength is λ.
Symbols: h is Planck's constant, and c is the speed of light.
Limitations: This expression only applies to *photons*.

The quantity hc has a conveniently sized and easily remembered value when we express it in units of electronvolts times nanometers:

$$hc \approx 1240 \, \text{eV} \cdot \text{nm} \qquad (Q3.8)$$

Electrons ejected by interaction with a single photon

Einstein's model further proposed that each electron was ejected from the metal as a result of a collision with (and subsequent absorption of) a *single* photon. The maximum kinetic energy K that an electron could have just after leaving the metal must therefore be

$$K = hf - W = \frac{hc}{\lambda} - W \qquad (Q3.9)$$

where $hc/\lambda = hf$ is the energy that the electron got from the photon and W is the **work function** for the metal, defined to be the minimum amount of energy required to liberate the electron from the metal (this roughly corresponds to the quantity we called E_{eject} in section Q3.3).

The photon model nicely explains results here

This model neatly explains *all* the experimental evidence listed in section Q3.4, as we shall see. The intensity of light is defined to be the amount of energy delivered to a unit area in a unit time. If the light has a fixed wavelength, the amount of energy carried by each photon is the same, so the light intensity will be directly proportional to the number of photons delivered per unit area per unit time. If each ejected electron is booted as a result of absorption of a single photon, then the number of ejected electrons should be directly proportional to the intensity of the light at a given frequency. This was observation 1 on our list of actual experimental results.

Since only a single photon is required to liberate an electron, electrons can start coming out of the plate as soon as photons arrive at the plate, even if the intensity of the light is so low that only a handful of photons arrive every second. This explains (without violating the conservation of energy) how electrons can be liberated instantly (observation 2) even at very low intensities.

Equation Q3.9 is also nicely consistent with the graph shown in figure Q3.2. For light of sufficiently high frequency f, the energy of the ejected electron depends linearly on f, with a slope h that is independent of the type of metal involved. On the other hand, if the frequency is so low that $hf < W$, then the absorption of a photon will not provide sufficient energy for the electron to escape the metal and no electrons will be observed. So the place at which the graph of K versus f intersects the horizontal axis is simply the point where $hf = W$, and this point should therefore depend on the type of metal used. This explains observations 4 and 6 on our list.

Exercise Q3X.6

How does the photon model explain observation 5, that the maximum kinetic energy of electrons ejected from a plate is *completely independent* of the intensity of the light falling on it (contrary to expectations)?

Exercise Q3X.7

Check that the value of hc quoted in equation Q3.8 is consistent with the values of h quoted below equation Q3.5.

Exercise Q3X.8

Green light has a wavelength of about 540 nm. What is the energy carried by a single photon in such light?

Einstein's model explains the experimental results from experiments on the photoelectric effect neatly and easily. Yet even after Millikan and others verified the predictions of this model, most physicists were very reluctant to embrace it. It was really not until 1923, when Arthur Compton verified that X-rays (which are EM waves with very short wavelengths) were scattered by electrons exactly as if they were tiny particles (see problem Q3A.1), that physicists began to really take the photon model of light seriously. Why was this resistance so strong?

The problem is that *Einstein's model is completely inconsistent with the results of experiments that illustrate the wave nature of light.* For example, no simple particle model is able to explain the behavior of light in the Young two-slit interference experiment, while the wave model does so quite admirably. Particularly when one recognizes that such an interference pattern can easily be displayed with a minimal amount of equipment, while the detailed measurements required to verify the photon model of light are hard to make (and were at the time subject to doubt), it is easy at second glance to forgive physicists who were reluctant to accept the photon model of light in the early decades of the 20th century.

Even the photon model implicitly recognizes the reality of the wave nature of light: the energy of a photon is expressed in terms of the frequency (or alternatively, the wavelength) of the light involved, concepts that would have no meaning in a strict particle model of light. We have to face the fact that neither a wave model nor a particle model is able to explain the behavior of light *completely*. The problem of reconciling these models lies at the very heart of quantum mechanics, and it will be the focus of our attention in chapters Q4 and Q5.

Photon model does not explain wave effects

Neither model adequately explains light by itself

Example Q3.1   Photoelectrons from Aluminum

Problem   When ultraviolet light of wavelength 250 nm falls on an aluminum surface, the measured maximum kinetic energy of ejected electrons is 0.88 eV. When light of wavelength 210 nm falls on the same plate, the maximum kinetic energy is 1.82 eV. Check that the value of hc implied by these data is indeed 1240 eV · nm, and find the value of the work function W for aluminum.

Translation   Let us define the following symbols: K_1 and λ_1 for the maximum kinetic energy and wavelength of the light in the first case, and K_2 and λ_2 for the same quantities in the second case.

Model   According to equation Q3.9, we have

$$K_1 = \frac{hc}{\lambda_1} - W \quad \text{and} \quad K_2 = \frac{hc}{\lambda_2} - W \qquad (Q3.10)$$

This provides the two equations in two unknowns that we need to solve the problem.

Solution   Subtracting the first of equations Q3.10 from the second, we get

$$K_2 - K_1 = \frac{hc}{\lambda_2} - \frac{hc}{\lambda_1} = hc\left(\frac{1}{\lambda_2} - \frac{1}{\lambda_1}\right) \qquad (Q3.11)$$

Solving this for hc, we find that

$$hc = \frac{K_2 - K_1}{1/\lambda_2 - 1/\lambda_1} = \frac{(1.82 - 0.88)\,\text{eV}}{(210\,\text{nm})^{-1} - (250\,\text{nm})^{-1}} = 1240\,\text{eV}\cdot\text{nm} \quad (Q3.12)$$

We can now find W from either equation in equation Q3.10. Choosing the first, we get

$$W = \frac{hc}{\lambda_1} - K_1 = \frac{1240\,\text{eV}\cdot\text{nm}}{250\,\text{nm}} - 0.88\,\text{eV} = 4.08\,\text{eV} \qquad (Q3.13)$$

The wavelength of light whose photon exactly equals this energy is

$$W = \frac{hc}{\lambda} \quad\Rightarrow\quad \lambda = \frac{hc}{W} = \frac{1240\,\text{eV}\cdot\text{nm}}{4.08\,\text{eV}} = 304\,\text{nm} \qquad (Q3.14)$$

Evaluation   The result in equation Q3.12 is as expected. The result in equation Q3.14 is well within the ultraviolet range of wavelengths, meaning that only ultraviolet photons with wavelengths shorter than this can eject electrons from an aluminum surface. Many common metals (such as copper, silver, lead, and iron) have work functions in this range. Lithium, sodium, and cesium are among the few metals whose electrons can be ejected by photons of visible light.

Example Q3.2   How Many Photons?

Problem   A 100-W incandescent lightbulb uses 100 W of electrical power but only radiates about 10 W to 15 W of actual visible light. Roughly how many visible photons per second hit the open pages of a typical hardcover book if the pages are about 2 m from the bulb and face it directly?

Model   We have to make a variety of estimations and approximations to solve this problem. To find the number of photons hitting the pages each second, we have to know the light energy hitting the pages per second and the energy per photon. We could compute the latter if we knew the wavelength of the light, but the yellow-white visible light emitted by a normal incandescent bulb is a mix of wavelengths. Let us estimate that the average wavelength of the visible light is about 590 nm, which is in the yellow region of the spectrum. This means that the average energy per photon is about

$$E_{\text{ph}} = \frac{hc}{\lambda} = \frac{1240\,\text{eV}\cdot\text{nm}}{590\,\text{nm}} = 2.1\,\text{eV}\left(\frac{1.6 \times 10^{-19}\,\text{J}}{1\,\text{eV}}\right) = 3.4 \times 10^{-19}\,\text{J} \quad (Q3.15)$$

To convert this to a number of photons, we need to know the energy per second that falls on the pages. Let's assume that the light energy from the bulb travels uniformly in all directions. Imagine a sphere centered on the lightbulb with a radius $R = 2$ m. Every second, 15 J or so of visible light energy crosses this sphere. If the light energy is spread uniformly over the sphere, then the intensity of the visible light at any point on the inner surface of the sphere will be

$$I = \frac{P_{\text{lamp}}}{4\pi R^2} \qquad \text{(Q3.16)}$$

where P_{lamp} is the power of the visible light emitted by the lamp (10 W to 15 W) and $4\pi R^2$ is the area of the inner surface of the sphere.

The book pages essentially lie on the inner surface of this imaginary sphere, so if they have area A, the energy per unit time P_{book} that falls on them will be

$$P_{\text{book}} = IA = P_{\text{lamp}} \frac{A}{4\pi R^2} \qquad \text{(Q3.17)}$$

Taking a typical hardcover book from my shelf and measuring it with a ruler, I find that the two facing pages form a rectangle about 18 cm tall and 24 cm wide, so $A = (0.18 \text{ m})(0.24 \text{ m}) = 0.058 \text{ m}^2$.

Finally, to find the number of photons per second falling on these pages, we have to divide the energy per second falling on them by the energy per photon:

$$\frac{\text{photons}}{\text{s}} = \frac{P_{\text{book}}}{E_{\text{ph}}} = \frac{P_{\text{lamp}}}{E_{\text{ph}}} \frac{A}{4\pi R^2} \qquad \text{(Q3.18)}$$

Since we now know all the quantities on the right side of equation Q3.18, we can calculate the number of photons per second.

Solution   Plugging in the numbers, we get

$$\frac{15 \text{ J/s}}{3.4 \times 10^{-19} \text{ J/photon}} \frac{0.058 \text{ m}^2}{4\pi (2.0 \text{ m})^2} = 5 \times 10^{16} \text{ photons/s} \qquad \text{(Q3.19)}$$

Evaluation   I have only quoted this result to one significant digit, because our approximations and estimations probably yield an uncertainty of more than one-third or about $\pm 2 \times 10^{16}$ photons/s. Still, this is a useful ballpark estimate.

Q3.6   Detecting Individual Photons

The experimental results described in section Q3.5 constitute rather indirect evidence of the particle nature of light. At the time of Millikan's verifying experiment in 1916, no more direct evidence was available. It is now possible, however, to build a detector capable of registering the absorption of a single photon. This is no small task: photons of visible light carry only a tiny amount of energy (roughly 2 eV $\approx 3 \times 10^{-19}$ J). This can be done, though, with a device known as a **photomultiplier**, whose design is illustrated schematically in figure Q3.3.

This device works as follows. A single photon strikes a very thin metal plate, called the **photocathode**. According to Einstein's explanation of the photoelectric effect, the absorption of this photon will liberate a single electron from the surface of the metal. A power supply or battery of some kind is used to set up a potential difference between the photocathode and another nearby metal plate (called a **dynode**), establishing an electric field between

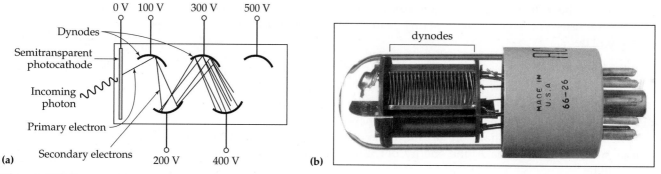

Figure Q3.3

(a) A schematic diagram of a photomultiplier tube. (b) A photograph of an actual photomultiplier tube.

these plates. The single electron liberated from the photocathode is accelerated toward the dynode by this electric field and strikes it with a substantial amount of kinetic energy (100 eV in the design above). The impact of this electron dislodges several electrons from the dynode. Another potential difference maintained between the first dynode and a nearby second dynode causes these secondary electrons to accelerate toward the third plate, strike it, liberating even more electrons, and so on. After several stages of this process, the number of electrons participating in the cascade becomes large enough that sensitive electronic circuitry can detect the tiny pulse of current that they represent. The circuitry then electronically amplifies this signal to the point that an obvious pulse can be registered on an oscilloscope or fed into an electronic counter.

I should point out that the fact that such a detector can register a single photon does not really provide evidence of the photon model of light independent of the evidence provided by the photoelectric effect. Note that my description of the operation of the device *assumed* that Einstein's description of the photoelectric effect was true. However, this device *does* operate as described, and at very low illumination levels, discrete pulses are in fact observed as individual photons hit the detector, giving *vivid* (even if not exactly independent) evidence of the particle nature of light.

TWO-MINUTE PROBLEMS

Q3T.1   In the experiment shown in figure Q3.1b, the final voltage difference between the plates is proportional to
A.   The rate at which electrons are ejected from the cathode
B.   The average kinetic energy of the ejected electrons
C.   The maximum kinetic energy of the ejected electrons
D.   Both A and C

Q3T.2   In the experiment shown in figure Q3.1b, which of the following possible results (if seen) regarding the final potential difference $\Delta\phi$ between the plates

would probably *not* be consistent with the wave model of light?
A. $\Delta\phi$ increases as the light's intensity increases.
B. $\Delta\phi$ is independent of the light's intensity.
C. $\Delta\phi$ varies as the light's wavelength changes.
D. $\Delta\phi$ increases as the rate of electron ejection increases.

Q3T.3   In the experiment shown in figure Q3.1a, which of the following possible results (if seen) about the rate of electron ejection would probably *not* be consistent with the wave model of light?
A.   The rate is zero for some time after light starts shining.

B. The rate increases as the light's intensity increases.
C. The rate varies as the light's wavelength changes.
D. The rate is zero if wavelength is above a certain value.

Q3T.4 A beam of light P has twice the wavelength but the same intensity as beam Q. The number of photons that hit a given area in a given time when it is illuminated by beam P is (A) twice, (B) the same, or (C) one-half of the number that hit when the area is illuminated by beam Q.

HOMEWORK PROBLEMS

Basic Skills

Q3B.1 Red light emitted by a standard helium-neon laser has a wavelength of about 633 nm. What is the energy of one photon of such red light?

Q3B.2 Photons from a certain light source are measured to have an energy of 3.5 eV. What is the wavelength associated with this light? Is this light visible?

Q3B.3 Photons from a certain light source are measured to have an energy of 0.62 eV. What is the wavelength associated with this light? Is this light visible?

Q3B.4 The value of W is about 4.24 eV for zinc. What is the maximum wavelength that light falling on the zinc can have if it is to be able to eject electrons? [For comparison, visible light has wavelengths ranging from about 700 nm (red) to 400 nm (violet).]

Q3B.5 The value of W is about 2.3 eV for lithium. What is the maximum wavelength that light falling on the lithium can have if it is to be able to eject electrons? [For comparison, visible light has wavelengths ranging from roughly 700 nm (red) to 400 nm (violet).]

Q3B.6 Verify that the intensity of light falling on a surface 1.0 m away from a lamp radiating 40 W of visible light is 3.2 W/m^2, as claimed in exercise Q3X.4.

Synthetic

Q3S.1 How does the photon model explain the observed result of photoelectric effect experiments indicating that the rate at which light ejects electrons is directly proportional to the intensity of the light?

Q3S.2 How does the photon model explain the observed result of photoelectric effect experiments that if the intensity of light falling on a metal plate is held fixed, the rate at which electrons are ejected from the metal decreases as the light's frequency increases?

Q3S.3 About how many photons per second are broadcasted by a FM radio station whose transmitter power is 10,000 W and whose frequency is 89.9 MHz?

Q3S.4 When sodium metal is illuminated with monochromatic light with a wavelength of 420 nm, the maximum potential difference developed between the plates in the experiment shown in figure Q3.1b is found to be 0.65 V; when the wavelength is 310 nm, this voltage is found to be 1.69 V. Check that these experimental results are consistent with a value of 1240 eV·nm for hc, and find the value of W for sodium.

Q3S.5 When cesium metal is illuminated with monochromatic light with a wavelength of 500 nm, the maximum potential difference developed between the plates in the experiment shown in figure Q3.1b is found to be 0.57 V; when the wavelength is 420 nm, this voltage is found to be 1.04 V. Check that these experimental results are consistent with a value of 1240 eV·nm for hc, and find the value of W for cesium.

Q3S.6 When iron is illuminated with ultraviolet light with a wavelength of 250 nm, the maximum potential developed between the plates in the experiment shown in figure Q3.1b is 0.46 V. From these data and the accepted value of hc, find the potential difference between the plates if the ultraviolet light wavelength is changed to 220 nm. Also find W for iron.

Q3S.7 When lead is illuminated with ultraviolet light with a wavelength of 250 nm, the maximum potential developed between the plates in the experiment shown in figure Q3.1b is 0.82 V. From these data and the accepted value of hc, find the potential difference between the plates if the wavelength is 215 nm. Also find W for lead.

Q3S.8 Imagine that you are standing and facing a 60-W incandescent lightbulb 100 m away. If the diameter of your pupils is about 2 mm, about how many photons enter your eye every second?

Rich-Context

Q3R.1 Experiments have shown that the nervous system of the human eye effectively takes about 30 "frames" per second (like a movie camera) and that when the eye is fully dark-adapted, it needs to receive only about 500 photons per frame from an object to register it. Our sun radiates a power of about 3.9×10^{26} W at all wavelengths, peaking in the yellow region of the spectrum (only about one-half of this energy is in the visible range, however). Estimate how far away a star like the sun could be and still be visible.

Advanced

Q3A.1   If we shine X-rays at a target, both the wave and photon models of light predict that the X-rays will be scattered in all directions by the target. The wave model, however, predicts that the scattered X-rays will have the *same* wavelength as that of the original waves, while the photon model predicts that the wavelength of the scattered X-rays will depend on angle in a specific way. Our purpose here is to derive this angle dependence.

Let us define our coordinate system so that the initial X-ray photons move in the $+x$ direction. Let the initial energy of the X-rays be E_0 and the mass of an electron be m. Imagine that a photon strikes the electron and scatters at an angle θ, while the electron scatters at an angle ϕ, as shown in figure Q3.4. Since the ratio of any particle's relativistic momentum p to its relativistic energy E is $p/E = v$, for a photon $p/E = 1$, so $p = E$ in units where $c = 1$ (such as the SR units discussed in unit R).

(a) Argue that conservation of 4-momentum implies that

$$\begin{bmatrix} m \\ 0 \\ 0 \\ 0 \end{bmatrix} + \begin{bmatrix} E_0 \\ E_0 \\ 0 \\ 0 \end{bmatrix} = \begin{bmatrix} E \\ E\cos\theta \\ E\sin\theta \\ 0 \end{bmatrix} + \begin{bmatrix} \sqrt{m^2 + p^2} \\ p\cos\phi \\ -p\sin\phi \\ 0 \end{bmatrix}$$

(Q3.20)

where E is the energy of the scattered photon and p is the relativistic momentum of the scattered electron.

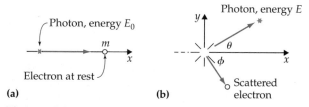

Figure Q3.4
Compton scattering (a) before and (b) after the interaction.

(b) Show that you can square and combine the middle two equations in equation Q3.20 to yield

$$E_0^2 - 2EE_0\cos\theta + E^2 = p^2 \qquad \text{(Q3.21)}$$

(c) Show, by isolating the square root and squaring both sides, the top equation implies that

$$2E_0 m + E_0^2 - 2E(m + E_0) + E^2 = p^2 \qquad \text{(Q3.22)}$$

(d) Show that by combining these equations to eliminate p, dividing both sides by $EE_0 m$, and using equation Q3.7, you get (in SI units)

$$\lambda - \lambda_0 = \frac{h}{mc}(1 - \cos\theta) \qquad \text{(Q3.23)}$$

(Remember that E and E_0 have units of kilograms in SR units and thus are equivalent to E/c^2 and when the energies are expressed in the SI unit of joules.) Arthur Compton verified that the scattered X-rays had exactly this angle dependence in 1923, and this is what finally convinced the community that the photon model was correct.

ANSWERS TO EXERCISES

Q3X.1   According to the discussion in the previous paragraphs, the maximum energy of the ejected electrons will be 3.32 eV $= (3.32 \text{ eV})(1.6 \times 10^{-19} \text{ J/eV})$ $= 5.3 \times 10^{-19}$ J.

Q3X.2   As long as there are *any* electrons that have enough kinetic energy to reach the anode, the charge difference between the plates (and thus the potential difference between them) will continue to increase. When the potential difference finally settles down to a certain value, it means that there are no electrons with a higher kinetic energy.

Q3X.3   The first: we want to measure the *rate* at which electrons are liberated as a function of intensity, and an *ammeter* measures the rate of electron flow.

Q3X.4   The rate at which each atom absorbs energy is

$$3.2\frac{\text{J/s}}{\text{m}^2}\left(\frac{10^{-20}\text{ m}^2}{\text{atom}}\right)\left(\frac{1\text{ eV}}{1.6 \times 10^{-19}\text{ J}}\right)$$

$$= 0.20\frac{\text{eV}}{\text{s}\cdot\text{atom}} \qquad \text{(Q3.24)}$$

If the electron must accumulate "a few electron-volts" of energy to be liberated, this will take 10 s or more.

Q3X.5   In order to liberate an electron within 10^{-8} s, we need to collect energy at the rate of maybe 2×10^8 eV/s, which is 10^9 times larger than the rate in equation Q3.24. This means that we need a collecting area 10^9 times as big; that is, we must collect the energy from a billion atoms to eject *one* electron! It is hard to see how this could happen.

Q3X.6   The energy of an ejected electron depends on the energy of the individual photon that ejected it, which is determined by the *wavelength* of the light. Higher intensities correspond to greater numbers of photons, but if λ is fixed, the energy of an *individual* photon does not change as the intensity increases.

Q3X.7   The calculation looks like this:

$$
\begin{aligned}
hc &= (6.63 \times 10^{-34}\, \cancel{J} \cdot \cancel{s}) \left(\frac{1\,\text{eV}}{1.6 \times 10^{-19}\, \cancel{J}} \right) \\
&\quad \times (3.0 \times 10^8\, \text{m}/\cancel{s}) \\
&= 1.24 \times 10^{-6}\, \text{eV} \cdot \cancel{m} \left(\frac{1\,\text{nm}}{10^{-9}\, \cancel{m}} \right) \\
&= 1240\, \text{eV} \cdot \text{nm}
\end{aligned}
\tag{Q3.25}
$$

Q3X.8   According to equation Q3.7, we have

$$
E = \frac{hc}{\lambda} = \frac{1240\,\text{eV} \cdot \cancel{nm}}{540\, \cancel{nm}} = 2.3\,\text{eV}
\tag{Q3.26}
$$

Q4

The Wave Nature of Matter

Chapter Overview

Introduction

In chapter Q3, we reviewed some of the most important evidence supporting a particle model of light. In this chapter, we will review some of the evidence supporting a wave model for entities we normally consider to be particles (e.g., electrons and protons). The unavoidable conclusion is that *all* forms of matter and energy have aspects that fit a wave model and aspects that fit a particle model, but that neither model is completely sufficient. We will begin developing a model that *is* sufficient in chapter Q5.

Section Q4.1: Subatomic Particles as Particles

There is abundant evidence of the particlelike character of electrons, protons, neutrons, and so on. These objects are individually countable and leave definite, well-defined tracks in particle detectors. A particle model of such objects seems so obvious that it is difficult to imagine that any other possibilities might exist.

Section Q4.2: The de Broglie Hypothesis

In spite of this, in 1923 Louis de Broglie proposed that such objects might, like photons, have an associated wavelength given by

$$p = \frac{h}{\lambda} \quad \text{or} \quad \lambda = \frac{h}{p} \qquad (Q4.4b)$$

Purpose: This equation specifies the relationship between the magnitude of a particle's (relativistic) momentum p and the effective wavelength λ of a beam of such particles.

Symbols: h is Planck's constant.

Limitations: This equation applies to both photons and other kinds of particles. It only applies, though, when the particles have a well-defined and constant value of p.

Some simple calculations show that this expression implies that the wavelengths of objects with nonzero rest mass (such as electrons or protons) are pretty short compared to the wavelengths of light, so wave effects will be difficult to observe.

Section Q4.3: Preparing an Electron Beam

The first step to testing de Broglie's hypothesis would be to prepare a beam of electrons with momenta having a uniform and well-defined magnitude p. Since $K = p^2/2m$ for nonrelativistic particles, *free* particles (i.e., particles subject to no significant external forces) with a well-defined kinetic energy K will also have a well-defined momentum magnitude p. The effective wavelength associated with particles having a definite kinetic energy is

$$\lambda = \frac{h}{\sqrt{2Km}} = \frac{hc}{\sqrt{2Kmc^2}} \qquad \text{(Q4.6)}$$

Purpose: This equation expresses the relationship between a free particle's kinetic energy K and its de Broglie wavelength λ.

Symbols: m is the particle's mass, c is the speed of light, and h is Planck's constant.

Limitations: The particle must be free (subject to negligible external forces), have nonzero rest mass m, and be nonrelativistic ($K \ll mc^2$).

This section describes how one can use an **electron gun** to create a **monoenergetic** beam of free electrons having a definite kinetic energy K.

Section Q4.4: The Davisson-Germer Experiment

C. H. Davisson and L. H. Germer (accidently) performed an experiment in 1925 in which they directed a monoenergetic electron beam at a nickel crystal. They found that the electrons reflecting off the regular lines of nickel atoms on the crystal's surface created an interference pattern that was nicely explained by de Broglie's hypothesis. This was the first recognized evidence that electrons sometimes behave as waves do.

Section Q4.5: Electron Interference

In the 1960s it became possible to make a pair of slits small enough to do two-slit interference experiments with electrons. Such experiments generate the classic two-slit interference pattern, exactly as if the electrons were simple waves.

Section Q4.6: Matter Waves

In recent years, interference effects consistent with the de Broglie model have also been observed in experiments involving beams of protons, neutrons, atoms, and even molecules. This firmly establishes that *all* forms of matter and energy have both wave and particle aspects. When the value of p for a particle is small (as is commonly the case for photons with wavelengths in the visible range and longer), the corresponding wavelengths are reasonably large and wave aspects (such as interference effects) dominate. When p is large (as is commonly the case for subatomic particles with nonzero rest mass), wavelengths can be very short and interference effects difficult to display, making the particle aspect dominant. The de Broglie wavelengths associated with objects of macroscopic mass are so short that observing any wave aspects at all is hopeless: for such objects, the particle aspect is completely dominant.

Q4.1   Subatomic Particles as Particles

Quantization of electric charge
and particle model of the atom

As early as 1913, experiments by the U.S. physicist Robert Millikan showed that small electric charges on oil drops were always integer multiples of $e = 1.6 \times 10^{-19}$ C. These results were taken by the physics community as evidence of the existence of particles (in this case, electrons) carrying this amount of charge that were either removed or added to the oil drop to give it its charge. Models of atoms developed from the turn of the century through the present are based on the assumption that every atom is constructed of an integer number of subatomic particles called *electrons, protons,* and *neutrons:* the periodic table of elements is fundamentally based on this particle model of the atom.

Particle tracks visible in bubble
chamber photographs

Since the middle of this century, it has been actually possible (using a device called a **bubble chamber**) to photograph the trajectories of individual electrons, protons, and other subatomic particles. Such a photograph is shown in figure Q4.1. A bubble chamber consists of a pressurized vat of liquid hydrogen whose temperature is just below hydrogen's boiling point. Energetic charged particles moving through the vat interact violently with the hydrogen atoms, often stripping off their electrons. When a liquid boils, it happens that bubbles preferentially form where the liquid has been disturbed in this way. So just before the photograph is taken, the ambient pressure in the chamber is reduced a bit, which lowers the boiling point of hydrogen just below its current temperature. As the hydrogen begins to boil, bubbles first form along the paths of any energetic charged particles that happen to be moving through it at the time.

It is hard to imagine more vivid evidence of the particle nature of electrons than a photograph like this: the track of each particle is sharply defined in space, providing evidence of the particle passing as vivid as footprints in the snow. Analogous bubble chamber photographs show tracks left by protons, muons, and other subatomic particles, firmly establishing their particle nature. Even though the photon's track is invisible (because the photon is uncharged), the photograph also offers compelling evidence for the particle

Figure Q4.1
A bubble chamber photograph showing a high-energy photon colliding with an electron at point *P*. The struck electron flies off to the right. In this case, some of the photon's energy is also converted to an electron/antielectron pair, whose tracks curl above and below the track of the struck electron. (The tracks are curved because the chamber is placed in a magnetic field: their curvature in this field allows the experimenters to determine the particles' momenta.) This diagram vividly illustrates the particle nature of these objects.

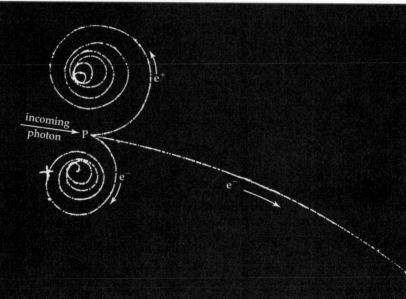

nature of light: *something* carries a lot of energy into the process and deposits it all essentially at point P, exactly as if an energetic collision between two particles were taking place there.

Physicists can also now see (and push around) individual atoms by using a scanning tunneling microscope, suspend individual atoms in a trap, and count individual electrons and photons by using sensitive detectors. Even as early as 1910, physicists could register and count individual helium nuclei ejected by radioactive atoms. In short, there is an abundance of evidence (both direct and indirect) for the particle nature of electrons, protons, neutrons, and nuclei. Indeed, there is so much evidence and it is so compelling (such as the evidence of the photograph) that any alternative model is hard to imagine. Yet, as we will see shortly, a particle model does *not* completely describe the behavior of such objects.

Other kinds of evidence

Q4.2   The de Broglie Hypothesis

In the photon model of light proposed by Einstein, each photon carries an energy that is related to the frequency f of the associated light wave as follows:

$$E = hf \qquad \text{(Q4.1)}$$

Link between energy and frequency

where h is Planck's constant ($hc = 1240\,\text{eV} \cdot \text{nm}$). Since light moves at speed c and the velocity of a traveling wave is equal to λf, this equation can be rewritten as

$$E = \frac{hc}{\lambda} \qquad \text{(Q4.2)}$$

Now, as we saw in unit R, the ratio of a particle's relativistic momentum p to its relativistic energy E is equal to its speed:

$$\frac{p}{E} = v \quad \text{(SR units: see unit R)} \qquad \Rightarrow \qquad \frac{pc}{E} = \frac{v}{c} \quad \text{(SI units)} \quad \text{(Q4.3)}$$

(Note that both sides of both equations are unitless in their appropriate unit systems.) If we apply this to a photon (whose speed is c by definition) and solve for its relativistic momentum p, we find that

$$pc = E\frac{c}{c} = \frac{hc}{\lambda}(1) \qquad \Rightarrow \qquad p = \frac{h}{\lambda} \qquad \text{(Q4.4}a\text{)}$$

The link between momentum and wavelength (the de Broglie relation)

In 1923, the French physicist Louis de Broglie proposed in his doctoral dissertation that a beam of *any* kind of particle that carries momentum and energy (electrons, protons, whatever) will (like light) exhibit wavelike behavior, and that the frequency f and the effective wavelength λ of the beam should be linked to the energy E and relativistic momentum p of each particle in the beam by equations Q4.1 and Q4.4, respectively. De Broglie proposed this hypothesis because the theory of relativity suggested to him that there should be no real distinction between matter and energy (and thus between photons and other particles). At the time, however, there was little evidence in support of this rather outrageous assertion, and while the faculty readers of de Broglie's thesis were impressed with the quality and care of his argument, they were very skeptical about the hypothesis itself (they passed it anyway, to de Broglie's relief). Equation Q4.4 is now called the **de Broglie relation** and the wavelength λ that appears in it the **de Broglie wavelength**

de Broglie's hypothesis

in his honor:

The de Broglie relation

$$p = \frac{h}{\lambda} \quad \text{or} \quad \lambda = \frac{h}{p} \tag{Q4.4b}$$

Purpose: This equation specifies the relationship between the magnitude of a particle's (relativistic) momentum p and the effective wavelength λ of a beam of such particles.
Symbols: h is Planck's constant.
Limitations: This equation applies to both photons and other kinds of particles. It only applies, though, when the particles have a well-defined and constant value of p.

For the moment, let us take this hypothesis seriously. If electrons exhibit wavelike behavior, how might we demonstrate it experimentally? The Young two-slit interference experiment clearly exposes the wave nature of light: might we do an analogous experiment with electrons?

Exercise Q4X.1

What kind of wavelengths are we talking about here? What is the effective wavelength of an electron moving at a speed of $0.01c$? This is slow enough that you can use the nonrelativistic expression $p \approx mv$ for the momentum. (The value of Planck's constant is $h = 6.63 \times 10^{-34}$ J · s, and the mass of an electron is 9.11×10^{-31} kg.) Express your answer in nanometers. How does this compare to the wavelength of visible light?

Q4.3   Preparing an Electron Beam

The first step in setting up a two-slit interference experiment involving light is to find a source of *monochromatic* light, that is, light having a well-defined wavelength λ. What would be the analogue of monochromatic light for electrons?

The basic point of using monochromatic light is that the interference pattern for waves having a sharply defined wavelength is clear and well defined. The positions of the bright spots of the interference pattern depends on the wavelength (see equation Q2.1); so if we were to attempt to create an interference pattern using a beam containing light with many wavelengths, the interference pattern would be fuzzy and not very well defined.

To get a clear interference pattern, we need a beam of particles all having the same p

If we want to get a clear interference pattern in a two-slit interference experiment with electrons, we want the *electrons* to have a well-defined wavelength as well. The de Broglie relation (equation Q4.4) implies that for electrons in a beam to have a well-defined wavelength, they should all have the same magnitude of momentum p. The momentum of a free (nonrelativistic) electron is related to its kinetic energy K by the equation

$$K = \frac{1}{2}mv^2 = \frac{m^2v^2}{2m} = \frac{p^2}{2m} \tag{Q4.5}$$

So if free electrons in a beam moving in a specific direction have a well-defined value of K, they will also have a well-defined value of p and thus λ. We describe such a beam as being a **monoenergetic** beam.

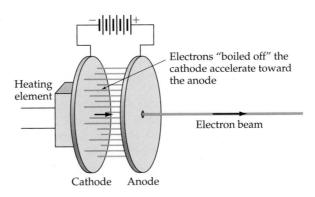

Heating element

Electrons "boiled off" the cathode accelerate toward the anode

Electron beam

Cathode   Anode

Figure Q4.2
A schematic diagram of an electron gun. The final kinetic energy of the electrons is determined by the potential difference between the plates: if the battery puts a potential difference of 100 V across the plates, each electron emerging from the hole in the anode will have a kinetic energy of 100 eV.

A description of an electron gun

One means of creating a monoenergetic beam of free electrons is shown in figure Q4.2. A metal plate (the *cathode* in the illustration) is heated with an electric current to a sufficiently high temperature to "boil off" some electrons from the metal (the average kinetic energy of the electrons in a metal increases as the temperature increases, and when the temperature is high enough, a significant fraction of electrons have enough energy to escape the metal entirely). A second metal plate with a hole in the center (the *anode* in the illustration) is placed near the cathode, and a battery is connected to the two plates so as to give the cathode a negative charge and the anode a positive charge. Electrons that have boiled off the cathode are repelled by it and attracted toward the anode, and are thus accelerated by the electric field between the plates. Most of these loose electrons collide with the anode plate and are absorbed by it, but some (by virtue of their momentum) happen to go through the hole in the anode. Once they are through the hole, their motion is essentially unaffected by the charged plates, since the electric field outside a pair of closely spaced charged plates is essentially zero. This device, called an **electron gun,** thus creates a beam of electrons that emerges from the hole in the anode.

The kinetic energy (and thus the momentum) of the electrons in the electron beam is determined by the potential difference that the battery sets up across the plates. An electron moving across a potential difference of 1 V experiences a change in potential energy (and thus kinetic energy) of 1 eV by definition. Therefore, a 100-V battery between the cathode and the anode will cause an electron moving between them to gain 100 eV of kinetic energy. Since the initial kinetic energy of electrons boiled off the cathode is very small (a few tenths of an electronvolt or so), the final kinetic energy of electrons emerging from the hole in the anode will be very nearly 100 eV.

What is the effective wavelength of such electrons, according to the de Broglie relation? Combining equations Q4.4 and Q4.5, we find that the wavelength associated with a nonrelativistic beam of electrons (or any particle with nonzero rest mass m) is related to the kinetic energy of each particle as follows:

$$\lambda = \frac{h}{\sqrt{2Km}} = \frac{hc}{\sqrt{2Kmc^2}} \qquad (Q4.6)$$

Wavelength as a function of kinetic energy

Purpose: This equation expresses the relationship between a free particle's kinetic energy K and its de Broglie wavelength λ.

Symbols: m is the particle's mass, c is the speed of light, and h is Planck's constant.

Limitations: The particle must be free (subject to negligible external forces), have nonzero rest mass m, and be nonrelativistic ($K \ll mc^2$).

Exercise Q4X.2

Verify equation Q4.6.

I personally find the second version of equation Q4.6 easier to use, because the values of hc (1240 eV · nm) and the electron's rest energy mc^2 (511,000 eV) are more manageable and easier to remember than h (6.63×10^{-34} J · s or 4.15×10^{-15} eV · s) and m (9.11×10^{-31} kg for an electron), and the wavelength comes out in the convenient unit of nanometers.

Example Q4.1   Wavelength of an Electron Beam

Problem   What is the de Broglie wavelength of a beam of electrons having kinetic energy $K = 100$ eV?

Solution   Using the second form of equation Q4.6, we get

$$\lambda = \frac{1240 \text{ eV} \cdot \text{nm}}{\sqrt{2(100 \text{ eV})(511,000 \text{ eV})}} = 0.12 \text{ nm} \qquad (Q4.7)$$

This wavelength is on the order of the size of a single atom. Now, equation Q3.1 for the two-slit interference pattern [$\theta_{nc} = \sin^{-1}(n\lambda/d) \approx n\lambda/d$ for small angles] implies that to get an interference pattern where the bright spots are separated by a few degrees of angle (i.e., about 0.03 rad or so), one would like to have the distance d between the slits be not much more than 30λ. This means that for an electron interference experiment, we would want to construct a mask with a pair of slits about 30 atoms' width apart. This is not going to be easy!

Exercise Q4X.3

Explain why equation Q4.6 only applies to nonrelativistic particles, pinpointing the step where we used the nonrelativistic approximation.

Exercise Q4X.4

One of your friends complains: "Equation Q4.6 links the kinetic energy K of a particle in a particle beam to the effective wavelength of the beam. Photons, being light, must have zero rest mass. If you plug $m = 0$ into equation Q4.6, $\lambda \to \infty$, which is absurd, since light has a finite wavelength. Therefore equation Q4.6 must be wrong." Pinpoint the error in this argument.

Q4.4   The Davisson-Germer Experiment

The difficulty of constructing a pair of slits for an electron interference experiment was one of the many reasons that physicists did not rush out to check the extravagant proposal of a French doctoral candidate. Even so, confirmation of de Broglie's hypothesis was not long in coming. In 1925, the U.S. physicists C. H. Davisson and L. H. Germer were the first to observe the effects of electron interference, rather by accident. The story is interesting.

Davisson and Germer were investigating the scattering of electrons from metal surfaces. These surfaces were enclosed in evacuated glass tubes so that the electrons could move freely without running into air molecules. One day, they accidentally broke a part of their vacuum system, and the hot nickel target oxidized when it came into contact with the air. In the process of trying to remove the oxidation from the target, they inadvertently converted the nickel from an aggregate of small nickel crystals to a few large crystals. When they mounted the cleaned target in a new tube and began firing electrons at it, they found that it did not behave at all as it had previously: now there appeared to be special angles where electron scattering was enhanced.

Davisson and Germer's fortuitous accident

Being a great physicist does not mean that you never break your expensive apparatus, nor does it mean that you never accidently modify your material in a fundamental way that you don't expect or understand. The difference between a good physicist and a mediocre one is more about the curiosity, tenacity, and ingenuity with which you examine unexpected results. Davisson and Germer could have thrown up their hands and then thrown out the apparently defective new tube, and if they had done so, their names would not now appear in the list of names familiar to physics students worldwide. Instead, their curiosity was aroused, and they set about to understand *why* they were seeing what they were.

Though they were not aware of de Broglie's work at the time that they initially observed the strange scattering effects, papers concerning the possibility of electron interference were soon brought to their attention, and with a series of careful experiments, they were able to solve their mystery. It turned out that the electrons were getting reflected from the regularly spaced planes of atoms in the nickel crystal; and at certain angles, the electron waves reflected from adjacent atomic planes interfered constructively with each other, and at other angles they interfered destructively. The effective wavelength that they inferred from the known spacing of atoms in nickel and the angles of constructive and destructive interference was consistent with the wavelength predicted by equation Q4.6 for the accelerating potential of their electron gun. The description of their experiments and their explanation was published in 1927 (Davisson and Germer, *Physical Review*, 30, p. 707).

What Davisson and Germer found

Example Q4.2   An Analysis of a Davisson-Germer Experiment

Problem   What follows is a slightly simplified analysis of one of Davisson and Germer's experiments. In this experiment, electrons with a kinetic energy of 54 eV were fired toward the crystal perpendicular to its face. A detector was mounted to receive electrons reflected at a variable angle θ from the direction of the incident beam (see figure Q4.3). The crystal was oriented so that the rows of atoms on the surface were perpendicular to the plane containing the incident and reflected electron beams. Imagine that the

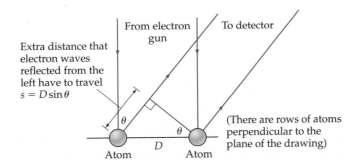

Extra distance that electron waves reflected from the left have to travel
$s = D \sin \theta$

From electron gun

To detector

(There are rows of atoms perpendicular to the plane of the drawing)

Atom D Atom

Figure Q4.3

A simplified model of how the electron beam in the Davisson-Germer interferes constructively with itself at certain scattering angles.

electron beams are reflected off the atomic rows on the surface as shown in the diagram: thus the atomic rows behave for the electron waves as parallel slits a distance D apart. The condition for constructive interference of these waves is that the extra distance s that a wave has to travel from one atomic row to the detector above that from the adjacent row to the right is equal to an integer number of wavelengths. Given that the spacing between atomic rows in nickel is $D = 0.215$ nm, what is the smallest nonzero angle for constructive interference?

Solution   According to equation Q4.6, the wavelength of a beam of electrons that each has a kinetic energy of $K = 54$ eV is

$$\lambda = \frac{hc}{\sqrt{2Kmc^2}} = \frac{1240 \text{ eV} \cdot \text{nm}}{\sqrt{2(54 \text{ eV})(511{,}000 \text{ eV})}} = 0.167 \text{ nm} \qquad (Q4.8)$$

The condition for constructive interference is that the extra distance s traveled by each reflected beam as we go left must be an integer number of wavelengths. As we can see in figure Q4.3, this extra distance is $s = D \sin \theta$, so the condition for constructive interference is

$$n\lambda = D \sin \theta \qquad \Rightarrow \qquad \theta = \sin^{-1} \frac{n\lambda}{D} \qquad (Q4.9)$$

where n is an integer. The smallest angle will be when $n = 1$:

$$\theta = \sin^{-1} \frac{\lambda}{D} = \sin^{-1} \frac{0.167 \text{ nm}}{0.215 \text{ nm}} = 0.889 = 51° \qquad (Q4.10)$$

This is exactly the angle where Davisson and Germer saw a bright spot in the electron scattering. Note that $n = 2$ yields a result for $n\lambda/D = \sin \theta$ that is greater than 1 and thus cannot correspond to any real angle. Therefore, this will be the only nonzero angle of constructive interference.

[The analysis in this problem is simplified because the electrons do not reflect from just the top layer of atoms but from a certain number of deeper layers as well. This makes the analysis more complicated, but it turns out when all is said and done that equation Q4.9 is *still* a necessary condition for constructive interference, though other conditions apply that make certain kinetic energies, including 54 eV, more effective than others for yielding sharp constructive interference. For a detailed discussion, see A. P. French and E. F. Taylor, *An Introduction to Quantum Physics*, Norton, New York, 1978, pp. 64–72.]

Exercise Q4X.5

If the electrons in the experiment described in example Q4.2 had a kinetic energy of 78 eV, what would be the smallest nonzero angle for constructive interference of the electron waves?

Q4.5   Electron Interference

Two-slit electron interference becomes possible

In the early 1960s it became possible to do actual two-slit interference experiments with electrons. The same techniques used to construct transistors and integrated circuits can be employed to create very tiny slits in a barrier. Even with these techniques, constructing slits 30 atoms apart is not possible; but clever techniques enabled experimenters to record and display the very narrow interference patterns produced by more widely separated slits.

Figure Q4.4 shows a photograph of a two-slit electron interference pattern produced in an experiment done by C. H. Jönsson (*Z. Phys.* **161**: 454, 1961). Jönsson created slits in copper foil about 0.5 μm in width and spaced 1 to 2 μm apart (1 μm = 10^{-6} m = 10^{-3} mm). He used a beam of electrons having a kinetic energy of 50 keV (corresponding to an effective de Broglie wavelength of about 0.005 nm) and displayed the interference pattern on a screen 35 cm from the slits. If you grind through the numbers (see problem Q4S.4), you will find that adjacent bright spots in the interference pattern are about 1 μm, which is far too small to see with the naked eye. To make the pattern visible, Jönsson employed a clever scheme of electrostatic lenses to magnify the image of the interference pattern. (A translation of Jönsson's paper appears in *American Journal of Physics*, **42:** 5, 1974.)

Figure Q4.4

An actual photograph of a two-slit electron interference pattern.

Q4.6   Matter Waves

In recent years, interference effects have also been observed in experiments involving beams of protons, neutrons, atoms, and even molecules. For example, in 1991, O. Carnal and J. Mlynek performed a virtual copy of Young's two-slit interference experiment except with a beam of helium atoms (O. Carnal and J. Mlynek, *Phys. Rev. Lett.*, **66:** 2689). They sent helium atoms with a de Broglie wavelength of 0.1 nm through a pair of slits in gold foil that were both 1 μm wide and separated by 8 μm. Figure Q4.5 shows the interference pattern obtained at a detector positioned 64 cm from the slits. The measured distance between interference maxima was 7.7 μm $\pm$ 1.0 μm, quite consistent with what one would expect from waves with that de Broglie wavelength (see problem Q4S.5).

Figure Q4.6 shows the results of an even more famous experiment. In November 1996, Wolfgang Ketterle and collaborators cooled about 2 million sodium atoms in an atom trap to a temperature of roughly 100 nK, which caused them to settle into a state called a *Bose-Einstein condensate* that has well-defined wave properties. Using laser beams, they cut and separated the cloud of supercooled sodium atoms into two separate globs, and then released the globs, allowing them to expand slowly into each other. When the globs began to overlap, the researchers could see bands of constructive and destructive interference, as shown in the figure. The distance between these

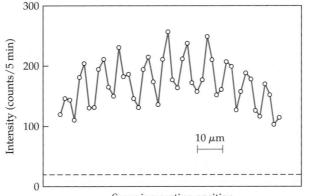

Figure Q4.5

An interference pattern created in a two-slit interference experiment involving a beam of helium atoms.

From O. Carnal and J. Mlyneck, Physical Review Letters, v. 66, p. 2691, 1991. Copyright 1991 by the American Physical Society.

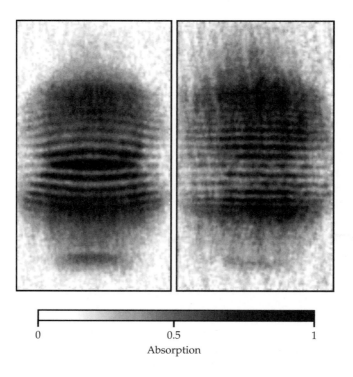

Figure Q4.6
The interference pattern created by merging clouds of supercooled sodium atoms. The two clouds were originally small dots about 40 μm apart: these photographs were taken during different experimental runs after the regions were allowed to expand freely for 40 ms. The observed distance between interference maxima was about 15 μm in the right-hand picture.

0   0.5   1

Absorption

interference bands was again completely consistent with the de Broglie model (see problem Q4R.3). (Ketterle shared the 2001 Nobel Prize for physics with Eric Cornell and Carl Weiman for his work related to Bose-Einstein condensates.)

All forms of matter-energy exhibit wavelike aspects

The conclusion is inescapable: though interference is difficult to observe for a beam of matter particles, such beams do indeed exhibit wavelike behavior in two-slit interference experiments and other analogous experiments. While the results of these experiments are impossible to explain with any simple particle model, they are easily explained if we assume that these particle beams are in fact comprised of waves whose wavelength is given by the de Broglie relation $p = h/\lambda$.

These results seem to stand in direct conflict with the experiments illustrating the particle nature of these physical objects, just as the results of the photoelectric effect stand in direct conflict with the Young two-slit interference experiment involving light. Bringing these paradoxical results into some harmony will be our main task in chapters Q5 and Q6.

There is a silver lining in these troublingly counterintuitive results: these experiments make it clear there is no physical distinction (in this regard) between the behavior of photons, electrons, or any other kind of matter-energy. De Broglie's equations $E = hf$ (equation Q4.1) and $p = h/\lambda$ (equation Q4.4) apply equally well to all forms of matter-energy. All exhibit wavelike characteristics in interference experiments and particlelike characteristics in experiments involving collisions (as shown in figure Q4.1) or other cases where energy and/or momentum is being transferred from one object to another (as in the photoelectric effect). The special theory of relativity (by asserting that the mass of an object is just another form of energy) had already blurred the distinction between matter and energy: this just drives home the point.

When does one aspect dominate?

Let me point out another important implication of these results. The wavelike character of an object becomes more apparent as its effective de Broglie wavelength increases and less apparent as its wavelength decreases.

It is much easier to display an interference pattern with visible light than it is to display it with electrons because the effective wavelength of visible light is roughly 10,000 times larger than the effective wavelength of electron beams that are easy to produce in the laboratory! Equation Q4.6 implies that objects having large kinetic energies and/or large masses will have shorter wavelengths, and thus will seem more like particles than objects having small masses and/or energies.

Let me illustrate. The particlelike nature of FM radio signals (electromagnetic radiation with $\lambda \approx 3$ m) is nearly impossible to demonstrate because the energy carried by a single photon is so tiny ($\approx 10^{-7}$ eV): it takes an astronomical number of such photons to do anything detectible. On the other hand, we can fairly easily demonstrate both the wave *and* particle aspects of visible light ($\lambda \approx 500$ nm). The wavelike character of gamma rays produced by nuclear reactions (electromagnetic radiation with $\lambda \approx 0.0005$ nm) is almost impossible to demonstrate because of their incredibly small wavelength, while their particlelike characteristics become abundantly clear (as we saw in the collision process illustrated in figure Q4.1).

This doesn't mean that high-energy gamma photons don't have *any* wavelike aspects or that the low-energy radio waves don't have *any* particlelike aspects. Electromagnetic waves *always* have both aspects. It is simply that at high energies, the particlelike aspects of light become more evident, and at low energies the wavelike aspects become more evident.

Exhibiting the wavelike characteristics of particles with nonzero rest mass (electrons, protons, and so on) is more challenging because these objects have short de Broglie wavelengths even at small energies. We have seen in this chapter that doing a two-slit interference experiment with electrons (the lightest known particle with nonzero rest mass) is daunting; doing such an experiment with more massive objects (e.g., protons, atoms, and molecules) is just barely possible at present. Thus, the particlelike character of these kinds of objects is *much* clearer than their wavelike aspects at ordinary energies.

The effective de Broglie wavelength of even very tiny "macroscopic" objects (from collections of several hundred atoms on upward) is generally smaller than the object itself, even if the object is moving with the minimum kinetic energy allowed by thermal effects (roughly 0.04 eV at room temperature). This means that it becomes so difficult to display interference effects involving such objects that the de Broglie wavelength of such an object really has no operational physical meaning. The particlelike nature of such objects is completely dominant.

The fact that objects that we experience directly with our senses lie within the latter category means that we are not prepared by daily living to deal with phenomena (such as visible light) that can display both particlelike and wavelike behavior. This makes it hard to intuitively understand such phenomena. Even so, we can still do some useful physics if we exert sufficient imagination and ingenuity. In chapter Q5, we will see that there is a way to bring both the wave and particle natures of matter-energy into a logically coherent (if strange) reconciliation.

In macroscopic objects, particle aspect completely dominates

Example Q4.3   How Mass Affects Wavelength

Problem   A beam of electrons and a beam of protons are accelerated from rest through the same potential difference. Which has the longer wavelength?

Solution   The de Broglie wavelength of a free particle of mass m and kinetic energy K is

$$\lambda = \frac{hc}{\sqrt{2Kmc^2}} \qquad (Q4.11)$$

If the electrons and protons are accelerated through the same potential difference, the magnitude of their change in electrical potential energy will be the same (since the magnitude of the charge on each is the same), so their final kinetic energy K will be the same. But since the mass of a proton is about 1836 times greater than that of the electron, the wavelength of the beam of protons with kinetic energy K will be about $\sqrt{1836} \approx 43$ times shorter than a beam of electrons with the same energy.

Example Q4.4   Kinetic Energy from Wavelength

Problem   What would be the kinetic energy of each electron in a beam of electrons having a de Broglie wavelength of 633 nm (the wavelength of light emitted by the common helium-neon laser)?

Solution   Solving equation Q4.6 for K, we get

$$\frac{(hc)^2}{\lambda^2} = 2mc^2 K \qquad \Rightarrow \qquad K = \frac{(hc)^2}{2mc^2\lambda^2} \qquad (Q4.12)$$

Plugging in numbers, we find that

$$K = \frac{(1240 \text{ eV} \cdot \text{nm})^2}{2(511{,}000 \text{ eV})(633 \text{ nm})^2} = 3.75 \times 10^{-6} \text{ eV} \qquad (Q4.13)$$

This is the kinetic energy of an electron accelerated from rest through a potential difference of 3.75 *microvolts* (μV). Since the basic thermal motion of electrons at room temperature will give them average kinetic energies about 10,000 times larger, one would have to actually cool this beam of electrons to a tiny fraction of a degree above absolute zero so that the electrons' random thermal motions would not totally swamp the tiny beam velocity we are trying to give them. It is therefore impractical to create an electron beam with such a wavelength.

Example Q4.5   The Wavelength of Macroscopic Particles

Problem   Consider particles of fine soot 100 nm in diameter, each containing roughly 10^9 carbon atoms. Imagine a beam of such particles moving at 1 mm/s. Carbon atoms turn out to have an average atomic mass of 12.011 u, where $1 \text{ u} = 1.66 \times 10^{-27}$ kg. What would the beam's de Broglie wavelength be?

Model and Translation   According to the de Broglie relation (equation Q4.4), the wavelength of a beam of particles having a given momentum $p = mv$ is

$$\lambda = \frac{h}{p} = \frac{h}{mv} \qquad (Q4.14)$$

In this case, we know that $m = 12.011$ u and $v = 1$ mm/s, so we can calculate λ.

Solution   Plugging the given numbers into this equation, we get

$$\lambda = \frac{6.63 \times 10^{-34}\,\text{J}\cdot\text{s}}{10^9 (12\,\text{u})(10^{-3}\,\text{m/s})} \left(\frac{1\,\text{u}}{1.66 \times 10^{-27}\,\text{kg}}\right) \left(\frac{1\,\text{kg}\cdot\text{m}^2/\text{s}^2}{1\,\text{J}}\right)$$

$$= 3 \times 10^{-14}\ \text{m} = 0.03\ \text{pm} \tag{Q4.15}$$

Evaluation   Since this is far smaller than the size of the soot particles involved, it would be very difficult to display an interference pattern for such particles. For example, imagine that we placed two slits 150 nm wide (just large enough to let the particles through) with centers about 300 nm apart (this would be at the extreme edge of modern microfabrication technology). The bright spots in the interference pattern would then be separated by an angle of $\theta \approx \lambda/d \approx 10^{-7}$ rad. To separate the maxima of the interference pattern by more than 1 μm (which would be barely large enough to see in a microscope), you would have to put a screen for collecting the soot particles about 10 m from the slits and make sure that the soot particles are completely undisturbed for the roughly 3 h that it would take them to travel from the slits to the screen. The whole experiment would have to be conducted at a temperature of a small fraction of a kelvin to make sure that the speed associated with the random thermal motion of the particles is much smaller than 1 mm/s. This and many other aspects make such an experiment impractical.

It is fair to note, however, that interference of objects as large as atoms and small molecules have been seen, in spite of enormous experimental difficulties. But the experimental challenges become more severe as the objects in question become more massive, and a point is reached long before we get to soot particles where the problems become insurmountable with current technology.

Note in example Q4.5 that because the velocity is given in meters per second and the mass of the particle is easier to compute in kilograms than in electronvolts, we are better off using the straight form of the de Broglie relation given in equation Q4.4 than the form given by equation Q4.6. The latter, on the other hand, is much more useful when both the kinetic and rest energies of the particles involved are expressed in electronvolts.

Exercise Q4X.6

"Thermal" neutrons are neutrons whose energy is roughly equal to the average kinetic energy that any object has at room temperature due to thermal effects (≈ 0.04 eV). Imagine that we prepare a beam of such neutrons. What will be the beam's effective wavelength? (*Hint:* the value of mc^2 for a neutron is 939.6 MeV.)

Exercise Q4X.7

What is the de Broglie wavelength of a "beam" of cars all having a mass of 1500 kg and speed of 65 mi/h?

TWO-MINUTE PROBLEMS

Q4T.1 Consider a beam of free particles with a certain speed v. If we double this speed, what happens to the beam's de Broglie wavelength?
- A. It increases by a factor of 2.
- B. It increases by a factor of $\sqrt{2}$.
- C. It remains the same.
- D. It decreases by a factor of $\sqrt{2}$.
- E. It decreases by a factor of 2.
- F. Something else happens to the wavelength (specify).

Q4T.2 Consider a beam of free particles that each have a certain (nonrelativistic) kinetic energy K. If we double this kinetic energy, what happens to the beam's de Broglie wavelength?
- A. It increases by a factor of 2.
- B. It increases by a factor of $\sqrt{2}$.
- C. It remains the same.
- D. It decreases by a factor of $\sqrt{2}$.
- E. It decreases by a factor of 2.
- F. Something else happens to the wavelength (specify).

Q4T.3 Imagine that in a Davisson-Germer type of experiment we shine a beam of electrons on a nickel crystal perpendicular to the crystal face and find that we get enhanced scattering at an angle of 50°. If we double the kinetic energy of the electron beam, the angle of enhanced scattering will (A) increase or (B) decrease by a factor of

- A. A bit less than $\sqrt{2}$
- B. Exactly $\sqrt{2}$
- C. A bit more than $\sqrt{2}$
- D. A bit less than 2
- E. Exactly 2
- F. More than 2
- T. Some other number (explain)

Q4T.4 Imagine that we shine a beam of electrons with a de Broglie wavelength of 0.1 nm through a pair of slits that have been miraculously constructed to be only 10 nm apart. What will be the order of magnitude of the distance between bright spots of the interference pattern displayed on a fluorescent screen placed 1 m from the slits?
- A. 1 cm
- B. 1 mm
- C. 0.1 mm
- D. 10 μm
- E. 1 μm
- F. Other (specify)

Q4T.5 The de Broglie wavelength of a beam of particles *must* be larger than each individual particle if we are to be able to display an interference pattern, true (T) or false (F)?

Q4T.6 If the value of h were bigger, it would be easier to display interference effects in macroscopic objects, T or F?

HOMEWORK PROBLEMS

Basic Skills

Q4B.1 Compute the de Broglie wavelength of an electron beam whose electrons each have a kinetic energy of 25 eV.

Q4B.2 Compute the de Broglie wavelength of an electron beam made up of electrons that each have a kinetic energy of 3.2 keV.

Q4B.3 Imagine that you want to use an electron gun to prepare an electron beam with a de Broglie wavelength of 1.0 nm. What potential difference would you set up between its plates?

Q4B.4 Imagine that you want to use an electron gun to prepare an electron beam with a de Broglie wavelength of 0.33 nm. What potential difference would you set up between its plates?

Q4B.5 Compute the de Broglie wavelength of a beam made up of neutrons that each have a kinetic energy of 25 eV. (The rest mass-energy mc^2 of a neutron is

939 MeV.) Why is this much smaller than the wavelength computed in problem Q4B.1?

Q4B.6 What would be the kinetic energy of electrons in a beam having a de Broglie wavelength of 350 nm (the wavelength of violet light)?

Q4B.7 Compute the de Broglie wavelength of a gamma-ray photon having an energy of 1.0 MeV.

Q4B.8 A baseball has a mass of 0.15 kg, and a major-league pitcher can deliver a ball with a speed of about 40 m/s ≈ 90 mi/h.
(a) Compute the de Broglie wavelength of a pitched baseball.
(b) Why do we not have to worry much about the wave aspects of such an object?

Synthetic

Q4S.1 In the Davisson-Germer experiment described in example Q4.2, what would be the smallest nonzero angle (relative to the direction of the original beam) where reflected electrons might constructively interfere if the kinetic energy of the electrons were 102 eV? Is there another possible angle of constructive interference?

Q4S.2   A beam of electrons is created by accelerating electrons from rest through a potential difference of 55 V.
   (a) What is the de Broglie wavelength of this batch of electrons? Express your result in nanometers.
   (b) Explain why it is not going to be easy to make two slits with a spacing that is roughly the same size as this wavelength. (*Hint:* the size of a typical atom is 0.1 nm.)
   (c) Find the de Broglie wavelength of a beam of protons instead of electrons accelerated through the same voltage difference ($mc^2 = 938$ MeV for a proton). Compare with your result in part (a). Is it going to be easier or harder to set up a two-slit interference experiment for protons?

Q4S.3   Verify that in the Jönsson experiment, electrons each having a kinetic energy of 50 keV going through slits placed 2.0 μm apart will produce an interference pattern on a screen 35 cm away having adjacent bright spots roughly 1 μm apart. [*Hint:* Treat the electrons as nonrelativistic (but see problem Q4S.9).]

Q4S.4   Imagine that we manage to create a pair of slits in a metal foil that are 100 nm apart, and imagine that we send a beam of 100-eV electrons through these slits and project them on a fluorescent screen 1.0 m away from the slits. What will be the approximate distance between bright spots in the interference pattern displayed on the screen?

Q4S.5   Consider the helium atom interference experiment discussed in section Q4.6. In this experiment, the detection screen was 64 cm from the slits.
   (a) If the beam of helium atoms has a wavelength of 0.103 nm, what is the approximate speed of an individual helium atom?
   (b) Calculate the theoretical distance between adjacent interference maxima on the detection screen and compare to the measured value.

Q4S.6   A spray gun creates a beam of identical water droplets, each of which has a diameter of about 10 μm and a mass of roughly 10^{-18} kg. (These are rough figures.)
   (a) If these drops move with a speed of 1.0 mm/s, what is the approximate de Broglie wavelength of this particle beam?
   (b) Why is there no practical hope of observing two-slit interference effects with such a particle beam?

Q4S.7   A buckeyball is a large molecule comprised of 60 carbon atoms arranged in a shape something

like a hollow sphere 0.71 nm in diameter. Imagine that we create a beam of buckeyballs all moving at the same speed v. What is the maximum value that v can have if the de Broglie wavelength of the buckeyball beam is to be at least 10 times the size of the buckeyball (so that we might actually be able to display interference of the buckeyballs)?

Q4S.8   The kinetic energy of a relativistic particle is defined (in SI units) to be

$$K = E - mc^2 \qquad (Q4.16)$$

where E is the particle's total relativistic energy. The relativistic link between mass, energy, and momentum is

$$(mc^2)^2 = E^2 - (pc)^2 \qquad (Q4.17a)$$
$$\Rightarrow \quad p^2c^2 = E^2 - (mc^2)^2 = (E - mc^2)(E + mc^2) \qquad (Q4.17b)$$

   (a) Combine these equations with the de Broglie relation to prove that the link between de Broglie wavelength and kinetic energy that is valid for both nonrelativistic and relativistic particles is given by

$$\lambda = \frac{hc}{\sqrt{K(K + 2mc^2)}} \qquad (Q4.18)$$

   (b) Also prove that this reduces to equation Q4.6 in the nonrelativistic limit.

Q4S.9   The 50-keV electrons in the Jönsson experiment are not really nonrelativistic (they are moving at about 40 percent the speed of light!).
   (a) Calculate the relativistically correct wavelength of a beam of such electrons, using the result of problem Q4S.8.
   (b) By about what percentage is the nonrelativistic calculation of the bright-spot spacing in problem Q4S.3 in error?

Q4S.10   An electron traveling at a speed of $0.60c$ has a kinetic energy of about how many electronvolts? What will be the de Broglie wavelength of an electron having this energy? (*Hint:* such an electron would be considered to be relativistic. See problem Q4S.8.)

Q4S.11   The 1961 Nobel Prize for physics was awarded to Robert Hofstadter for experimental work involving scattering of 20-GeV electrons from atomic nuclei.
   (a) What is the de Broglie wavelength of a beam of 20-GeV electrons? (*Hint:* see problem Q4S.8.)
   (b) How does this wavelength compare to the typical size of an atomic nucleus, which is about 10^{-15} m? (It is impossible to examine objects with any kind of beam whose wavelength is

much larger than the object in question: the beam will diffract around the object rather than reflect from it and will thus not produce a sharp image.)

(c) Does it much matter whether the "20 GeV" here refers to the total or just the relativistic kinetic energy of the electrons?

Q4S.12 Why can't we build an electron gun to produce an electron beam whose kinetic energy is accurately 0.1 eV or even 1 eV? (Such a gun would be desirable for producing electron beams with long de Broglie wavelengths.) List some of the difficulties that would have to be addressed in building such a low-energy electron gun, and explain. [*Hints:* The cathode has to be heated red hot to "boil off" an appreciable number of electrons. As we've seen in unit C, thermal kinetic energy of anything at a given temperature T is roughly $(3/2)k_B T$, where $k_B = 1.38 \times 10^{-23}$ J/K = Boltzmann's constant. Thermal effects are not the only problem here, though. What might be some others?]

Rich-Context

Q4R.1   (a) Verify that a photon of an FM radio signal with a frequency of 100 MHz has an energy of about 4×10^{-7} eV.

(b) In a typical radio receiver, the antenna collects enough photons to push a current averaging (very roughly) 1 μA through a potential difference of (very roughly) 10 mV. Roughly how many photons does the antenna have to absorb every second to drive this motion of electrons? (*Hint:* Find the *power* that has to be supplied to move the electrons as described, using concepts from unit E.)

Q4R.2   In his famous book *Mr. Tomkins in Wonderland*, the physicist George Gamow imagined a trip to a "quantum jungle" where the value of Planck's constant h was 1.0 J $\cdot$ s instead of its real value of 6.63×10^{-34} J $\cdot$ s. Imagine that while exploring in this quantum jungle, you disturb a community of jungle bats residing in a ruined temple. Imagine that a "beam" of identical bats with a mass of 0.5 kg flies at 6 m/s through two temple doors 3 m apart and into a flat, large courtyard beyond. Where could you stand in the courtyard to avoid being struck by any bats?

Q4R.3   Consider the interfering sodium atoms experiment described in section Q4.6. Assume that initially the atoms were confined to two small spots a distance $d \approx 44$ μm apart. After the confinement is turned off, the spots expand due to random variations in the atom velocities. Assume that a given atom's momentum is conserved as the spots expand. Figure Q4.6 shows the situation after the spots have expanded for a time $t \approx 40$ ms. Consider the interference pattern near a point P halfway between the initial positions of the two spots.

(a) Carefully argue that the speed of any atoms passing P at the time of the photograph must be $v = \frac{1}{2}d/t$, no matter which of the two spots they originally came from.

(b) The de Broglie hypothesis implies that atoms flowing in each spot form waves. Figure Q1.9 shows that waves of equal wavelength moving in opposite directions create a standing wave. Figure Q4.6 displays the *intensity* of atom interference, which amounts to the time-averaged square of the standing-wave wave function. Use this information to predict the distance between peaks of the interference pattern in the neighborhood of point P according to the de Broglie model, and compare with the observed spacing of about 15 μm. (*Hint:* If you get a result that is off by a simple multiple, you might note that the spacing between peaks of the interference pattern is *not* the same as the wavelength of the standing wave. Why not?)

Advanced

Q4A.1   Show that taken together, the two de Broglie equations $E = hf$ and $p = h/\lambda$ imply that the velocity of the crests of a beam's de Broglie waves is *greater* than that of light for any beam of particles with nonzero rest mass m moving at a speed $v < c$! Find a formula for that velocity (do *not* assume that the particles are nonrelativistic), and show that the velocity reduces to c as the particles' speed approaches c. [This faster-than-light speed does not violate the theory of relativity, because it turns out (for reasons that would take too long to explain here) that *information* actually travels along a wave at the speed $d\omega/dk$, which turns out to be simply equal to v in this case. For a full discussion, see sections 2-1 and 2-2 in A. French and E. F. Taylor, *An Introduction to Quantum Physics*, Norton, New York, 1978.

ANSWERS TO EXERCISES

Q4X.1   The wavelength is about

$$\lambda = \frac{h}{mv} = \frac{(6.63 \times 10^{-34}\ \cancel{J} \cdot s)[(1\ \cancel{kg} \cdot m^2/s^2)/1\ \cancel{J}]}{(9.11 \times 10^{-31}\ \cancel{kg})(0.01)(3.0 \times 10^8\ \cancel{m/s})}$$

$$= 2.4 \times 10^{-10}\ \text{m} = 0.24\ \text{nm} \qquad (Q4.19)$$

Q4X.2   Combining equations Q4.4 and Q4.5, we get

$$K = \frac{(h/\lambda)^2}{2m} = \frac{h^2}{2m\lambda^2} \qquad (Q4.20)$$

Solving for λ, we get

$$\lambda^2 = \frac{h^2}{2mK} \quad \Rightarrow \quad \lambda = \frac{h}{\sqrt{2mK}} \qquad \text{(Q4.21)}$$

We can obtain the second version of this equation simply by multiplying the top and bottom of the fraction by c.

Q4X.3 Equation Q4.6 is based on equation Q4.5, which is the nonrelativistic expression for the kinetic energy.

Q4X.4 Equation Q4.6 is valid only when particles are nonrelativistic. Photons move at the speed of light and are thus extremely relativistic, so the formula does not apply to them. For a generalization of equation Q4.6 that applies to all particles, relativistic or not, see problem Q4S.8.

Q4X.5 The wavelength of electrons with $K = 78$ eV is

$$\lambda = \frac{hc}{\sqrt{2mc^2 K}} = \frac{1240\,\text{eV} \cdot \text{nm}}{\sqrt{2(511,000\,\text{eV})(78\,\text{eV})}} = 0.139 \text{ nm}$$

$$\text{(Q4.22)}$$

The angle of constructive interference for $n = 1$ (according to equation Q4.9) is thus

$$\theta = \sin^{-1}\frac{\lambda}{D} = \sin^{-1}\frac{0.139\,\text{nm}}{0.215\,\text{nm}} = 40° \qquad \text{(Q4.23)}$$

Q4X.6 If the neutrons all had the same energy 0.04 eV, then their wavelength would be

$$\lambda = \frac{hc}{\sqrt{2mc^2 K}} = \frac{1240\,\text{eV} \cdot \text{nm}}{\sqrt{2(940,000,000\,\text{eV})(0.04\,\text{eV})}}$$

$$= 0.143 \text{ nm} \qquad \text{(Q4.24)}$$

This is about the same order as magnitude as the electron wavelengths in the Davisson-Germer experiment, so we might be able to see interference effects by shining such a beam on a crystal. Unfortunately, the individual neutrons in a beam of thermal neutrons will not necessarily have exactly the same speed and thus wavelength, so any interference pattern they create will be somewhat blurred.

Q4X.7 Since 65 mi/h $\approx$ 30 m/s, the "wavelength" of this hypothetical beam would be about

$$\lambda = \frac{h}{mv} = \frac{6.63 \times 10^{-34}\,\text{J} \cdot \text{s}}{(1500\,\text{kg})(30\,\text{m/s})}\left(\frac{1\,\text{kg} \cdot \text{m}^2/\text{s}^2}{1\,\text{J}}\right)$$

$$= 1.5 \times 10^{-38} \text{ m} \qquad \text{(Q4.25)}$$

which is about 100 billion trillion times smaller than an atomic nucleus. Such a ridiculously small wavelength has no testable implications and thus no physical meaning.

Q5

The Quantum Facts of Life

Chapter Overview

Introduction

This chapter launches the quantum theory subdivision by examining some experiments that clearly display the physical facts that a quantum theory must explain.

Section Q5.1: Particle or Wave?

Since all forms of matter and energy display both wave and particle aspects, we will call any entity small enough to exhibit quantum behavior (e.g., a photon, electron, proton, or atom) a **quanton.** This helps us avoid the prejudicial term *particle* and reminds us that all quantons behave similarly.

Section Q5.2: Single-Quanton Interference

This section discusses a two-slit interference experiment in which (1) we replace the display screen with a detector array capable of counting individual quantons and (2) we reduce the intensity so that only one quanton goes through the apparatus at a time. We find that individual quantons appear at apparently random locations in the detector array, but that the two-slit interference pattern emerges as a *statistical* description of the behavior of many quantons.

Section Q5.3: Implications

This behavior cannot be due to interactions *between* quantons, because only one quanton goes through the apparatus at a time. Therefore, each quanton must *individually* contribute to the interference pattern. But since the pattern's characteristics depend on the separation and width of the slits, each individual quanton must "know" about *both* slits. A wave has no trouble "knowing" about both slits, because different parts of the wave go through each slit, but it is hard to imagine a particle model that would be consistent with this fact. Yet no simple wave model can explain why the detectors register quantons individually. *Both* the wave and particle models are inadequate.

Section Q5.4: Desperately Seeking Trajectories

To test the idea that quantons might go through both slits at once, imagine placing near each slit a proximity detector that can register a quanton passing through that slit. When we do such an experiment, we find that the detectors never register a quanton going through both slits at once, but also that the two-slit interference pattern is *gone!* The very existence of the pattern depends on *not* asking embarrassing questions about quanton trajectories in such an experiment.

Section Q5.5: Spin Experiments

Electrons have a vectorlike property called **spin** whose behavior displays the same issues in an even simpler form. Any experiment we perform to determine the projection of an electron's spin on a given directed axis finds that the spin is either completely aligned with or completely antialigned with (opposite to) that axis. A

Stern-Gerlach (SG) device performs such an experiment: it accepts an electron and sends it out the plus output channel if the electron's spin is aligned with the stated axis or the minus output channel if the spin is antialigned with that axis. Experiments involving sequences of SG devices display the following aspects of spin behavior.

1. *We can generally only make probabilistic predictions about an individual electron's behavior.* This is analogous to our inability to make more than probabilistic predictions in the single-quanton two-slit interference experiment. In the spin case, we find specifically that an electron's spin that is determined to be aligned with a given axis will have a 50/50 chance of being determined to be either aligned with or antialigned with an axis perpendicular to the first. If the angle between the first and second axes is θ, the probabilities are, respectively,

$$\text{Pr(aligned)} = \cos^2 \tfrac{1}{2}\theta \qquad \text{and} \qquad \text{Pr(antialigned)} = \sin^2 \tfrac{1}{2}\theta \qquad (Q5.1)$$

2. *SG devices are self-consistent.* If we use an SG device to determine an electron's spin projection along a given axis and then immediately check the result with a second device, we get the same result.

3. *An SG device changes the electron's spin state, but only if we actually know the result.* For example, if one SG device determines that an electron's spin is aligned with the $+z$ direction, and a second device determines that the same electron's spin is aligned with the $+y$ direction, then a third SG device shows that electron's spin is no longer necessarily aligned with the $+z$ direction: its spin state was *changed* by the second device. Strangely, though, if we remix the output beams from the second SG device so that we *never really know* the result of its determination, then the third device tells us that the electron's spin is still aligned with the $+z$ direction! This is analogous to the way that the two-slit interference pattern changes depending on whether we know or do not know which slits the quantons went through.

Section Q5.6: Complex Numbers

The model of quantum mechanics is most elegantly expressed using *complex numbers*. A **complex number** is a number of the form $c = a + ib$, where a and b are ordinary real numbers and $i \equiv \sqrt{-1}$. We can add and multiply complex numbers as we would binomials. The **complex conjugate** and **absolute square** of a complex number are, respectively,

$$c^* \equiv a - ib \qquad (Q5.5a)$$

$$|c|^2 \equiv c^*c = (a - ib)(a + ib) = a^2 + b^2 \qquad (Q5.5b)$$

We define

$$e^{i\theta} \equiv \cos\theta + i\sin\theta \qquad (Q5.7)$$

where θ is real. As the notation suggests, this function behaves as if it were an exponential function [for example, $e^{i0} = 1$, $e^{i\theta_1}e^{i\theta_2} = e^{i(\theta_1+\theta_2)}$, $d(e^{i\theta})/d\theta = ie^{i\theta}$]. Note however, that

$$e^{-i\theta} = \cos\theta - i\sin\theta = (e^{i\theta})^* \qquad \text{and} \qquad |e^{i\theta}|^2 = 1 \qquad (Q5.9)$$

The mathematical model of quantum mechanics specifically involves vectors with various numbers of complex components. The traditional abstract notation for such a complex vector is $|u\rangle$. We define the **inner product** of two vectors $|u\rangle = [u_1, u_2, \ldots]$ and $|w\rangle = [w_1, w_2, \ldots]$ having the same number of components to be

$$\langle u \mid w \rangle = u_1^* w_1 + u_2^* w_2 + \cdots \qquad (Q5.11)$$

This is a generalization of the idea of the dot product of ordinary vectors. Two vectors are **orthogonal** if $\langle u \mid w \rangle = 0$ and $|u\rangle$ is **normalized** if $\langle u \mid u \rangle = |u_1|^2 + |u_2|^2 + \cdots = 1$.

Q5.1   Particle or Wave?

In chapter Q2, we saw that Young's two-slit interference experiment and the phenomenon of diffraction clearly support the wave model of light. In chapter Q3, we saw that the photoelectric effect supports a particle model of light. Many experiments support the particle nature of objects such as electrons, protons, neutrons, and atoms, but in chapter Q4, we saw that such objects can produce an interference pattern in a two-slit interference experiment just as if they were waves! Any successful theory is clearly going to have to somehow unify these very different aspects of "particle" behavior.

Definition of quanton

Indeed, because all forms of matter/energy (photons, electrons, protons, neutrons, atoms, and so on) exhibit both wave *and* particle aspects, it will help in what follows to have a word that describes them all without calling them *particles*, which prejudices the mind toward the particle model. In what follows, we will describe these objects as **quantons,** a word coined by the French quantum physicists Jean-Marc Lévy-Leblond and Françoise Balibar. Using this whimsical term actually has three very serious purposes: (1) It enables us to talk about photons, electrons, protons, neutrons, and so on in the abstract while avoiding the word *particle*. (2) It reminds us that *all* these quantons behave in the same fundamental way, exhibiting both wave and particle aspects. (3) It emphasizes that we will be focusing our attention in what follows on the strange "quantum" behavior of these objects. We can use *quanton* to refer to any object (even complicated structures such as atoms and molecules) that is an entity small and light enough to exhibit quantum (non-newtonian) behavior.

An experiment that manifests wave and particle aspects simultaneously

One approach to learning more about how we might unify the particle and wave aspects of quantons is to find an experiment that exhibits *both* aspects simultaneously. The two-slit interference experiments considered in chapters Q2 and Q4 can do this—*if* we replace whatever we were using as a display screen with an array of detectors capable of registering and counting single quantons. Furthermore, let us reduce the intensity of the beam going through the slits so that only a single quanton is in flight between the beam source and the detector array at any given time! Such an experiment should vividly illustrate whatever relationship exists between the wave and particle aspects of quantons.

Counting quantons

Of course, this presumes that we can actually detect and count individual quantons. In chapter Q3, we discussed how we can count individual photons with a photomultiplier tube. Similar tricks and techniques have been developed in recent years for detecting many types of quantons in a wide range of circumstances. How these devices work is not really crucial: my main point here is that one *can* build devices capable of registering individual quantons. In what follows, I will treat such a detector as simply a "black box" that absorbs a single quanton and produces an electrical signal in response.

Q5.2   Single-Quanton Interference

In this section we will discuss a possible single-quanton interference experiment. For the sake of concreteness I will describe the experiment as we would carry it out using *photons*, but (in principle at least) we could do the same kind of experiment with *any* kind of quanton.

Low-intensity two-slit experiment with photons

Consider the experiment illustrated in figure Q5.1. Monochromatic light from a laser goes through a set of semitransparent filters that reduce the

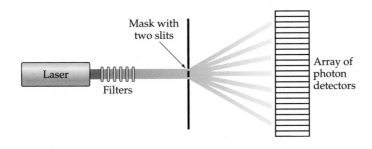

Figure Q5.1
The setup for a low-intensity two-slit interference experiment.

beam's intensity by absorbing most of its light. The light that gets through the filters goes through a double slit and ends up at an array of photon detectors. The photon detectors are connected to a computer that records when and where each photon hits the array.

Imagine that we use the filters to set the intensity so that there is only a 1/1000 chance that there is a photon in flight between the filters and the detector array at any given time. Even though the laser produces photons at a huge rate (on the order of 10^{15} photons per second for a typical laser), you can show that if each filter transmits only 5 percent of the light falling on it, it only takes a handful of filters in series to reduce the intensity by this much. Such an intensity means that there is essentially no possibility that photons going through the apparatus will be able to influence each other: 99.9 percent of the photons going through the apparatus will do so entirely *alone*.

Exercise Q5X.1

Assume that we have a laser that emits about 3×10^{15} photons per second. If the distance between the filters and the detector array is about 3 m, then it will take a photon about 10^{-8} s to cover this distance. If we arrange things so that an average of one photon every 10^{-5} s (i.e., an average of 10^5 photons per second) gets through the last filter, then there will be only a 1/1000 chance that a photon is in flight at any given time. So we want to reduce the 3×10^{15} photons per second emitted by the laser to roughly 10^5 photons per second emerging from the last filter. Show that a series of eight filters very nearly gives the required reduction.

The wave and particle models of light predict very different outcomes for this experiment. The wave model predicts that the superposition of light waves from each slit will result in an interference pattern at the plane of the detectors: detectors located where the waves constructively interfere will therefore be bathed in more intense light than detectors located where the waves destructively interfere. But the wave model also predicts that the detectors will *not* register discrete pulses of energy, but rather that the gently oscillating waves will smoothly and slowly deposit energy into each detector at a rate that depends on the detector's position in the interference pattern.

The particle model of light, on the other hand, predicts that the detectors will register discrete bursts corresponding to the absorption of individual photons. But any simple particle model of light predicts that there should be two bright spots at the detector array (one behind each slit) and no interference pattern (because only *waves* interfere with each other).

When we actually do this experiment, this is what happens. The detectors *do* register discrete detection events, consistent with the particle model. Rather than observing the events to be confined to two small regions directly

Predictions of the wave and particle models

What really happens

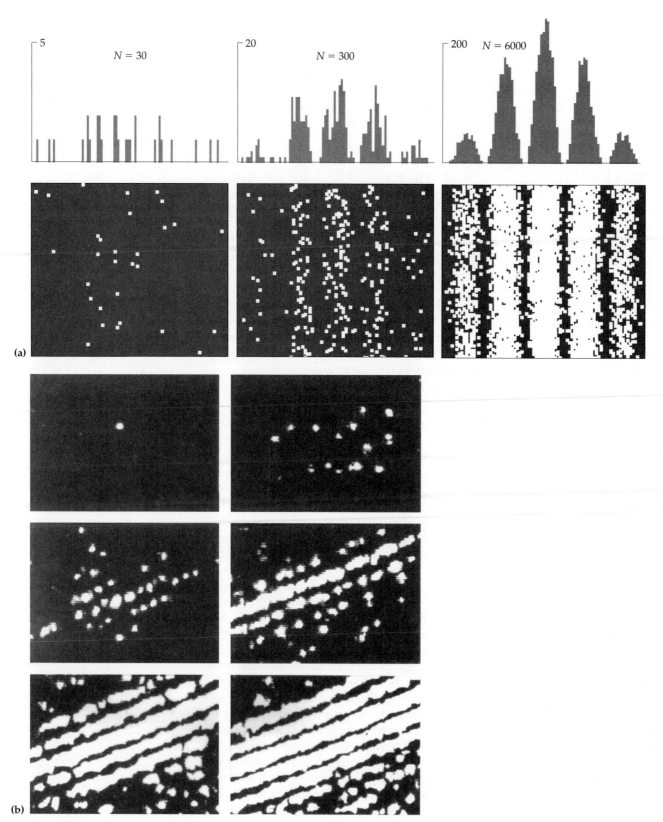

Figure Q5.2

(a) Simulated data for a low-intensity two-slit interference experiment. In each of the three cases, the bottom box shows what the individual hits might look like displayed on a screen. The top graph shows a histogram displaying the number of hits as a function of displacement perpendicular to the beam direction. (b) An actual picture of a developing double-slit electron interference pattern.

From P. G. Merli, G. F. Missiroli, and G. Pozzi, *Journal of American Physics*, Vol. 44, No. 3, March 1976, Fig. 1, pg. 306.

behind each slit, though, the photon detection events seem at first to be spread *randomly* over a much larger area of the detector array than we would expect from the particle model. After we count thousands of photons, though, it becomes plain that photons seem to prefer certain detector locations and to avoid others. After we have recorded millions of detection events, we can see that a typical two-slit interference pattern emerges as the *statistical* outcome of a huge number of what look like individually random photon detection events (see figure Q5.2).[†]

Thus, it seems to be impossible to predict which detector in the array will receive any *given* photon: that appears to be random. *Statistically*, though, we see that detectors at positions corresponding to the bright spots of the standard two-slit interference pattern are more likely to register photons than those that happen to be at dark spots. It turns out that the mathematical *probability* that a given photon will arrive at a given detector is exactly proportional to the intensity of the interference pattern at that detector that is predicted by the *wave* model of light.

We see, therefore, that the particle model accurately describes the *interaction of each individual photon with a detector* in this experiment, while the wave model describes *the statistical distribution* of the photons collectively.

Exercise Q5X.2

Can we use a two-slit interference experiment to measure the wavelength of a single photon? Explain how or why not.

Q5.3   Implications

The behavior of quantons in this experiment is clearly *not* newtonian. The first thing to notice is the apparently unpredictable behavior of individual quantons. Instead of having the nice, predictable trajectories that we expect of projectiles in newtonian mechanics, the quantons instead appear to be scattered by the slits in individually unpredictable directions.

The second thing to notice is that the interference pattern characterizes not the behavior of individual quantons but the *statistical* behavior of many quantons. Even though the individual quantons *appear* to behave unpredictably, their behavior obeys a deep underlying order that is not apparent until we have registered many quantons. Moreover, this order is definitely *not* what any simple particle model would predict. What possible mechanism could impose this order?

One possible mechanism is to postulate some kind of interaction *between* quantons: perhaps quantons emerging from one slit exert some kind of force on those emerging from the other slit, and vice versa, and this interaction somehow diverts them into beams that give rise to the interference pattern. Unfortunately, this explanation doesn't work. Remember that we arranged the experiment so that there was only 1/1000 chance that a single photon was in flight at any given time. This means that during the flight time of any given photon there is only a 1/1000 chance that another photon is also in flight. The chance that both are near enough to each other to significantly

Quanton behavior is not newtonian!

Each individual quanton must "know" about both slits

[†]The computer program Interference, which is freely available on the *Six Ideas* website, simulates single-quanton one-slit and two-slit interference experiments.

affect each other's motion at the slits is far smaller, so only a tiny fraction of photons arriving at the detectors could have possibly been influenced by another photon. People have done this experiment over a wide range of intensities, and the hard truth is that the interference pattern produced by a million quantons of any type going through two slits is the same whether a million quantons per second or one quanton per second goes through the slits.

We are thus compelled against our will to understand that this interference pattern is generated by *individual* quantons, one at a time, and not by interacting quantons. But this means that each individual quanton must "know" something about which points on the detector array to prefer or to avoid. This leads to a serious conceptual problem with *any* kind of particle model because the size of the interference pattern depends on the *separation* of the two slits. If individual quantons are to generate this pattern, each individual quanton must somehow "know" about the slit separation, so that it can know what positions in the interference pattern it should prefer or avoid. But how can a *particle* know about two slits separated by a significant distance? A real "particle" can only go through one slit or the other, and there is no obvious mechanism that would allow it to even know about the *presence* of the slit that it doesn't go through, much less its size, distance, or other characteristics. A *wave*, on the other hand, does know about both slits because different parts of the wave *do* go through each slit. But if the quanton is really a wave, as this argument seems to require, then why is it always registered by the detector as a particle?

Neither the particle model nor the wave model is adequate at all

This starkly illustrates the complete inadequacy of either the particle or the wave model to explain the observed behavior of quantons in this experiment. A particle model cannot explain where the interference pattern comes from. A wave model cannot explain why the quantons always are detected as if they were particles. Neither model can explain the seemingly random behavior of individual quantons. **Classical models** of particle or wave behavior simply break down when we attempt to apply them to this situation.

You may justifiably feel that the quanton behavior described here does not make sense. Quantons in this context simply do not behave like *anything* that we are familiar with in daily life. Because there is nothing in daily life upon which we can build an analogy, we cannot "make sense" of these results by comparing them to something familiar. This is one of the basic problems of quantum mechanics (and 20th-century physics in general): it is simply something that we will have to deal with.

Q5.4   Desperately Seeking Trajectories

If we are unwilling as yet to give up the idea that quantons really are particles, then one of the questions that we might want to answer about the two-slit interference experiment concerns exactly what kinds of trajectories quantons follow in this experiment. Knowing something about these trajectories might tell us something about how quantons are able to respond to both slits.

Two-slit experiment with proximity detectors

Therefore, let us modify our two-slit interference experiment by adding a *proximity detector* near each slit. Unlike the detectors in the detector array far beyond the slit (which are triggered when they absorb a quanton), imagine that these proximity detectors are triggered simply by proximity to the quanton as it passes by. (If the quanton is charged, e.g., the proximity detector might detect the quanton's electric or magnetic field.) These detectors are arranged so that a quanton passing through slit *A* triggers detector *A* and an electron passing through slit *B* triggers detector *B*. Such proximity detectors

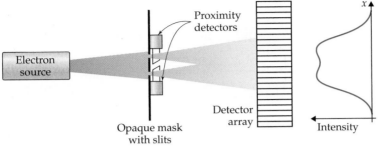

Figure Q5.3
The outcome of a two-slit interference experiment with proximity detectors near each slit. Note that there are two overlapping single-slit patterns, but there is no interference.

allow us to "view" each quanton's trajectory at the most critical place in its journey between the source and the final detector array, and they represent a necessary first step to determining the trajectories in greater detail.

People in fact have done more realistic versions of the idealized experiment I have just described. The results superficially seem to support the idea that a quanton has a well-defined trajectory, because only one proximity detector ever fires at a time, indicating that the quantons do not seem to go through both slits at once. But the horrible thing is that when we look for the interference pattern at the detector array after many electrons have passed through the slits, *it is no longer there!* Instead, we see the pattern that we would expect for ordinary particles going through two slits (see figure Q5.3). This is not a problem with the design of the proximity detectors: different detector designs yield the same results (assuming that newtonian mechanics would predict that the effect of the detector on the quanton was negligible).

So it seems that one *cannot* measure a quanton's trajectory in this experiment without radically affecting the results of the experiment. The very *existence* of the interference pattern depends on our *not* asking even the most rudimentary questions about the quanton's trajectory! We can meaningfully talk about the trajectory of a baseball or planet because we know from experience that such an object's trajectory does not seem to depend on whether we have been trying to observe it or not. However, it is simply *not meaningful* to talk about a quanton's trajectory in a two-slit interference experiment. Not only does the concept of a trajectory lead to logical absurdities (such as a particle whose trajectory goes through one slit somehow "sensing" the other slit), but it seems that at least in this experiment we cannot measure *anything* about this hypothetical trajectory without screwing up the results. If we cannot measure a quanton's trajectory in this experiment and the idea doesn't help us in understanding the experiment's outcome, then is the idea worth believing?

There is no way around it: the behavior described here is *very* bizarre. No slight modification of newtonian mechanics is going to explain this behavior: what we need is a radically new way to think about quantons. In chapter Q6, I will begin to outline the theory of **quantum mechanics,** which provides such a fresh perspective.

> Trying to examine quanton trajectories kills the interference pattern!

> We need a new approach to mechanics

Q5.5   Spin Experiments

The two-slit interference experiment, however, has certain complexities that obscure the simplicity and logic of quantum mechanics. Experiments that explore a phenomenon called *spin* display exactly the same basic issues but in a much simpler form. In chapter Q6, we will use this phenomenon as a "toy problem" to display in a straightforward way how we can solve quantum mechanical problems. This will provide a strong foundation for discussing the more complicated problem of two-slit interference later in that chapter.

> Introduction to *spin*

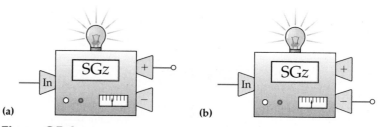

Figure Q5.4
A Stern-Gerlach device that determines the projection of an electron's spin on the +z direction. Diagram (a) shows the device registering an electron as having spin aligned with the z axis; diagram (b) shows the device registering an electron as having a spin antialigned with that direction.

We will discuss the physics of spin and its applications in much greater detail in chapter Q8. It is enough for the present to know that electrons have a vectorlike property called **spin** (which shares many properties of the classical angular momentum vector expressing rotation about an axis) and that if we do an experiment to determine the projection of a given electron's spin on any directed axis, we always find the spin to be either totally *aligned* with that axis (that is, it points in the *same direction* as that axis), or totally *antialigned* with the axis (meaning that it points in the *opposite* direction), never anything in between. This is strange and nonnewtonian behavior, but it is an experimental fact.

We can determine this alignment with an experimental apparatus we will call a **Stern-Gerlach, or SG,** device. We will explore how this device works in greater detail in chapter Q8, but for now we will treat this device as simply being a black box that accepts an electron through an input port at one end and emits it from the output channel labeled plus if electron's spin is aligned with the axis in question or the channel labeled minus if the spin is antialigned with that axis (see figure Q5.4). The label on the box indicates the direction along which the device determines the spin projection. For example, an SGx device determines the spin's projection S_x in the +x direction, an SGy device determines the spin's projection S_y in the y direction, an SGz device determines the spin's projection S_z in the z direction, and an SGθ device determines the spin's projection S_θ on an axis in the yz plane that makes an angle of θ with the z axis.

In what follows, I will describe a set of three different experiments that display (in this particular context) the same strange behaviors that troubled us in the two-slit interference experiment. I want to make it very clear that in spite of the way that I have abstracted and simplified these experiments, realistic versions of these experiments have been performed and electrons *really do* behave exactly as I am about to describe.

We can only make statistical predictions about electron spin behavior

Consider the experiment shown in figure Q5.5a. In this experiment, the first SG device ensures that all electrons entering the second device have spins aligned with the z direction. The second device then determines the spin projection of these electrons on the y direction. We find experimentally we *cannot* predict whether the second device will determine that a given electron's spin is aligned or antialigned with the y direction: each electron seems to choose an output channel at random. However, we find after taking many measurements that the *probability* that the electron will come out either of the two channels is $\frac{1}{2}$. The same probabilities apply if the second device is an SGx device as in figure Q5.5b; if we have an SGx device followed by an SGy device, as in figure Q5.5c; or if we reverse the order of devices in any of these pictures (see figure Q5.5d). This illustrates that, just as we found in the

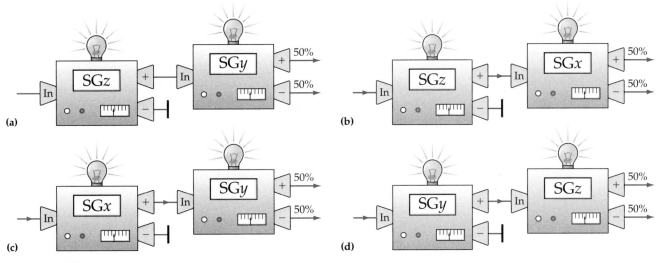

Figure Q5.5
When we determine an electron's spin projection on one axis and then immediately determine its spin projection on a perpendicular axis, we always find a 50 percent probability that the SG device will determine it to have a spin aligned with or antialigned with that second axis.

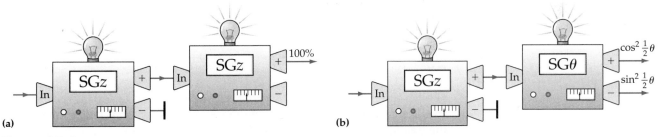

Figure Q5.6
(a) Two identical experiments in succession yield consistent results. (b) If we determine an electron's spin projection on one axis and then immediately determine its projection on an axis tilted at an angle θ relative to the first, the probabilities of the two possible outcomes change smoothly as θ increases.

two-slit interference experiment, we cannot predict the behavior of individual electrons, but we can make *statistical* predictions about electron behavior.

It might seem, however, that our SG devices are just *arbitrarily* sorting these electrons into the two output channels. The experiment shown in figure Q5.6a dispels this notion. If the first SGz device determines that a given electron has its spin aligned with the +z direction, the second device will *always* concur with that result. We see, therefore, that the SG devices really do determine an electron's spin in a self-consistent way. Figure Q5.6b shows that if the second device determines the spin projection along an axis lying in the yz plane that makes an angle of θ with the z axis, then the probabilities that an electron determined to have spin aligned with the +z axis by the first device will be determined by the second to have spin aligned with or antialigned with the tilted axis are

$$\Pr(\text{aligned}) = \cos^2 \tfrac{1}{2}\theta \quad \text{and} \quad \Pr(\text{antialigned}) = \sin^2 \tfrac{1}{2}\theta \quad \text{(Q5.1)}$$

respectively. Note that these probabilities match the probabilities for the case shown in figure Q5.6a if $\theta = 0$ and match those shown in figure Q5.5a if $\theta = 90°$. We see that there is clearly an underlying order in these probability predictions, but we can still only predict probabilities (unless we exactly repeat an experiment as in figure Q5.6a).

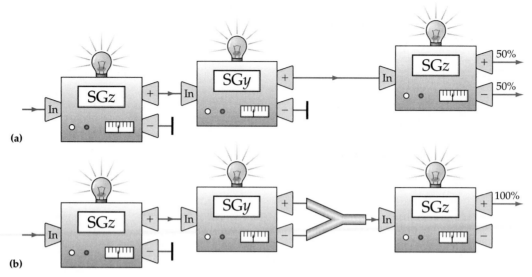

Figure Q5.7

(a) This figure illustrates that determining an electron's spin affects its state. The intervening determination of projection on the y axis means that electrons entering the third device no longer necessarily have their spin aligned with the z axis. (b) On the other hand, if we recombine electrons from the second device so that we cannot determine which path the electron followed, the result is the same as if the second device were not there!

Exercise Q5X.3

Show that we really do get the probabilities shown in figure Q5.5a if $\theta = 90°$ (i.e., if the tilted axis of the second device in figure Q5.6b is really the same as the y axis).

Determining spin alignment changes an electron's spin state!

Finally, consider the case shown in figure Q5.7a. Here the electrons have been determined by the first device to have spin aligned with the $+z$ direction. The second device determines that one-half of these electrons also have spin aligned with the $+y$ direction. We find that the third device determines that one-half of these electrons have spin aligned with the $+z$ direction and one-half have spin antialigned with that direction. This is in spite of the fact that *every electron going into the third device was originally determined by the first device to have spin* aligned with the $+z$ direction!

This illustrates something very important and nonintuitive about quantum mechanics. In newtonian mechanics, we can in principle *measure* the value of some property of a system without affecting that property or any other property of the system. But in quantum mechanics, doing an experiment that determines the value of some quanton property often actually *changes* the quanton's state. In this case, determining the electron's spin alignment with the y axis causes it to "forget" its original alignment with the z axis. This is analogous to the way that putting proximity detectors near the slits in the two-slit interference experiment causes the quantons to "forget" that they were supposed to be contributing to a two-slit interference pattern.

On the other hand, if we remix the electrons emerging from the second device so that we have *no way of knowing* which electron followed which path through that device, then we get the result shown in figure Q5.7b. Now we see that the electrons *do* "remember" their original spin orientation! This is directly analogous to the way that we get an interference pattern in the two-slit experiment only if we have *no way of knowing* which slit a given quanton went through. This is an intrinsic feature of quantum phenomena that has nothing to do with the specific designs of the SG devices or proximity detectors.

These are the quantum facts of life. They seem impossible (even patently absurd!), but this *is* the way that the universe works at the microscopic level. Our job here is not to complain but to explain. Our minds rebel at the very notion, though, because *nothing* at the macroscopic level of our daily experience behaves even remotely like this.

The quantum facts of life

But, if you step back and think about it calmly, there is a kind of (twisted) logic to this behavior. Our goal in chapter Q6 will be to find this logic and put it on a firmer mathematical foundation.

Exercise Q5X.4

The drawing below shows a sequence of SG devices. Use analogies to the cases discussed in this section and logic to determine the probabilities that an electron entering the final device will leave through its plus and minus channels, respectively.

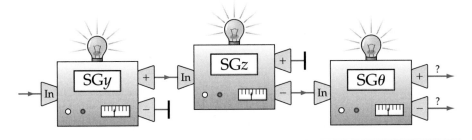

Q5.6   Complex Numbers

The model of quantum mechanics we will explore in chapter Q6 is most elegantly expressed using *complex numbers.* In this section, I would like to review some things about complex numbers so that we will be fully prepared for the next chapter.

A **complex number** is essentially a clever way of compressing two real numbers into a single quantity. Complex numbers are written in the form

The definition of a complex number

$$c = a + ib \tag{Q5.2}$$

where c is the complex number, a and b are ordinary real numbers, and

$$i \equiv \sqrt{-1} \tag{Q5.3}$$

We call the quantity a the **real part** of c and the quantity b the **imaginary part** of c. The quantity ib is considered an **imaginary number,** because $(ib)^2 = i^2 b^2 = -b^2$, which is a negative number. Because negative numbers do not have square roots according to the standard rules of real arithmetic, we call any quantity whose square is negative an imaginary number. Note that we can consider a real number to be a complex number whose imaginary part is zero. Similarly, we can have complex numbers that are "purely imaginary," that is, whose real part is zero.

We define addition and multiplication of complex numbers as follows:

Operations with complex numbers

$$c_1 + c_2 \equiv (a_1 + ib_1) + (a_2 + ib_2) \equiv (a_1 + a_2) + i(b_1 + b_2) \tag{Q5.4a}$$

$$c_1 c_2 \equiv (a_1 + ib_1)(a_2 + ib_2) \equiv a_1 a_2 - b_1 b_2 + i(a_2 b_1 + a_1 b_2) \tag{Q5.4b}$$

Note that the rules for these operations are pretty much as one would expect: simply treat the complex number as if it were a binomial. Division by a complex number is hard to define, but division of a complex number by a real number is the same as multiplying that complex number by the inverse of the real number.

Complex conjugate and absolute square

Given a complex number $c = a + ib$, we define its **complex conjugate** c^* and its **absolute square** $|c|^2$ as follows:

$$c^* \equiv a - ib \qquad (Q5.5a)$$

$$|c|^2 \equiv c^*c = (a - ib)(a + ib) = a^2 + b^2 \qquad (Q5.5b)$$

Note that the absolute square of a complex number is always a nonnegative purely real number. The complex conjugate has the useful following properties:

$$(c^*)^* = c \qquad (Q5.6a)$$

$$(c_1 + c_2)^* = c_1^* + c_2^* \qquad (Q5.6b)$$

$$(c_1 c_2)^* = c_1^* c_2^* \qquad (Q5.6c)$$

$$c^* = \begin{cases} c & \text{if } c \text{ is real} \\ -c & \text{if } c \text{ is imaginary} \end{cases} \qquad (Q5.6d)$$

We will find both the complex conjugate and the absolute square *very* useful in chapter Q6.

Exercise Q5X.5

Calculate the following complex quantities.

a. $(1 + i) + (2 - 3i)$
b. $1 + (-1 - 2i)$
c. $(1 + i)(2 - 3i)$
d. $3i(2 - 5i)$
e. $[3i(2 - 5i)]^*$
f. $|-5 + 12i|^2$

The function $e^{i\theta}$

The function $e^{i\theta}$ is *defined* as follows.

$$e^{i\theta} \equiv \cos\theta + i\sin\theta \qquad (Q5.7)$$

For the moment, do not worry about the notation on the left: simply take this equation as a *definition*. This function is basically a clever way of combining both a cosine and a sine into a single function, and it is useful in many contexts where we want to describe oscillations or waves mathematically (you can see why it might be useful in quantum mechanics!).

Using this definition, one can easily prove the following properties of this function:

$$e^{i0} = 1 \qquad (Q5.8a)$$

$$e^{i\theta_1} e^{i\theta_2} = e^{i(\theta_1 + \theta_2)} \qquad \text{(even if } \theta_1 \text{ or } \theta_2 \text{ is negative)} \qquad (Q5.8b)$$

$$\frac{d}{d\theta}(e^{i\theta}) = ie^{i\theta} \qquad (Q5.8c)$$

These properties are exactly what we would expect if $e^{i\theta}$ were a genuine exponential function whose exponent contains a constant i. *This* is why we use the notation $e^{i\theta}$ for this function; it is a mnemonic device to remind us of some of our function's most important properties.

Exercise Q5X.6

Verify equations Q5.8. (*Hint:* You will need to use some trigonometric identities to prove equation Q5.8*b*.)

We will also find useful the following special properties of the function $e^{i\theta}$ that are *not* really analogous to the exponential of a real number:

$$|e^{i\theta}|^2 = 1 \qquad\qquad (Q5.9a)$$

$$e^{-i\theta} \equiv e^{i(-\theta)} = \cos\theta - i\sin\theta = (e^{i\theta})^* \qquad\qquad (Q5.9b)$$

$$e^{i\theta} + e^{-i\theta} = 2\cos\theta \qquad\qquad (Q5.9c)$$

$$e^{i\theta} - e^{-i\theta} = 2i\sin\theta \qquad\qquad (Q5.9d)$$

(assuming that θ is real). Note that $|e^{i\theta}|^2$ does not grow as θ grows, unlike the exponential of a real number.

The mathematical model of quantum mechanics specifically uses column vectors which have varying numbers of complex components. The conventional way to represent such a vector in quantum mechanics is to enclose a symbol on the left with a vertical line and on the right with an angle bracket as follows:

$$|\psi\rangle = \begin{bmatrix} \psi_1 \\ \psi_2 \\ \vdots \end{bmatrix} \qquad\qquad (Q5.10a)$$

where $\psi_1, \psi_2, \ldots$ are complex numbers. The reversed symbol refers to a vector whose components are the complex conjugates of the original vector's components:

$$\langle\psi| = \begin{bmatrix} \psi_1^* \\ \psi_2^* \\ \vdots \end{bmatrix} \qquad\qquad (Q5.10b)$$

If two vectors $|u\rangle$ and $|w\rangle$ have the same number of components, then we can define their **inner product** $\langle u \mid w\rangle$ as follows:

$$\text{If} \qquad |u\rangle = \begin{bmatrix} u_1 \\ u_2 \\ \vdots \end{bmatrix} \qquad \text{and} \qquad |w\rangle = \begin{bmatrix} w_1 \\ w_2 \\ \vdots \end{bmatrix}$$

$$\text{Then} \qquad \langle u \mid w\rangle \equiv \begin{bmatrix} u_1^* \\ u_2^* \\ \vdots \end{bmatrix} \cdot \begin{bmatrix} w_1 \\ w_2 \\ \vdots \end{bmatrix} = u_1^* w_1 + u_2^* w_2 + \cdots \qquad (Q5.11)$$

Note that this is just like the component definition of the dot product of ordinary vectors except that we use the complex conjugate of the first vector's components in the products and the result is a complex scalar instead of being a real scalar. Think of the inner product as simply being a generalized dot product.

We say that two complex vectors are **orthogonal** if their inner product is zero. The squared magnitude of a vector (by analogy to the dot product) is the inner product of that vector with itself:

$$[\text{mag}(|u\rangle)]^2 \equiv \langle u \mid u\rangle = u_1^* u_1 + u_2^* u_2 + \cdots = |u_1|^2 + |u_2|^2 + \cdots \quad (Q5.12)$$

Vectors with complex components

Note that this magnitude is always a nonnegative real number. We say that the vector is **normalized** if its magnitude is equal to 1.

Exercise Q5X.7

Consider the two-component complex vector $|\psi\rangle = [a, ia]$, where a is a real number. What value must a have if the vector is normalized?

TWO-MINUTE PROBLEMS

Q5T.1 Consider the experimental evidence that we have discussed in chapters Q1 through Q4. Classify the following experimental results according to this scheme: These results

A.   Are consistent with a pure wave model for the quantons in question

B.   Are consistent with a pure particle model for the quantons in question

C.   Are consistent with either model (the results do not distinguish between models)

D.   Cannot be explained by either model alone

(a) When light goes through a slit, the beam broadens somewhat, and if it is projected on a screen, one can see bright and dark fringes on the slit image.

(b) When light shines on a metal surface, electrons are ejected from the surface.

(c) The number of electrons ejected from a metal surface depends directly on the intensity of the light shining on that metal surface.

(d) The maximum kinetic energy of electrons ejected from a metal surface depends on the frequency of the light shining on the surface, and not on the intensity.

(e) When light shines on a surface, at least a few electrons seem to get ejected virtually as soon as the surface is illuminated, even if the light is extremely dim.

(f) An electron beam shining on a nickel crystal is preferentially reflected in certain directions instead of being scattered uniformly in all directions.

(g) Light that goes through two narrow, closely spaced openings in an opaque barrier and then falls on a screen creates a pattern of multiple bright spots on the screen.

(h) When dim light falls on a photomultiplier, the photomultiplier produces discrete electric pulses at random times.

(i) A rapidly moving electron leaves a track in a bubble chamber photograph.

(j) Photons sent one at a time through two slits appear individually to hit a detector array at random positions, but statistically they fall into a standard two-slit interference pattern.

Q5T.2 Figure Q5.8 displays a sample interference pattern and five interference patterns labeled A through E. Each row of the table below describes a set of parameters that will create one of the displayed patterns (the first row shows the parameters that display the "Sample" pattern). The table rows are organized so that each set of parameters differs from the previous one by a change in only one value. The Detectors column refers to whether there are proximity detectors operating at the slits. Your task is to write the appropriate letter in the rightmost column of the table. You might use some of the letters more than once. (Thanks to David Tanenbaum and Jason Evans for developing this exercise.)

Case	Wavelength	Slit Width	Slit Separation	Detectors?	Pattern
Sample	5 nm	3 μm	10 μm	No	Sample
(a)	5 nm	6 μm	10 μm	No	_____
(b)	5 nm	6 μm	20 μm	No	_____
(c)	10 nm	6 μm	20 μm	No	_____
(d)	10 nm	6 μm	20 μm	Yes	_____
(e)	10 nm	6 μm	10 μm	Yes	_____

Q5T.3 The drawing below shows a sequence of Stern-Gerlach devices. By analogy to the cases discussed in the chapter, what do you think are the probabilities that an electron entering the last device will come out of the plus and minus channels of that device?

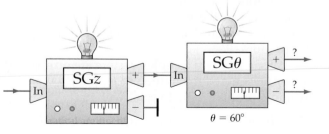

$\theta = 60°$

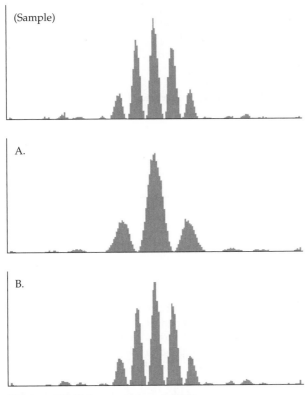

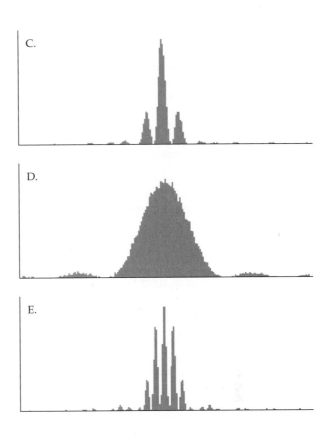

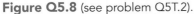

Figure Q5.8 (see problem Q5T.2).

A.   1 and 0, respectively
B. $\frac{3}{4}$ and $\frac{1}{4}$, respectively
C. $\frac{1}{2}$ for both channels
D. $\frac{1}{4}$ and $\frac{3}{4}$, respectively
E.   0 and 1, respectively
F.   Some other probabilities (specify)

Q5T.4 The drawing below shows a sequence of Stern-Gerlach devices. By analogy to the cases discussed in the chapter, what do you think are the probabilities that an electron entering the last device will

come out of the plus and minus channels of this device?

A.   1 and 0, respectively
B. $\frac{3}{4}$ and $\frac{1}{4}$, respectively
C. $\frac{1}{2}$ for both channels
D. $\frac{1}{4}$ and $\frac{3}{4}$, respectively
E.   0 and 1, respectively
F.   Some other probabilities (specify)

Q5T.5 The drawing below shows a sequence of Stern-Gerlach devices. By analogy to the cases discussed in this chapter, what do you think are the probabilities that an electron entering the last device will come out of the plus and minus channels of this device?

A.   1 and 0, respectively
B. $\frac{3}{4}$ and $\frac{1}{4}$, respectively
C. $\frac{1}{2}$ for both channels
D. $\frac{1}{4}$ and $\frac{3}{4}$, respectively
E.   0 and 1, respectively
F.   Some other probabilities (specify)

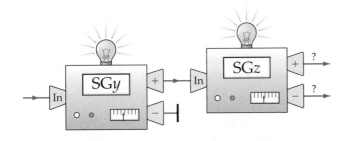

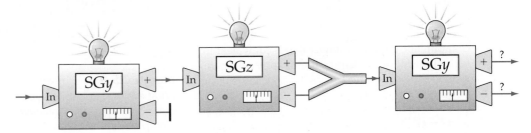

Q5T.6 $e^{-i\theta} = 1/e^{i\theta}$, true (T) or false (F)?

Q5T.7 The inner product of two complex vectors is a real number, T or F? The magnitude of a complex vector is a real and nonnegative number, T or F?

Q5T.8 Consider the two-component complex vector $|\psi\rangle = [a, ib]$, where a and b are real numbers. Which of the following vectors are orthogonal to this vector?

A. $|\psi_A\rangle = [a, ib]$
B. $|\psi_B\rangle = [ia, b]$
C. $|\psi_C\rangle = [ib, a]$
D. $|\psi_D\rangle = [b, -ia]$
E. $|\psi_E\rangle = [ib, -a]$
F. C and D
T. Some other vector (specify)

HOMEWORK PROBLEMS

Basic Skills

Q5B.1 A standard laboratory helium-neon laser produces about 1 mW (that is, 0.001 J/s) of power in the form of light at a wavelength of 633 nm. How many photons per second does this laser produce?

Q5B.2 A certain argon laser produces about 5 mW of power in the form of light at a wavelength of 514 nm. How many photons per second does this laser produce?

Q5B.3 Imagine that you have a supply of filters that pass 15 percent of the light incident on them. How many such filters would you have to use in series to reduce the intensity of a laser beam by at least a factor of 3×10^{10}? (Be sure to show your work.)

Q5B.4 Consider the situation shown in figure Q5.6b. If the axis of the SGθ device is oriented at an angle of 45°, what is the probability that an electron entering the device will leave through the plus channel? The minus channel?

Q5B.5 Consider the situation shown in figure Q5.6b. At what angle would the SGθ device have to be oriented if the probabilities that an electron entering the device left from the plus and minus channels are $\frac{1}{3}$ and $\frac{2}{3}$, respectively?

Q5B.6 Compute the complex number that is equivalent to each of the following expressions. Express your answer in the form $a + bi$.
(a) $(3 + 5i) + (-2 + i)$
(b) $2i(-3 + i)$
(c) $(1.2 + 0.5i)(1.2 + 0.5i)$
(d) $(2 + 3i)^*$
(e) $[(1 - i)i]^*$
(f) $|\frac{3}{4} + \frac{5}{4}i|^2$

Q5B.7 Compute the complex number that is equivalent to each of the following expressions. Express your answer in the form $a + bi$.
(a) $(2 - i) + (-3 + i)$
(b) $(6 + 3i)(1 - i)$
(c) $(2 + i)^*(3 - i)$

(d) $(e^{-i\pi})^*$
(e) $|-3 - 2i|^2$
(f) $|\frac{3}{5} - \frac{4}{5}i|^2$

Q5B.8 Verify equations Q5.9.

Q5B.9 Compute the inner product $\langle u | w \rangle$ of the following two-component complex vectors.
(a) $|u\rangle = [1, -i]$ and $|w\rangle = [2i, 3]$
(b) $|u\rangle = [1, -2]$ and $|w\rangle = [i, -5]$
(c) $|u\rangle = [1 + i, -2 + i]$ and $|w\rangle = [i, 2 - i]$

Q5B.10 Find the value of a that normalizes the following two-component complex vectors.
(a) $|\psi\rangle = [a, -2ia]$
(b) $|\psi\rangle = [a(1 + i), ai]$
(c) $|\psi\rangle = [ae^{i\theta}, ae^{-i\theta}]$

Synthetic

Q5S.1 Imagine that you want to run a single-quanton two-slit interference experiment with electrons. If the distance between your electron gun and the electron detectors is about 1.2 m, your electrons have a kinetic energy of about 50 eV, and you want there to be only a 1/10,000 chance that an electron is in flight at any instant, what should the current represented by the electron beam be (in amperes)?

Q5S.2 Figure Q5.9 shows a sequence of Stern-Gerlach devices. Use analogies to the cases discussed in

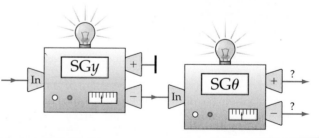

Figure Q5.9
What are the probabilities of the two possible results of the final spin determination experiment?
(See problem Q5S.2.)

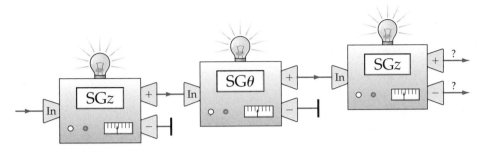

Figure Q5.10
What are the probabilities of the two possible results of the final spin determination experiment? (See problem Q5S.3.)

section Q5.5 and logic to determine the probabilities that an electron entering the final SGθ device will leave through the plus and minus channels, respectively. Express your answer in terms of θ and explain your reasoning.

Q5S.3 Figure Q5.10 shows a sequence of Stern-Gerlach devices. Use analogies to the cases discussed in section Q5.5 and logic to determine the probabilities that an electron entering the final SGz device will leave through the plus and minus channels, respectively. Express your answer in terms of θ and explain your reasoning. (*Hints:* What, if anything, does the first device do? What would the probabilities be if θ were 0°? 180°?)

Q5S.4 Prove the properties of the complex conjugate listed in equations Q5.6.

Q5S.5 Prove the following statements about the inner product of two complex vectors with the same arbitrary number of components.
(a) $\langle u \mid w \rangle = \langle w \mid u \rangle^*$
(b) $|\langle u \mid w \rangle|^2 = |\langle w \mid u \rangle|^2$

Q5S.6 The vectors

$$|\psi_a\rangle = \begin{bmatrix} 1 \\ 0 \end{bmatrix} \quad \text{and} \quad |\psi_b\rangle = \begin{bmatrix} \sqrt{\tfrac{1}{2}} \\ \sqrt{\tfrac{1}{2}} \end{bmatrix} \quad \text{(Q5.13)}$$

are normalized and satisfy $|\langle \psi_a \mid \psi_b \rangle|^2 = \tfrac{1}{2}$. Find a *third* normalized vector $|\psi_c\rangle$ such that $|\langle \psi_a \mid \psi_c \rangle|^2 = \tfrac{1}{2}$ and $|\langle \psi_b \mid \psi_c \rangle|^2 = \tfrac{1}{2}$. Explain your method. (Even if you start with trial and error, see if you can work backward to find a logical method for finding this vector.)

Rich-Context

Q5R.1 Figure Q5.11 shows a sequence of Stern-Gerlach devices. Use analogies to the cases discussed in section Q5.5 and logic to determine the probabilities that an electron entering the final SG device will leave through the plus and minus channels. Express your answer in terms of θ, and explain your reasoning.

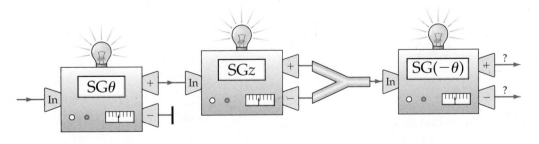

Figure Q5.11
What are the probabilities of the two possible results of the final spin determination experiment? (See problem Q5R.1.)

ANSWERS TO EXERCISES

Q5X.1 Let the original intensity of the light be I_0. After the first filter the intensity is $0.05I_0$, after the second filter it is $(0.05)(0.05I_0)$, and so on. So the intensity after the eighth filter is

$$(0.05)^8 I_0 = (3.9 \times 10^{-11}) I_0 \qquad (Q5.14)$$

Since the intensity of a monochromatic light beam is proportional to the number of photons per second flowing, if we have photons flowing at a rate of 3×10^{15} photons per second out of the front end of the laser, the rate of photons flowing out of the final filter is

$$(3.9 \times 10^{-11})(3.0 \times 10^{15} \text{ photons/s})$$
$$= 120{,}000 \text{ photons/s} \qquad (Q5.15)$$

Q5X.2 In a two-slit interference experiment, we determine the wavelength by analyzing the distance between bright spots in the interference pattern (see chapter Q2). Since we cannot tell anything about the interference pattern from where a *single* photon lands in the detector array, we *cannot* use the experiment to measure the wavelength of a single photon. If a large number of indistinguishable photons from the same source together create a clear interference pattern corresponding to a sharply defined wavelength, then we might be able to *infer* that each photon in the set has the same wavelength and calculate the value of the wavelength; but this does not quite amount to measuring that specific photon's wavelength.

Q5X.3 If $\theta = 90°$, the SGθ device would be oriented in the same way as an SGy device. If $\theta = 90°$, the general probability formulas yield

$$\Pr(\text{aligned}) = \cos^2 \left[\tfrac{1}{2}(90°) \right] = \cos^2 45°$$

$$= \left(\frac{\sqrt{2}}{2} \right)^2 = \frac{1}{2} \qquad (Q5.16a)$$

$$\Pr(\text{antialigned}) = \sin^2 \left[\tfrac{1}{2}(90°) \right] = \sin^2 45°$$

$$= \left(\frac{\sqrt{2}}{2} \right)^2 = \frac{1}{2} \qquad (Q5.16b)$$

These results are consistent with the probabilities stated in figure Q5.5a.

Q5X.4 As we noted in the case illustrated in figure Q5.7a, when an electron goes through an SG device, it changes its state so that it "forgets" any previous determinations of its spin. Therefore, the presence of the first device is irrelevant. This means that the situation in the drawing is much like the situation shown in figure Q5.6b, except that we are sending electrons whose spins are aligned with the $-z$ direction into the SGθ device instead of electrons with spin up. Notice that if $\theta = 0$, the last device becomes essentially another SGz device; so as we discussed in conjunction with figure Q5.6a, the probabilities should be 0 and 1 for the up and down channels of the final device, respectively. As we change θ, we expect the probabilities to vary much as they did before, except that the probabilities should be 0 and 1 for the up and down channels, respectively, when $\theta = 0$ instead of 1 and 0. This strongly suggests that the probabilities should be

$$\Pr(\text{aligned}) = \sin^2 \tfrac{1}{2}\theta \qquad \Pr(\text{antialigned}) = \cos^2 \tfrac{1}{2}\theta$$
$$(Q5.17)$$

Alternatively, we could arrive at the same conclusion by noting that if we rotate our coordinate system 180° around the x axis, we get *exactly* the situation shown in figure Q5.6a, except that the angle θ' between the new z axis and the axis of the SGθ device corresponds to $\theta + 180°$. So the probabilities for emerging from the up and down channels should be

$$\Pr(\text{aligned}) = \cos^2 \tfrac{1}{2}\theta' = \left[\cos \tfrac{1}{2}(\theta + 180°) \right]^2$$
$$= \left[\cos \left(\tfrac{1}{2} + 90° \right) \right]^2$$
$$= \left(\cos \tfrac{1}{2}\theta \cos 90° - \sin \tfrac{1}{2}\theta \sin 90° \right)^2$$
$$= \sin^2 \tfrac{1}{2}\theta \qquad (Q5.18a)$$

$$\Pr(\text{antialigned}) = \sin^2 \tfrac{1}{2}\theta' = \left[\sin \left(\tfrac{1}{2}\theta + 90° \right) \right]^2$$
$$= \left[\sin \tfrac{1}{2}\theta \cos 90° + \cos \tfrac{1}{2}\theta \sin 90° \right]^2$$
$$= \cos^2 \tfrac{1}{2}\theta \qquad (Q5.18b)$$

where I have used the trigonometric identities $\cos(A + B) = \cos A \cos B - \sin A \sin B$ and $\sin(A + B) = \sin A \cos B + \cos A \sin B$.

Q5X.5 The results are as follows:

a. $(1 + i) + (2 - 3i) = (1 + 2) + (1 - 3)i = 3 - 2i$
$$(Q5.19a)$$

b. $1 + (-1 - 2i) = (1 - 1) - 2i = -2i \qquad (Q5.19b)$

c. $(1 + i)(2 - 3i) = 2 + 2i - 3i - 3i^2$
$$= 2 - i + 3 = 5 - i \qquad (Q5.19c)$$

d. $3i(2 - 5i) = 6i - 15i^2 = 15 + 6i \qquad (Q5.19d)$

e. $[3i(2 - 5i)]^* = (15 + 6i)^* = 15 - 6i \qquad (Q5.19e)$

f. $|-5 + 12i|^2 \equiv (-5)^2 + 12^2$
$$= 25 + 144 = 169 \qquad (Q5.19f)$$

Q5X.6 The proofs are as follows:

$$e^{i0} \equiv \cos 0 + i \sin 0 = 1 + 0 = 1 \qquad (Q5.20a)$$

$$e^{i\theta_1} e^{i\theta_2} \equiv (\cos\theta_1 + i \sin\theta_1)(\cos\theta_2 + i \sin\theta_2)$$

$$= \cos\theta_1 \cos\theta_2 + i(\sin\theta_1 \cos\theta_2 + \cos\theta_1 \sin\theta_2)$$
$$+ i^2 \sin\theta_1 \sin\theta_2$$

$$= \cos\theta_1 \cos\theta_2 - \sin\theta_1 \sin\theta_2 + i(\sin\theta_1 \cos\theta_2$$
$$+ \cos\theta_1 \sin\theta_2)$$

$$= \cos(\theta_1 + \theta_2) + i \sin(\theta_1 + \theta_2) \equiv e^{i(\theta_1 + \theta_2)}$$
$$(Q5.20b)$$

In the next-to-last step I used the sum-of-angles trigonometric identities.

$$\frac{d}{d\theta}(e^{i\theta}) \equiv \frac{d}{d\theta}(\cos\theta + i \sin\theta)$$

$$= -\sin\theta + i \cos\theta$$

$$= i(i \sin\theta + \cos\theta) = i e^{i\theta} \qquad (Q5.20c)$$

Q5X.7 Saying that the vector is normalized means that

$$1 = \langle \psi \mid \psi \rangle = |\psi_1|^2 + |\psi_2|^2 = |a|^2 + |ia|^2$$

$$= a^2 + (ia)^*(ia)$$

$$= a^2 + (-ia)(ia) = a^2 - i^2 a = 2a^2$$

$$\Rightarrow \quad a^2 = \tfrac{1}{2} \quad \Rightarrow \quad a = \pm\sqrt{\tfrac{1}{2}} \qquad (Q5.21)$$

Q6

The Wavefunction

Chapter Overview

Introduction
This core chapter uses the experimental results described in chapter Q5 as a springboard for presenting the fundamental mathematical model of quantum mechanics. This chapter provides important background information about quantum wavefunctions that we will use throughout the rest of the unit.

Section Q6.1: The Game of Quantum Mechanics
Because quantons behave like nothing we are familiar with, the quantum mechanics model *directly* links physical phenomena to mathematical objects and equations, without intervening conceptual analogies to motivate the mathematics. This is more palatable if we treat quantum theory as a *game* with simple but apparently arbitrary rules.

Section Q6.2: The Rules
Quantons are the playing pieces. The goal of the game is to predict the outcomes of experiments we do to determine the values of a quanton's **observables,** quantities such as the components of its spin orientation, its position, its momentum, or its energy. We can divide observables into two subsets that we can usually address independently, a **spin subset** and a **spatial subset.**

The quantum-mechanics game has six basic rules:

1. **The State Vector rule.** The state of a quanton at a given time is described using a *normalized state vector* $|\psi\rangle$ having a certain number of complex components. In the context of a certain subset of observables, a quanton's state vector has as many components as there are possible values for any one of the subset's basic observables.

2. **The Eigenvector rule.** For every possible numerical value that an observable might have, there is an associated normalized state vector, which we call that value's **eigenvector.** (See table Q6.1 for a list of spin eigenvectors.)

3. **The Collapse rule.** When we perform any experiment to determine the value of one of a quanton's observables, the experiment *determines* the observable's value by "collapsing" the quanton's state to a randomly selected one of that observable's eigenvectors, and yields the observable value corresponding to that eigenvector.

4. **The Outcome Probability rule.** In such an experiment, the probability of any given result is the *absolute square of the inner product* of the quanton's original state vector and the result's eigenvector. (*Note:* Self-consistency then implies that a given observable's eigenvectors must be orthogonal.)

5. **The Superposition rule.** Consider an experiment in which we take a quanton in an initial state $|\psi_0\rangle$ and determine some observable whose possible values are $a, b, c, \ldots$ and whose corresponding eigenvectors are $|a\rangle, |b\rangle, |c\rangle, \ldots$. If we recombine the outgoing quantons so that it is *completely impossible* to tell whether the quanton was determined to have value a or b, then the state $|\psi_f\rangle$ of quantons in the combined final beam is an appropriately weighted superposition of eigenvectors $|a\rangle$ and $|b\rangle$.

6. **The Time-Evolution rule.** If we know a quanton's state vector $|\psi(0)\rangle$ at $t = 0$, we can determine its state vector $|\psi(t)\rangle$ at time t by (a) writing $|\psi(0)\rangle$ as a superposition of the quanton's energy eigenvectors $|E_1\rangle, |E_2\rangle, \ldots$ and (b) multiplying each energy eigenvector in this sum by the factor $e^{-iE_nt/\hbar}$, where E_n is the energy value associated with that eigenvector and $\hbar = h/2\pi$.

The section describes these rules in greater detail and presents many examples involving the spin subset of observables.

Section Q6.3:   The Wavefunction

The spatial subset of observables involves the quanton's position x, its momentum p_x, and any observables that in newtonian mechanics we would calculate from these. The state vectors and eigenvectors needed to describe experiments in this subset have an infinite number of components, but it turns out that we can describe any such vector using a **wavefunction** $\psi(x)$. Such a wavefunction has the following simple probability interpretation:

$$\Pr(x_0 \le x \le x_1) = \int_{x_0}^{x_1} |\psi(x)|^2\, dx \qquad \text{in the limit that } dx \to 0 \quad \text{(Q6.16)}$$

Purpose: This equation describes how to compute the probability that a quanton whose wavefunction is $\psi(x)$ would be determined to have an x-position between x_0 and x_1 if we do an experiment to determine the quanton's position.

Limitations: This equation assumes that the quanton's wavefunction is normalized and that the quanton moves only along the x axis.

Note: This implies that if $\psi(x) \approx$ constant over an interval of width Δx centered on x_0, then the probability of finding the quanton in that range is $\approx |\psi(x_0)|^2 \Delta x$.

A wavefunction is normalized if

$$1 = \int_{-\infty}^{\infty} |\psi(x)|^2\, dx \qquad\qquad \text{(Q6.17)}$$

The position eigenfunction for a given position value x_0 is a spike that is zero everywhere except at x_0.

Section Q6.4:   Explaining the Two-Slit Experiment

The momentum eigenfunction corresponding to a momentum value is

$$\psi_{p_0}(x) = Ae^{ip_0x/\hbar} \qquad\qquad \text{(Q6.20)}$$

which is an oscillating function with amplitude A and a wavelength equal to the de Broglie wavelength. One can show that a free particle's time-dependent wavefunction is a traveling wave very much like those we have studied before. We can generalize this to two-dimensional waves to show that the quantum game rules correctly predict that quantons will most likely be found at positions corresponding to maxima of the two-slit interference pattern, and that the pattern goes away if we use proximity detectors to determine which slit the quanton went through.

Section Q6.5:   The Collapse of the Wavefunction

This section discusses what the collapse rule implies about the behavior of wavefunctions, and why it is both absolutely necessary and extremely troubling. You can think of an experiment as determining an observable's value by essentially removing parts of the wavefunction that are inconsistent with a given value so that the observable now has a well-defined value to report.

Q6.1   The Game of Quantum Mechanics

In chapters Q2 through Q5, we have discussed experiments that underline the need for a fundamentally new theory of mechanics. The purpose of this chapter is to describe the basic structure of the new theory of mechanics that arose during the decades between 1920 and 1950 to address the issues raised by these experiments. Physicists call this theory (for historical reasons) **quantum mechanics.** Before we dig into the theory, I want to make some general comments about what to expect.

The layers of a normal theory

We can think of most physical theories as having three discernible layers (see figure Q6.1a). The first and deepest layer consists of the *basic experimental results* that the theory is supposed to explain. The purpose of chapters Q2 through Q5 has been to construct this layer. The second layer is some kind of *conceptual analogy* that helps us think about the results. The analogy is often to something familiar (e.g., water waves as an analogy for light waves, or spinning tops as an analogy for electrons). The best conceptual analogies not only empower our intuition but also help motivate the third layer, which is a *mathematical model.* Such a model links features of the conceptual analogy with mathematical quantities and operations. For example, in newtonian mechanics, the mathematical model links the conceptual idea of the "position of a particle" with a position vector $\vec{r}$. We can use the mathematical model to make quantitative predictions because the equations of the model in a certain sense mimic what happens in the physical world.

The conceptual layer is missing in quantum mechanics

Now, the fundamental problem with quantum mechanics is that physics at the microscopic scale is *completely unlike anything that we are familiar with* in daily life. This makes it difficult to find a satisfying conceptual analogy! The result is that quantum mechanics often describes how physical systems behave by drawing an essentially *direct* connection to the mathematical behavior of certain equations, skipping the conceptual analogy layer (see figure Q6.1b). As we lay out the structure of quantum mechanics in what follows, you will probably experience the lack of a conceptual analogy as a feeling that the mathematical model is arbitrary and unmotivated. You are not alone! I think that most thoughtful physicists are troubled by the seemingly absurd and unmotivated nature of the model. Yet the mathematical

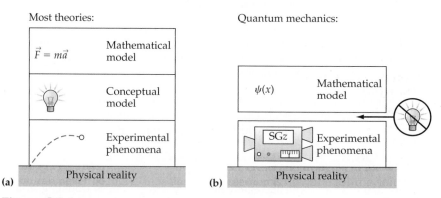

Figure Q6.1

(a) Most physical theories involve three layers of interpretation: a listing of important experimental results, a conceptual interpretation of those results, and a mathematical model whose equations mimic the behavior of the universe.
(b) Quantum mechanics seems strange partly because the lack of everyday analogies means that there is essentially no conceptual model: the mathematical model is linked more directly to the phenomena.

model of quantum mechanics has stood the test of time as being a model that *works* very well in a very broad range of circumstances.

I recommend that, instead of stressing out about the lack of a conceptual model, you think of the ideas and equations that follow as being a kind of *game* with somewhat arbitrary but simple and self-consistent rules. The rules of the "Quantum Mechanics Game" are really no more difficult to *learn* than the rules of chess or "Jeopardy," and the game is actually rather fun to play. The miracle is that this crazy game is actually a useful and powerful tool for describing physics at the microscopic level!

Think of quantum mechanics as a game

Q6.2   The Rules

I will begin with some basic information about the playing pieces. Every quanton has a certain number of **intrinsic characteristics,** such as its mass, charge, and spin. These characteristics never change while the quanton exists. For example, an electron *always* has a mass of 9.11×10^{-31} kg, a charge of -1.6×10^{-19} C, a certain spin magnitude, and so on. A complete list of such intrinsic characteristics essentially *defines* the quanton we call an electron.

Background information about observables

Every quanton also has some associated *variable* quantities (or **observables**), such as its position in space, its momentum, its energy, and the projection of its spin on a given axis. The goal of the quantum mechanics game is to predict the results of experiments that determine the values of these observables.

Observables can be usefully grouped into subsets of related observables. For example, the components S_x, S_y, S_z of an electron's spin orientation are the basic observables of the **spin subset** of observables. The subset also includes any observables that in newtonian mechanics we could calculate from these basic observables. For example, if an electron's energy depends only on its spin orientation, then its energy would be a part of this subset.

If a quanton moves in one spatial dimension, its position x and its x-momentum p_x are the basic observables of the **spatial subset** of observables. In newtonian mechanics we could calculate a quanton's total mechanical energy $E \equiv K + V(x) = p_x^2/2m + V(x)$ from x and p_x, so E is another observable in this subset, along with a quanton's angular momentum components L_x, L_y, L_z due to its motion about a point, and so on.

These are the only two subsets of observables that we will consider in this text. In many situations, observables in these subsets are independent (for example, an electron's spin orientation is usually completely independent of its position or momentum and vice versa). This means that we can generally focus on the behavior of observables in one subset while ignoring other.

The game of quantum mechanics has six basic rules. I will state each rule in general terms, give an example of how the rule applies to the spin subset of observables (we will discuss the trickier spatial subset in section Q6.3), and provide some general commentary about what the rule means.

1. **The State Vector rule.** The state of a quanton at a given time is described by using a *normalized state vector* $|\psi\rangle$ having a certain number of complex components. In the context of a certain subset of observables, a quanton's state vector has as many components as there are possible values for any one of the subset's basic observables. Even so, the components of this vector do *not* correspond to the values of that or any other observable.

The state vector rule

Example   In the spin subset, because we always find that for electrons the spin is either aligned with or antialigned with a given axis, each of the basic spin observables S_x, S_y, and S_z has only two possible values: $+s$ or $-s$ (where s is related to the magnitude of the electron's spin). Therefore the complex vector describing a electron's spin state will have two complex components ψ_1 and ψ_2:

$$|\psi\rangle = \begin{bmatrix} \psi_1 \\ \psi_2 \end{bmatrix} \qquad\qquad (Q6.1)$$

Since this vector is supposed to be normalized, we must have $|\psi_1|^2 + |\psi_2|^2 = 1$.

Commentary In newtonian mechanics, we would describe a quanton's spin state by specifying the actual *values* of its spin's vector components S_x, S_y, and S_z. We do *not* do this kind of thing in quantum mechanics. As we will see, the state vector instead provides cleverly encoded probability information about *possible* outcomes if we measure S_x, S_y, or S_z, but is agnostic about the particular values of these observables. A quantum state vector describes future *possibilities*, but only rarely *determines* anything.

The eigenvector rule

2. **The Eigenvector rule.** For every possible numerical value that an observable might have, there is an associated normalized state vector, which we call that value's **eigenvector.** (Conversely, the value corresponding to a given eigenvector is called that vector's **eigenvalue.**)

Example Table Q6.1 specifies the (conventional) eigenvectors for the electron spin observables S_x, S_y, S_z, and S_θ (the last is the spin component along an axis in the yz plane that makes an angle of θ with the z axis). The italic letter inside the symbol for each eigenvector specifies the axis along which the spin is being determined and the plus or minus sign corresponds to the sign of the result.

Exercise Q6X.1

Check that each of the two vectors $|+\theta\rangle$ and $|-\theta\rangle$ in the rightmost column of table Q6.1 is normalized. Show also that they are orthogonal to each other for all values of θ.

Commentary *Eigen* is a German root word meaning "characteristic." These eigenvectors do not just come out of the blue: in an upper-level quantum mechanics course you would learn how to *calculate* such vectors from more basic principles (see problems Q6A.1 and Q6A.2 for a taste of this). However, the arguments are subtle, and the mathematics required is beyond the level of this course. I will provide you with any eigenvectors that you need.

Table Q6.1   Table of spin eigenvectors

Observable:	S_x	S_y	S_z	S_θ
Value: $+s$	$\|+x\rangle = \begin{bmatrix} \sqrt{1/2} \\ \sqrt{1/2} \end{bmatrix}$	$\|+y\rangle = \begin{bmatrix} \sqrt{1/2} \\ i\sqrt{1/2} \end{bmatrix}$	$\|+z\rangle = \begin{bmatrix} 1 \\ 0 \end{bmatrix}$	$\|+\theta\rangle = \begin{bmatrix} \cos\frac{1}{2}\theta \\ i\sin\frac{1}{2}\theta \end{bmatrix}$
Value: $-s$	$\|-x\rangle = \begin{bmatrix} \sqrt{1/2} \\ -\sqrt{1/2} \end{bmatrix}$	$\|-y\rangle = \begin{bmatrix} i\sqrt{1/2} \\ \sqrt{1/2} \end{bmatrix}$	$\|-z\rangle = \begin{bmatrix} 0 \\ 1 \end{bmatrix}$	$\|-\theta\rangle = \begin{bmatrix} i\sin\frac{1}{2}\theta \\ \cos\frac{1}{2}\theta \end{bmatrix}$

3. **The Collapse rule.** When we perform any experiment to determine the value of one of a quanton's observables, the experiment *determines* the observable's value by "collapsing" the quanton's state to a randomly selected one of that observable's eigenvectors, and yields the observable value corresponding to that eigenvector.

The collapse rule

Example   Assume that an electron has an arbitrary initial state $|\psi\rangle = [\psi_1, \psi_2]$. If we send that electron through an SGz device and it emerges from the plus channel, its state as it emerges is $|+z\rangle = [1, 0]$. If it emerges from the device's minus channel, its state is $|-z\rangle = [0, 1]$.

Commentary   This rule is simple and straightforward, and yet so strange and unbelievable that it remains one of the biggest stumbling blocks in the whole theory for physicists. I will discuss this rule in greater depth in section Q6.5. At present, let me simply note that this rule, in conjunction with the next, is *absolutely essential* for explaining why determinations of observable values are self-consistent. It also tells us that we can *prepare* a quanton in a well-defined state by determining the value of one of its observables.

4. **The Outcome Probability rule.** In an experiment that determines the value of an observable, the probability of any given result (i.e., the probability that the quanton's state will collapse to that result's eigenvector) is the *absolute square of the inner product* of the quanton's original state and the result's eigenvector.

The outcome probability rule

Example   Consider the situation shown in figure Q6.2. Any electrons entering the second Stern-Gerlach device have an initial state of $|+z\rangle = [1, 0]$. The probability that an electron will emerge from the second device's plus channel is therefore

$$|\langle +z \mid +\theta\rangle|^2 = \left|\begin{bmatrix} 1 \\ 0 \end{bmatrix}^* \cdot \begin{bmatrix} \cos\frac{1}{2}\theta \\ i\sin\frac{1}{2}\theta \end{bmatrix}\right|^2 = \left|1^*\left(\cos\frac{1}{2}\theta\right) + 0^*\left(i\sin\frac{1}{2}\theta\right)\right|^2 = \cos^2\frac{1}{2}\theta$$

(Q6.2)

This is consistent with the empirical results discussed in chapter Q5.

Exercise Q6X.2

Show that the probability that the electron will emerge from the minus channel is $\sin^2\frac{1}{2}\theta$.

Comment   This rule describes how we decode the probability information hidden in a quantum state by using eigenvectors essentially as decoding keys. Note that taking the *absolute square* of the inner product ensures that the probability of a given result is always real and positive. Since all our vectors are *normalized*, this probability will also never be greater than 1 (indeed, this is the *point* of normalizing the vectors).

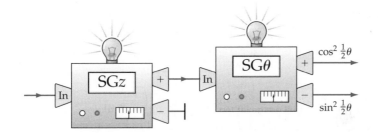

Figure Q6.2

We can use the outcome probability rule to calculate the probabilities of the two possible spin results in this experiment. (The SGθ device determines the projection of the electron's spin along an axis in the *yz* plane that makes an angle of θ with the *z* axis.)

Figure Q6.3

The collapse and outcome probability rules imply that the spin determinations in this experiment will be self-consistent.

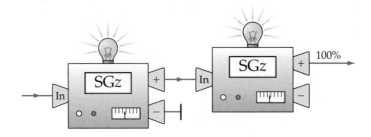

Note that this rule, in conjunction with the collapse rule, ensures that all determination experiments are *self-consistent*. For example, consider the situation shown in figure Q6.3. We know from the collapse rule that the spin state of any electron *entering* the second device is $|+z\rangle = [1, 0]$. The probability that the electron emerges from the plus channel of that device is therefore $|\langle +z \,|+z\rangle|^2 = |1|^2 = 1$, meaning that the second device will corroborate the determination of the first. If the collapse rule were *not* true, the quanton's spin state upon entering the second device might not be $|+z\rangle = [1, 0]$ and the second device would not duplicate with certainty the result of the first.

A necessary corollary of self-consistency is that *the set of eigenvectors corresponding to different possible values of any given observable will be mutually orthogonal*. For example, if the first device in figure Q6.3 collapses the electron's spin state to $|+z\rangle$ and the second device finds that the probability of duplicating that result is 1, then the probability of coming out the minus channel had better be zero, which will only be true if $\langle +z \,|-z\rangle = 0$.

The superposition rule

5. **The Superposition rule.** Consider an experiment in which we take quantons in an initial state $|\psi_0\rangle$ and determine some observable whose possible values are $a, b, c, \ldots$ and whose corresponding eigenvectors are $|a\rangle, |b\rangle, |c\rangle, \ldots$. If we take the subset of quantons determined to have values a or b and recombine them so that it is *completely impossible* to tell which value a given quanton was determined to have, then the state $|\psi_{rc}\rangle$ of the recombined quantons is the superposition

$$|\psi_{rc}\rangle = c_a|a\rangle + c_b|b\rangle \qquad \text{where} \quad c_a = A\langle a\,|\,\psi_0\rangle \quad \text{and} \quad c_b = A\langle b\,|\,\psi_0\rangle$$
$$\text{(Q6.3)}$$

and A is chosen so that $|\psi_{rc}\rangle$ is normalized.

Example   Consider the situation shown in figure Q6.4 (which duplicates figure Q5.7b). The spin state of electrons going into the SGy device is $|\psi_0\rangle = |+z\rangle = [1, 0]$. According to this rule, the state of an electron in the recombined beam going into the third device is

$$|\psi_{rc}\rangle = A\langle +y\,|\,\psi_0\rangle|+y\rangle + A\langle -y\,|\,\psi_0\rangle|-y\rangle$$

$$= A\left(\begin{bmatrix}\sqrt{\tfrac{1}{2}}\\ i\sqrt{\tfrac{1}{2}}\end{bmatrix}^* \cdot \begin{bmatrix}1\\0\end{bmatrix}\right)\begin{bmatrix}\sqrt{\tfrac{1}{2}}\\ i\sqrt{\tfrac{1}{2}}\end{bmatrix} + A\left(\begin{bmatrix}i\sqrt{\tfrac{1}{2}}\\ \sqrt{\tfrac{1}{2}}\end{bmatrix}^* \cdot \begin{bmatrix}1\\0\end{bmatrix}\right)\begin{bmatrix}i\sqrt{\tfrac{1}{2}}\\ \sqrt{\tfrac{1}{2}}\end{bmatrix}$$

$$= \left(A\sqrt{\tfrac{1}{2}}^* + 0\right)\begin{bmatrix}\sqrt{\tfrac{1}{2}}\\ i\sqrt{\tfrac{1}{2}}\end{bmatrix} + \left(A\left[i\sqrt{\tfrac{1}{2}}\right]^* + 0\right)\begin{bmatrix}i\sqrt{\tfrac{1}{2}}\\ \sqrt{\tfrac{1}{2}}\end{bmatrix} \qquad \text{(Q6.4)}$$

$$= A\sqrt{\tfrac{1}{2}}\begin{bmatrix}\sqrt{\tfrac{1}{2}}\\ i\sqrt{\tfrac{1}{2}}\end{bmatrix} - iA\sqrt{\tfrac{1}{2}}\begin{bmatrix}i\sqrt{\tfrac{1}{2}}\\ \sqrt{\tfrac{1}{2}}\end{bmatrix} = A\begin{bmatrix}\tfrac{1}{2}\\ \tfrac{i}{2}\end{bmatrix} + A\begin{bmatrix}\tfrac{1}{2}\\ \tfrac{-i}{2}\end{bmatrix} = \begin{bmatrix}A\\0\end{bmatrix}$$

which is normalized if $A = 1$. This explains why *all* these electrons emerge from the plus channel of the third device:

$$\text{Pr(aligned)} = |\langle \psi_{rc} \,|+z\rangle|^2 = \left|\begin{bmatrix} 1^* \\ 0^* \end{bmatrix} \cdot \begin{bmatrix} 1 \\ 0 \end{bmatrix}\right|^2 = |1 + 0|^2 = 1 \quad (\text{Q6.5}a)$$

$$\text{Pr(antialigned)} = |\langle \psi_{rc} \,|-z\rangle|^2 = \left|\begin{bmatrix} 1^* \\ 0^* \end{bmatrix} \cdot \begin{bmatrix} 0 \\ 1 \end{bmatrix}\right|^2 = |0 + 0|^2 = 0 \quad (\text{Q6.5}b)$$

as we found empirically in chapter Q5. On the other hand, if we block the electrons emerging from the second device's minus channel, so that this path is not a possibility, then the state of electrons going into the third device is simply $|+y\rangle$ and the probabilities become

$$\text{Pr(aligned)} = |\langle +y\,|+z\rangle|^2 = \left|\begin{bmatrix} \sqrt{\tfrac{1}{2}} \\ i\sqrt{\tfrac{1}{2}} \end{bmatrix}^* \cdot \begin{bmatrix} 1 \\ 0 \end{bmatrix}\right|^2 = \left|1\sqrt{\tfrac{1}{2}} + 0\right|^2 = \frac{1}{2} \quad (\text{Q6.6}a)$$

$$\text{Pr(antialigned)} = |\langle +y\,|-z\rangle|^2 = \left|\begin{bmatrix} \sqrt{\tfrac{1}{2}} \\ i\sqrt{\tfrac{1}{2}} \end{bmatrix}^* \cdot \begin{bmatrix} 0 \\ 1 \end{bmatrix}\right|^2 = \left|0 - i\sqrt{\tfrac{1}{2}}\right|^2 = \frac{1}{2}$$

$$(\text{Q6.6}b)$$

This is again consistent with what we found in chapter Q5.

Comment   We see that this rule is *essential* for correctly describing the observation displayed in figures Q5.7b and Q6.4. We will see in section Q6.4 that this rule also explains the two-slit interference probability pattern. The more believable hypothesis—recombining the electrons in figure Q6.4 should simply mean that one-half of the electrons entering the third device have state $|+y\rangle$ and one-half have state $|-y\rangle$—yields the same probability predictions as in equations Q6.6 (see problem Q6S.1) and so is inconsistent with experiment.

6.   **The Time-Evolution rule.** It turns out that *any* quanton's state $|\psi\rangle$ can be written as a superposition of energy eigenvectors $|E_1\rangle, |E_2\rangle, \ldots$. If at time $t = 0$ the quanton's state is

The time-evolution rule

$$|\psi(0)\rangle = c_1|E_1\rangle + c_2|E_2\rangle + \cdots \quad (\text{Q6.7}a)$$

then (assuming we leave the quanton alone and do not try to determine any of its observables) at a later time t, the quanton's state will be

$$|\psi(t)\rangle = c_1 e^{-iE_1 t/\hbar}|E_1\rangle + c_2 e^{-iE_2 t/\hbar}|E_2\rangle + \cdots \quad (\text{Q6.7}b)$$

where $\hbar \equiv h/2\pi$ (h is Planck's constant) and $E_1, E_2, \ldots$ are the energy values associated with the eigenvectors $|E_1\rangle, |E_2\rangle, \ldots$, respectively.

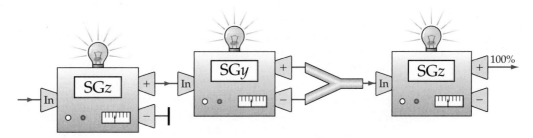

Figure Q6.4
The superposition rule explains the outcome of this experiment, which we first saw in figure Q5.7b.

Example   Assume that an electron's energy in a given situation is $+E_0$ if the electron's spin is aligned with the z axis and $-E_0$ if it is antialigned with that axis. The energy eigenvectors in this case will be $|+z\rangle = [1, 0]$ and $|-z\rangle = [0, 1]$, the same as the eigenvectors of S_z. Let us also assume that the quanton's initial state at $t = 0$ is $|\psi(0)\rangle = |+x\rangle = [\sqrt{1/2}, \sqrt{1/2}]$. We can write this initial state as a superposition of energy eigenvectors as follows:

$$|\psi(0)\rangle = \begin{bmatrix} \sqrt{\frac{1}{2}} \\ \sqrt{\frac{1}{2}} \end{bmatrix} = \sqrt{\frac{1}{2}} \begin{bmatrix} 1 \\ 0 \end{bmatrix} + \sqrt{\frac{1}{2}} \begin{bmatrix} 0 \\ 1 \end{bmatrix} \qquad (Q6.8)$$

According to this rule, then, the electron's state at time t would be

$$|\psi(t)\rangle = \sqrt{\tfrac{1}{2}} e^{-iE_0 t/\hbar} \begin{bmatrix} 1 \\ 0 \end{bmatrix} + \sqrt{\tfrac{1}{2}} e^{-i(-E_0)t/\hbar} \begin{bmatrix} 0 \\ 1 \end{bmatrix} = \begin{bmatrix} \sqrt{\tfrac{1}{2}} e^{-iE_0 t/\hbar} \\ \sqrt{\tfrac{1}{2}} e^{+iE_0 t/\hbar} \end{bmatrix} \qquad (Q6.9)$$

Exercise Q6X.3

Calculate $|\psi(t)\rangle$ if $|\psi(0)\rangle = [0, 1]$.

Commentary   This is essentially Newton's second law for quantum mechanics: it tells us how to calculate a quanton's future state in terms of its present state. Note that since all the energy eigenvectors are normalized (see the eigenvector rule) and are mutually orthogonal (see the commentary to outcome probability), they can play the same role for quantum state vectors that the mutually orthogonal unit vectors $\hat{x}$, $\hat{y}$, and $\hat{z}$ play for ordinary three-dimensonal vectors: since we can express any ordinary vector as a sum of the $\hat{x}$, $\hat{y}$, and $\hat{z}$ vectors multiplied by various constants, we ought to be able to express any quantum state in terms of a sum of energy eigenvectors multiplied by constants.

Special relativity is based on a *single* easily stated and very plausible postulate, as we saw in unit R. The game of quantum mechanics, in contrast, involves no less than six very strange and completely implausible rules. However, these rules *really do work*, as we will see!

Q6.3   The Wavefunction

Let us turn our attention to what these rules imply for the spatial subset of observables, whose basic observables are position x and momentum p_x (for quantons moving in one dimension). This subset is more challenging than the spin subset primarily because there are an infinite number of possible position coordinates that a quanton might have. Therefore, the first rule implies that a quanton's state vector must have an infinite number of components!

Describing infinite vectors by using functions

This is less of a problem than it seems, though, because we already know how to deal with a mathematical idea that is essentially equivalent to an infinite vector. Imagine that I wanted to communicate to you an endless sequence of numerical values. The *straightforward* (but dunderheaded) way to do this would be simply to start at some point and read the values to you in sequence. However, this would take (literally) forever! But if I can find a way to express the numbers on the list as a *function* $f(x)$ of some parameter x that tells where we are on the list, then all that I have to do is to describe the function [for example, $f(x) = A \sin kx$] and then *you* can compute whatever numerical values on the list you need. This is what we will do to describe our infinite vectors.

This trick would not work if the vectors of interest were badly behaved (e.g., the vector's components have values that just randomly jump around or something like that). The universe is kind, however: the infinite vectors we need to consider can all be described reasonably well by functions.

As a first step in the process of matching infinite vectors to functions, let us imagine that a quanton can only exist at specific discrete positions x_i separated by some tiny but constant step dx (we will later take $dx \to 0$). In such a situation, all quantum state vectors and eigenvectors in the spatial subset of variables will have one component for every possible value of x. It turns out that in such a case, we can define the eigenvector corresponding to the ith position observable value x_i to have a 1 in the ith position and a 0s everywhere else. The eigenvectors for possible adjacent position values x_{i-1}, x_i, x_{i+1} of the position observable therefore look like this:

Position eigenvectors

$$|x_{i-1}\rangle = \begin{bmatrix} \vdots \\ 1 \\ 0 \\ 0 \\ \vdots \end{bmatrix} \qquad |x_i\rangle = \begin{bmatrix} \vdots \\ 0 \\ 1 \\ 0 \\ \vdots \end{bmatrix} \qquad |x_{i+1}\rangle = \begin{bmatrix} \vdots \\ 0 \\ 0 \\ 1 \\ \vdots \end{bmatrix} \qquad \text{etc.} \qquad \text{(Q6.10)}$$

These eigenvectors are all normalized and orthogonal, as eigenvectors should be.

Now consider a general state vector $|\psi\rangle$. Note that if

A function that describes the vector $|\psi\rangle$

$$|\psi\rangle = \begin{bmatrix} \vdots \\ \psi_{i-1} \\ \psi_i \\ \psi_{i+1} \\ \vdots \end{bmatrix} \qquad \text{then} \qquad \psi_i = \begin{bmatrix} \vdots \\ 0^* \\ 1^* \\ 0^* \\ \vdots \end{bmatrix} \cdot \begin{bmatrix} \vdots \\ \psi_{i-1} \\ \psi_i \\ \psi_{i+1} \\ \vdots \end{bmatrix} = \langle x_i \mid \psi \rangle \qquad \text{(Q6.11)}$$

So a function $\overline{\psi}(x)$ such that $\overline{\psi}(x_i) \equiv \psi_i = \langle x_i \mid \psi \rangle$ completely describes the infinite vector. (We will see the point of the overbar notation shortly.)

This function has a simple interpretation. If we do an experiment to determine a quanton's position, then according to the outcome probability rule, the probability that a quanton in a state $|\psi\rangle$ will be found to have position x_i is

$$\Pr(x_i) = |\langle x_i \mid \psi \rangle|^2 = |\overline{\psi}(x_i)|^2 \qquad \text{(Q6.12)}$$

Therefore the function $\overline{\psi}(x)$ simply describes the square root of the probability that we would determine the quanton's position to be any arbitrary position x.

Now let us see what happens if we take the limit as $dx \to 0$. Note that as the step size between available quanton positions decreases, the number of possible positions *increases*, and the probability that one will determine the quanton to be at any *given* position will decrease to zero. Therefore our function $\overline{\psi}(x)$ will also go to 0 everywhere in this limit. This is *not* helpful.

Taking the limit as $dx \to 0$

There is a way around this problem. Consider a fixed range of values Δx that is much larger than dx but still small enough that $\overline{\psi}(x)$ is essentially the same for all points in the range. The probability that the quanton would be found in this fixed physical range should be a constant value even as $dx \to 0$. Since the number of points within the fixed range is proportional to $1/dx$, this means that the probability $|\overline{\psi}(x_i)|^2$ of finding the quanton at any specific point within that range should decrease in direct proportion to dx

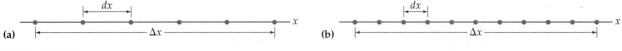

Figure Q6.5

(a) The dots in these diagram represent discrete position points within a range of fixed width Δx. (b) As we take the point separation dx to zero, the number of points increases in proportion to $1/dx$. But the total probability of finding the quanton in the fixed range Δx should be unchanged. This implies that the probability of finding the quanton at a given discrete point must decrease in proportion to dx.

(see figure Q6.5). Therefore we should find that

$$\lim_{dx \to 0} \left[\frac{1}{dx} |\overline{\psi}(x_i)|^2 \right] = \text{constant} \tag{Q6.13}$$

Definition of the *wavefunction* $\psi(x)$

With this in mind, let us define a function $\psi(x)$ (*without* the overbar) to be

$$\psi(x) \equiv \frac{\overline{\psi}(x)}{\sqrt{dx}} \tag{Q6.14}$$

This function still represents the quantum state vector but has a well-defined (i.e., nonzero) value in the limit that $dx \to 0$. If we rescale all state vectors and eigenvectors in this way, we can describe them *all* by using well-defined functions. When the rescaled function describes the quanton's state, we call it the quanton's **wavefunction;** a rescaled function that describes an eigenvector we call an **eigenfunction.**

The probability interpretation of the wavefunction

Note that if $\psi(x)$ represents a quanton's wavefunction, the probability that an experiment to determine the quanton's position yields a specific value x_0 is

$$\Pr(x_0) = |\overline{\psi}(x_0)|^2 = \left| \psi(x_0) \sqrt{dx} \right|^2 = |\psi(x_0)|^2 \, dx \tag{Q6.15}$$

The probability of being within a range of points from $x_0 \to x_1$ is the sum over all probabilities within that range,

$$\Pr(x_0 \le x \le x_1) = \sum_{\text{range}} |\overline{\psi}(x)|^2 = \sum_{\text{range}} \left| \psi(x)\sqrt{dx} \right|^2 = \sum_{\text{range}} |\psi(x)|^2 \, dx$$

$$\Rightarrow \quad \Pr(x_0 \le x \le x_1) = \int_{x_0}^{x_1} |\psi(x)|^2 \, dx \qquad \text{in limit that } dx \to 0$$

$$\tag{Q6.16}$$

Purpose: This equation describes how to compute the probability that a quanton whose wavefunction is $\psi(x)$ will be determined to have an x-position between x_0 and x_1 if we do an experiment to determine the quanton's position.

Limitations: This equation assumes that the quanton's wavefunction is normalized and that the quanton moves only along the x axis.

Normalization means that

$$1 = \langle \psi \mid \psi \rangle = \sum_{\text{all } x} |\overline{\psi}(x)|^2 = \sum_{\text{all } x} |\psi(x)|^2 \, dx = \int_{-\infty}^{\infty} |\psi(x)|^2 \, dx \tag{Q6.17}$$

in the limit that $dx \to 0$. Note that according to equation Q6.16, this essentially says that the probability of finding the quanton at *some* position between $-\infty$ and $+\infty$ is 1.

Note also that if $\psi(x)$ is approximately constant over some range of width Δx centered on x_0 (i.e., from $x_1 = x_0 - \frac{1}{2}\Delta x$ to $x_2 = x_0 + \frac{1}{2}\Delta x$), the probability that we will determine the quanton's position to be in that range is

$$\Pr(x_0 \leq x \leq x_1) = \int_{x_1}^{x_2} |\psi(x)|^2 \, dx \approx |\psi(x_0)|^2 \int_{x_1}^{x_2} dx \approx |\psi(x_0)|^2 \, \Delta x \quad \text{(Q6.18)}$$

Equation Q6.17 has an interesting and important consequence: *not all functions $\psi(x)$ can be valid quanton wavefunctions.* For equation Q6.17 to make any sense, the total area under the curve of $|\psi(x)|^2$ must be 1. As long as the area under the curve of a given function's absolute square is finite, we can rescale it so that equation Q6.17 is true. But if that area is infinite, the function cannot possibly be a valid quantum wavefunction.

Not all functions can be wavefunctions

Example Q6.1   Qualitative Probability Estimation

Problem   Imagine that the quanton has the wavefunction $\psi(x)$ shown in figure Q6.6a. If we do an experiment to locate the quanton, what result is most likely? Least likely?

Solution   The probability of a given result is proportional to the *square* of the wavefunction at that point. Figure Q6.1b shows qualitatively what $|\psi(x)|^2$ looks like. The probability is greatest where the *absolute value* of $\psi(x)$ is largest, so the most likely result is position x_B, even though $\psi(x_B)$ is actually negative. The smallest possible probability is zero, whose square root is zero, so the quanton is *least* likely to be found where the wavefunction crosses the x axis (at x_A): in fact, the quanton will *never* be found there.

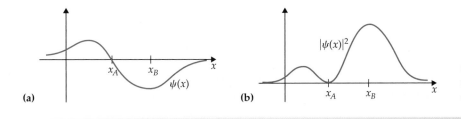

(a)   (b)

Figure Q6.6
(a) An example wavefunction.
(b) The square of the same wavefunction.

Example Q6.2   Quantitative Estimation

Problem   Imagine that at a certain time, a quanton's wavefunction is as shown in figure Q6.7a (a real wavefunction will never be this "boxy," but let's pretend). If we do an experiment that locates the quanton at this time, what is the probability that the result will be to the right of the origin?

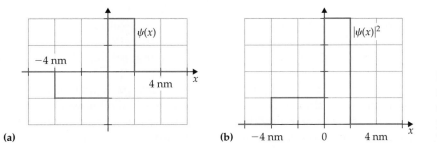

(a)   (b) $-4\,\text{nm}$ 0 $4\,\text{nm}$

Figure Q6.7
(a) The graph of a hypothetical quantum wavefunction. (b) A graph of the absolute square of that same quantum wavefunction.

Solution   Figure Q6.7b shows a graph of the square of the wavefunction. It doesn't matter what the units of the vertical axis are: if the height of $\psi(x)$ is 1 unit, then the height of $|\psi(x)|^2$ is 1 unit squared; if the height of $\psi(x)$ is 2 units, the height of $|\psi(x)|^2$ is 4 units squared, and so on. Figure Q6.7b makes it clear that there are four units squared of area under the graph of $|\psi(x)|^2$ to the right of $x = 0$ and six units squared under the whole graph [assuming that $\psi(x) = 0$ for points not in the diagram], so the probability that the position result will turn out to be to the right of 0 is $\frac{4}{6}$, or $\frac{2}{3}$.

Exercise Q6X.4

What is the probability that the position experiment yields a result within ± 1 nm of the origin for the wavefunction in this case?

How to calculate a dot product using functions

You can show that the inner product of two wavefunctions is

$$\langle u \,|\, w \rangle = \int_{-\infty}^{\infty} [u(x)]^* w(x)\, dx \qquad \text{(Q6.19)}$$

Such inner products are useful for calculating probabilities in experiments that determine observables *other* than position. Fortunately, we will not need to calculate inner products much in this text.

Exercise Q6X.5

Verify equation Q6.19.

Q6.4   Explaining the Two-Slit Experiment

Momentum eigenfunctions

It turns out that the eigenfunction for a given value p_0 of the momentum observable p_x is

$$\psi_{p_0}(x) = A e^{i p_0 x / \hbar} \qquad \text{(Q6.20)}$$

where $\hbar = h/2\pi$ and A is a constant chosen to normalize the function. Note that since $e^{i\theta} \equiv \cos\theta + i\sin\theta$, this function is an oscillating complex wave with amplitude A. This function undergoes a complete oscillation when $p_0 x / \hbar$ increases by 2π. Therefore the wavelength λ of this function is such that

$$2\pi = \frac{p_0}{\hbar}(x + \lambda) - \frac{p_0}{\hbar}x = \frac{p_0 \lambda}{\hbar}$$

$$\Rightarrow \quad \lambda = 2\pi \frac{\hbar}{p_0} = \frac{2\pi(h/2\pi)}{p_0} = \frac{h}{p_0} \qquad \text{(Q6.21)}$$

Therefore, if we determine the momentum of a set of quantons and select only those that have momentum p_0, the wavefunction of those quantons will be an oscillating wave whose wavelength is that predicted by the de Broglie relation!

How momentum eigenfunctions depend on time

The energy of a free particle (i.e., a particle participating in no external interactions) is simply $E = p^2/2m$. Therefore, if we know a free quanton's momentum with certainty (which is the same thing as saying that the quanton's wavefunction is a momentum eigenfunction), then we also know the quanton's energy with certainty. This means that a momentum eigenfunction is also an energy eigenfunction for a free particle. The time-evolution rule tells us that if the quanton's wavefunction at time $t = 0$ is given by equation Q6.20, then its wavefunction at time t is

$$\psi_{p_0}(x, t) = Ae^{ip_0x/\hbar}e^{-iEt/\hbar} = Ae^{i(p_0x-Et)/\hbar} \qquad (Q6.22)$$

If we identify $k \equiv p_0/\hbar$ and $\omega \equiv E/\hbar$, then this wave has the form

$$\psi_{p_0}(x, t) = Ae^{i(kx-\omega t)} = A\cos(kx - \omega t) - iA\sin(kx - \omega t) \qquad (Q6.23)$$

which is the combination of a real and imaginary traveling wave. *Therefore, free quantons whose momentum we have determined have wavefunctions that are simple traveling waves!*

It is not too big a stretch to imagine that in a two-dimensional situation such as the two-slit interference experiment, a quanton whose momentum has a definite magnitude p_0 after it emerges from a tiny slit has a wavefunction described by the circular wave

$$\psi_{p_0}(r, t) = A(r)e^{i(p_0r-Et)/\hbar} \qquad (Q6.24)$$

where r is the distance between the slit and the point where we are evaluating the wave and A is an amplitude that depends on r. (I am assuming here that the slit is so small compared to the quanton's wavelength that the wave is essentially diffracted equally in all directions. The amplitude A depends on r because the wave weakens as it gets farther from the slit.) If we have *two* slits and we have no idea which slit the quanton went through, then the superposition rule essentially implies that the wavefunction of the quanton beyond the slits should be the superposed function

$$\psi(r_1, r_2, t) = A(r_1)e^{i(p_0r_1-Et)/\hbar} + A(r_2)e^{i(p_0r_2-Et)/\hbar} \qquad (Q6.25a)$$

where r_1 and r_2 are the distances to each of the two slits, respectively, as shown in figure Q6.8. [Here I am assuming that the slits are symmetrically arranged so that the dot products appearing in equation Q6.3 are the same, and that we choose the magnitude of $A(r)$ so that the mixed state is normalized.] Now, if the detector array is far from the slits, then $r_1 \approx r_2$ at any point on the array, so this becomes essentially

$$\psi(r_1, r_2, t) \approx A(r)\left(e^{i(p_0r_1-Et)/\hbar} + e^{i(p_0r_2-Et)/\hbar}\right) \qquad (Q6.25b)$$

where $r = \frac{1}{2}(r_1 + r_2)$. (We do *not* want to replace the r_1 and r_2 in the exponentials, though, for reasons that will become clear shortly.)

The detector array essentially performs an experiment that determines the quanton's position. The probability of getting a result that is a distance r_1 from one slit and r_2 from the other slit is therefore proportional to

$$|\psi(r_1, r_2, t)|^2 = |A(r)|^2\left(e^{i(p_0r_1-Et)/\hbar} + e^{i(p_0r_2-Et)/\hbar}\right)^*\left(e^{i(p_0r_1-Et)/\hbar} + e^{i(p_0r_2-Et)/\hbar}\right)$$

$$= |A(r)|^2\left(e^{-i(p_0r_1-Et)/\hbar} + e^{-i(p_0r_2-Et)/\hbar}\right)\left(e^{i(p_0r_1-Et)/\hbar} + e^{i(p_0r_2-Et)/\hbar}\right)$$
$$\qquad (Q6.26)$$

Wavefunctions in the two-slit experiment

The probability of observing a quanton at a given position

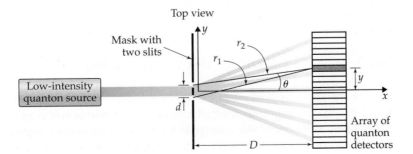

Figure Q6.8

Quantities needed to calculate the absolute square of a quanton's wavefunction at the detector marked in black.

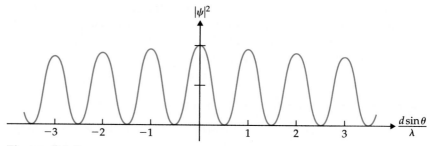

Figure Q6.9
A plot of the function shown in equation Q6.28. This function expresses the probability of finding the quanton at a deflection angle θ from its original direction of travel. Note that since $y = D \tan\theta \approx D\theta$ for small angles, $y \propto \theta \approx \sin\theta$, meaning that the horizontal axis is proportional to the y position for small angles.

With a bit of calculation (using results from section Q5.6), you can show that this reduces to

$$|\psi(r_1, r_2, t)|^2 = |A(r)|^2 \{2 + 2\cos[p_0(r_1 - r_2)/\hbar]\} \qquad (Q6.27)$$

Exercise Q6X.6

Verify equation Q6.27.

Now, $p_0/\hbar = 2\pi p_0/h = 2\pi/\lambda$, and we saw in chapter Q2 that $r_1 - r_2 \approx d\sin\theta$, where θ is as defined in figure Q6.8. Therefore

$$|\psi(r_1, r_2, t)|^2 = |A(r)|^2 \left[2 + 2\cos\left(2\pi \frac{d\sin\theta}{\lambda}\right) \right] \qquad (Q6.28)$$

A plot of this function appears in figure Q6.9. Note that this function has peaks whenever $d\sin\theta/\lambda$ is an integer, meaning where $n\lambda = d\sin\theta$, the condition for constructive interference that we found in that chapter. Therefore, the quantons are most likely to arrive at exactly the positions of constructive interference predicted by the wave model!

What happens when we use proximity detectors

What happens when we put a proximity detector near to each slit? According to the collapse postulate, if a proximity detector registers that the quanton passed through slit 1, its wavefunction thereafter is given simply by $\psi_{p_0}(r_1, t) = A(r)e^{i(p_0 r_1 - Et)/\hbar}$. The probability that this quanton will arrive at a given position on the detector array is proportional to

$$\left|\psi_{p_0}(r_1, t)\right|^2 = |A(r)|^2 \left|e^{i(p_0 r_1 - Et)/\hbar}\right|^2 = |A(r)|^2 \cdot 1 = |A(r)|^2 \qquad (Q6.29)$$

which is just a smooth function. This applies to a quanton going through slit 2 also, so *all* quantons end up being smoothly distributed along the array, displaying no interference pattern. This is completely consistent with the results described in chapter Q5.

Q6.5   The Collapse of the Wavefunction

One of the most important but troubling features of the quantum model is that the collapse rule implies that whenever we perform an experiment that determines any observable, the quanton's wavefunction *changes* in a sudden and irreversible way so as to be consistent with the result actually obtained.

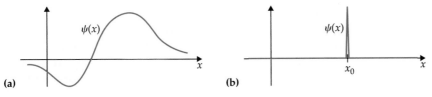

Figure Q6.10
(a) If we do an experiment to localize a quanton having this wavefunction, a number of different results are possible. (b) During the experiment, the quanton's wavefunction collapses to a position eigenfunction something like this, which is only nonzero where we found the quanton's position to be.

Let us consider how this all works when the observable is x-position. Imagine that at a certain time, the quanton has a wavefunction $\psi(x)$ that is nonzero at many positions (see figure Q6.10a). If we do an experiment to determine the quanton's position, there are therefore a variety of possible results, each with its corresponding probability. But when the experiment is over, there is only *one* result (say, x_0) and the quanton's wavefunction is now the position eigenfunction

$$\psi_{x_0}(x) = \frac{\langle x \mid x_0 \rangle}{\sqrt{dx}} \tag{Q6.30}$$

which essentially has a large spike at $x = x_0$ and is zero everywhere else (see figure Q6.10b). This collapse is essential for self-consistency, because such a wavefunction is clearly the only wavefunction that will yield zero probability of finding the quanton anywhere else if we immediately determine its position again.

The collapse of the wavefunction is what enables quantons to leave tracks that look like trajectories in a bubble chamber photograph. Each bubble is essentially the result of a microscopic experiment that locates the quanton at a certain instant of time in the liquid hydrogen. Because the quanton's wavefunction collapses after the first bubble, subsequent bubbles mark out positions consistent with the first, and thus mark out a consistent track across the bubble chamber.

It is sometimes said that this collapse is due to the interaction between the experimental apparatus and the quanton. This is correct, but it is also somewhat misleading if taken too literally. Empirically, we find that the collapse does not depend on the *design* of the experimental apparatus used. Indeed, recent experiments have shown that even performing an experiment on one of two linked quantons that allows us to *infer* the value of an observable for the other collapses the other's wavefunction, even when it is *completely impossible* (from a newtonian point of view) for the apparatus to influence the other quanton at all! Wavefunction collapse is a built-in and necessary part of the quantum mechanics *model*, not a characteristic of the apparatus.

Another way of looking at this is the following. A quanton with a wavefunction like that shown in figure Q6.9a does not *have* a well-defined position. It has a well-defined *wavefunction*, but the wavefunction talks about a number of possible positions with associated probabilities. An experiment that determines a quanton's position *literally* "localizes" that quanton by forcing its wavefunction to become a position eigenfunction similar to the one shown in figure Q6.9b. Only *after* the collapse process does the quanton have a well-defined position that the experiment can report! Thus *any* experiment that determines a quanton's position must first collapse its wavefunction to a position eigenfunction before it can deliver a meaningful result.

When we locate a quanton, its wavefunction collapses

Collapse is basic feature of the quantum model

Determining a quanton's position actually *defines* its position

Even so, this rule still bothers people very much, and physicists have spilled a lot of ink trying to decide *what* causes the collapse, *when* it actually occurs, what it implies about the physical nature of the wavefunction, and whether it violates relativity (the consensus in this case is that it does not). Except for the last, all these questions remain disturbingly unresolved.

TWO-MINUTE PROBLEMS

Q6T.1 Imagine that a set of electrons all have a spin state of $|\psi\rangle = [1, 0]$. If we send these electrons into an SGx device, what is the probability that a given electron will be found to have its spin antialigned with the +x direction?

    A.   0
    B. $\frac{1}{4}$
    C. $\frac{1}{2}$
    D. $\sqrt{1/2}$
    E.   1
    F.   Some other value (specify)

Q6T.2 Imagine that a set of electrons all have a spin state of $|\psi\rangle = [\sqrt{1/2}, \sqrt{1/2}]$. If we send these electrons through an SGx device, what is the probability that a given electron will be found to have its spin antialigned with the +x direction?

    A.   0
    B. $\frac{1}{4}$
    C. $\frac{1}{2}$
    D. $\sqrt{1/2}$
    E.   1
    F.   Some other value (specify)

Q6T.3 Imagine that an electron with a spin state of $|\psi\rangle = [\sqrt{1/2}, \sqrt{1/2}]$ goes into an SGz device and comes out the minus channel of that device. The electron's spin state is now

    A. $[\sqrt{1/2}, \sqrt{1/2}]$
    B. $[-\sqrt{1/2}, -\sqrt{1/2}]$
    C. $[1, 0]$
    D. $[0, 1]$
    E. $[\sqrt{1/2}\,e^{-iE_0t/\hbar}, \sqrt{1/2}\,e^{iE_0t/\hbar}]$
    F.   Some other state vector (specify)

Q6T.4 Imagine that a quanton's energy is zero, irrespective of its spin orientation. Its spin state vector will not change with time, true (T) or false (F)?

Q6T.5 Imagine that at a certain time a quanton has the wavefunction shown. If we were to perform an experiment to locate the quanton, what would be the most likely result or results?

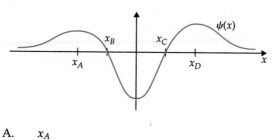

    A. x_A
    B. x_B
    C. x_C
    D. x_D
    E. $x = 0$
    F. x_B and x_C

Q6T.6 Consider the situation described in problem Q6T.5. If we were to perform an experiment to locate the quanton, what would be the least likely result or results? (Select from the answers in problem Q6T.5.)

Q6T.7 Imagine that at a certain time, a quanton has the (unrealistic) wavefunction shown. If we perform an experiment to locate the quanton, what is the probability that the result will be ≥ 1 nm? (Assume that the wavefunction is zero outside the picture.)

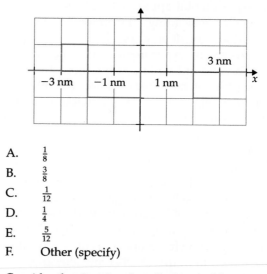

    A. $\frac{1}{8}$
    B. $\frac{3}{8}$
    C. $\frac{1}{12}$
    D. $\frac{1}{4}$
    E. $\frac{5}{12}$
    F.   Other (specify)

Q6T.8 Consider the situation described in problem Q6T.7. If we perform an experiment to locate the quanton, what is the probability that the result will be within the range $-2\,\text{nm} \le x \le 0$? (Assume that the wavefunction is zero at all points outside the picture.)

A. $-\frac{1}{4}$

B. $-\frac{1}{6}$

C. $\frac{1}{4}$

D. $\frac{1}{6}$

E. $\frac{1}{3}$

F. Other (specify)

Q6T.9 Imagine that at a certain time, a quanton has a well-defined wavelength of 0.15 nm. At that time we perform an experiment to locate the quanton, and we get the result (in a certain reference frame) $x =$ 5.26 nm to a high degree of precision. What can we say about the wavelength of the quanton's wavefunction after the experiment?

A. It is still 0.15 nm.

B. It is now 5.26 nm.

C. It is now not well defined.

D. Other (specify).

Q6T.10 Consider the function $\psi(x) = Ax^2 + B$, where A and B are constants. Is this function a possible wavefunction for a quanton under the right circumstances?

A. Yes

B. Yes, but only for certain values of A and B

C. Yes, but only if A and B are complex numbers

D. Yes, if B is zero

E. No.

Q6T.11 A quanton *really* has a position at all times, but quantum mechanics doesn't allow us to *predict* that position with certainty at future times as we could in newtonian mechanics, T or F?

HOMEWORK PROBLEMS

Basic Skills

Q6B.1 Consider the sequence of Stern-Gerlach devices shown.

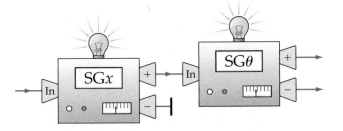

(a) What is the probability that an electron emerging from the minus channel of the first device will end up in the minus channel of the second device? Express your answer in terms of θ.

(b) What is the quanton's spin state vector after leaving the second device?

Q6B.2 Consider the sequence of Stern-Gerlach devices shown.

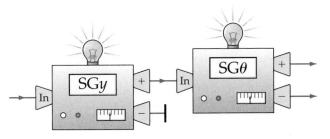

(a) What is the probability that an electron emerging from the minus channel of the first device

will end up in the minus channel of the second device? Express your answer in terms of θ.

(b) What is the quanton's spin state vector after leaving the second device?

Q6B.3 Imagine that a given time a quanton has the (unrealistic) wavefunction shown. If we perform an experiment to locate the quanton that time, what is the probability that the result will be greater than zero? Explain your reasoning. [Assume that $\psi(x) = 0$ everywhere outside the region shown. Surprisingly, the size of the vertical and horizontal scales is not relevant.]

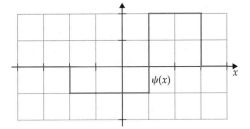

Q6B.4 Imagine that at a given time a quanton has the (highly artificial) wavefunction shown. If we perform an experiment to locate the quanton that

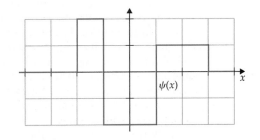

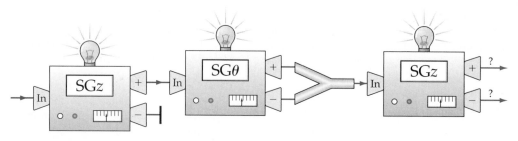

Figure Q6.11

What is the state of a quanton going into the third SG device? (See problem Q6S.2.)

time, what is the probability that the result will be greater than zero? Explain your reasoning. [Assume that $\psi(x) = 0$ everywhere outside the region shown. Surprisingly, the vertical and horizontal scales are not relevant.]

Q6B.5 Could the function $\psi(x) = Ax + B$ (where A and B are real constants) possibly be a valid wavefunction for a quanton under the right circumstances? Why or why not?

Q6B.6 Could the function $\psi(x) = A/x$ (where A is a real constant) possibly be a valid wavefunction for a quanton under the right circumstances? Why or why not?

Q6B.7 A quanton's wavefunction $\psi(x)$ at a certain time passes through zero at a certain position x_0. If we do an experiment to locate the quanton at that time, is the probability of getting the result x_0 less than (or equal to) the probability of getting any other result, or does this depend on the shape of the wavefunction at other positions? Explain your response.

Synthetic

Q6S.1 Assume that the superposition postulate is *not* true, and that in the situation shown in figure Q6.4, one-half of the electrons going into the final SG device have state $|+y\rangle$ and one-half have state $|-y\rangle$. (This is a much more *reasonable* assumption than the superposition postulate.)
  (a) What is the probability that an electron with spin state $|+y\rangle$ will be determined by the third device to have its spin aligned with the $+z$ direction? Antialigned with that direction?
  (b) What is the probability that an electron with spin state $|-y\rangle$ will be determined by the third device to have its spin aligned with the $+z$ direction? Antialigned with that direction?
  (c) Therefore, what is the probability that an arbitrary electron entering the third device will be determined by that device to have its spin aligned with the $+z$ direction? Antialigned with that direction?
  (d) Does this result contradict what is observed, as reported in chapter Q5? Explain.

Q6S.2 Consider the sequence of Stern-Gerlach devices shown in figure Q6.11. Prove that according to the superposition principle, the state of electrons in the recombined beam entering the third device is $|\psi\rangle = [1, 0]$ independent of the angle θ.

Q6S.3 Imagine that an electron has energy $+E_0$ when its spin is aligned with the $+x$ direction and $-E_0$ when its spin is antialigned with that direction. Assume that the electron's spin state at time $t = 0$ is $|\psi(0)\rangle = [1, 0]$. Find its spin state $|\psi(t)\rangle$ at time t. (*Hint:* You can simplify your result by using equations Q5.9c and Q5.9d.)

Q6S.4 Imagine that an electron has energy $+E_0$ when its spin is aligned with the $+y$ direction and $-E_0$ when its spin is antialigned with that direction. Assume that the electron's spin state at time $t = 0$ is $|\psi(0)\rangle = [0, 1]$. Find its spin state $|\psi(t)\rangle$ at time t.

Q6S.5 Consider the state vectors $|\psi\rangle$ and $|\psi'\rangle = e^{i\theta}|\psi\rangle$, where θ is an arbitrary real constant.
  (a) Prove that these vectors will yield the *same* prediction for the probability of collapse to any arbitrary final eigenvector $|\psi_f\rangle$ in an arbitrary experiment. (This means that these state vectors are really physically equivalent.)
  (b) Find values of θ such that $e^{i\theta} = -1$, $e^{i\theta} = +i$, and $e^{i\theta} = -i$. (This means that multiplying a state vector by -1 or $\pm i$ does not change the physical meaning of that vector.)

Q6S.6 Can we determine the units of the wavefunction $\psi(x)$ from, say, equation Q6.16? If so, what are the units? If not, why not?

Q6S.7 Near a certain position x_A, a quanton's wavefunction $\psi(x_A) \approx -0.3$ in some units. Near a certain position x_B, a quanton's wavefunction $\psi(x_B) \approx +0.12$ in the same units. If we do an experiment to locate the quanton, how many times more likely is it that the result is in a small range near x_A than in a range twice as large near x_B? Please explain.

Q6S.8 Imagine that a quanton's wavefunction at a given time is $\psi(x) = A\sin(3\pi x/L)$, for $0 \le x \le L$ and $\psi(x) = 0$ everywhere else, where A is whatever constant normalizes the wavefunction and L is a constant with units of length. If at this time we do an experiment to locate the quanton, what is the probability that the result will be between $x = \frac{1}{3}L$ and $x = \frac{2}{3}L$? Explain. (*Hint:* Draw a graph of the wavefunction.)

Q6S.9   Imagine that a quanton's wavefunction at a given time is $\psi(x) = Ae^{-(x/a)^2}$, where A is an unspecified constant and $a = 1.5$ nm. According to my table of integrals,

$$\int_{-\infty}^{\infty} \left(e^{-(x/a)^2}\right)^2 dx = a\sqrt{\frac{\pi}{2}} \qquad \text{(Q6.31)}$$

(a) Show that we must have $A = [2/\pi a^2]^{1/4}$ if $\psi(x)$ is to be normalized.

(b) If we were to perform an experiment to locate the quanton at this time, what would be the probability of a result within ± 0.1 nm of the origin (approximately)? (*Hint:* Note that 0.1 nm is pretty small compared to the range over which the exponential varies significantly.)

Q6S.10   Imagine that a quanton's wavefunction at a given time is $\psi(x) = A[1 + (a/x)^2]^{-1}$, where A is an unspecified constant and $a = 4.0$ nm. Note that according to my table of integrals

$$\int \frac{dx}{[1 + (x/a)^2]^2} = \frac{x/2}{1 + (x/a)^2} + \frac{a}{2}\tan^{-1}\frac{x}{a} \qquad \text{(Q6.32)}$$

(a) Show that we must have $A = \sqrt{2/\pi a}$ if $\psi(x)$ is to be normalized.

(b) If we were to perform an experiment to determine the quanton's location at this time, what would be the probability of a result between $x = 0$ and $x = 8.0$ nm?

Q6S.11   Imagine that you put a die in a cup, jostle the cup around violently, and then place the cup on a table with your hand over the top so we can't see the die. You can now perform an experiment to determine the die's value by looking at it and reading the number. Before we did this, there were six possible, equally likely results. After you remove your hand, there is only one possibility—the one you see. In *some* ways this is like an experiment that determines a quanton's position. Before we do the experiment, the quanton has many possible positions. After the experiment, it has just *one* position—the one we measured.

(a) However, we suspect that the die has *really* settled down to the number shown *before* we look; we just don't know what the number is. Why do we believe this? What physical evidence do we have (if any) that the die has really settled down to a choice before we look at it?

(b) On the other hand, in the quantum model, we believe that the quanton really does *not* have a position before we determine that position.

Why do we believe this? What physical evidence do we have that the quanton does *not* have a well-defined position before we localize it? (*Hint:* Consider a low-intensity two-slit interference experiment.)

Rich-Context

Q6R.1   Imagine that you have a box that emits quantons that have a definite but unknown spin state. If we run quantons from this box through an SGz device, we find that 20 percent of the electrons come out the plus channel and 80 percent from the minus channel. If we run quantons from the same box through an SGx device, we find that 50 percent of the electrons come out of each channel. If we run quantons from the box through an SGy device, we find that 90 percent come out the plus channel and 10 percent out of the minus channel. Find a quantum state vector for quantons emerging from the box that is consistent with these data. Make sure that your vector is normalized! (We cannot determine the state completely from experimental data because of the issue discussed in problem Q6S.5. Do not worry about this: just find *any* vector that is normalized and consistent with the data.)

Q6R.2   Imagine that you have a box that emits quantons which have a definite but unknown spin state. If we run quantons from this box through an SGz device, we find that 4/13 of the electrons come out the plus channel and 9/13 from the minus channel. If we run quantons from the same box through an SGx device, we find that 1/26 of the electrons come out of the plus channel and 25/26 out of the minus channel. If we run quantons from the box through an SGy device, we find that 1/2 come out of each channel. Find a quantum state vector for quantons emerging from the box that is consistent with these data. Make sure that your vector is normalized! (We cannot determine the state completely from experimental data because of the issue discussed in problem Q6S.5. Do not worry about this: just find *any* vector that is normalized and consistent with the data.)

Advanced

Q6A.1   Here is an example of one way to determine eigenvectors. In the spin experiments discussed in chapter Q5, we found that a given electron coming out of either channel of an SG device has a 50/50 chance of coming out either channel of a second SG device as long as the two devices correspond to perpendicular axes. We also know that the two eigenvectors corresponding to any one of the three spin observables S_x, S_y, or S_z must be normalized and orthogonal.

(a) Let us arbitrarily pick the eigenvectors of one of the observables (the conventional choice is S_z) to be $[1, 0]$ and $[0, 1]$. The experimental results described and the probability rule imply that the absolute square of the dot product of either one these vectors and *any* eigenvector corresponding to another observable must be $\frac{1}{2}$. Using this information, find a pair of normalized but orthogonal vectors that have *real* components and each has the top component positive.

(b) The pair of vectors that you found in part (a) is conventionally taken to be the vectors for the S_x observable. Now find a *second* pair of normalized but orthogonal vectors satisfying the experimental constraints and the conventional constraint that both components are positive and one component is real. These will be the eigenvectors for S_y.

Comment   We are free to choose both components real and the top component positive in the case of the S_x vectors, and both components positive and one component real in the case of the S_y vectors, because of the issue discussed in problem Q6S.5. We can take *any* pair of vectors satisfying the experimental constraints and put them into the form specified by these additional constraints by multiplying the vectors by $e^{i\theta}$ with an appropriate choice of θ. Therefore, the eigenvectors displayed in table Q6.1 are *possible* choices for the spin eigenvectors that are consistent with experiment, but other physically equivalent choices can be generated by multiplying the individual vectors by $e^{i\theta}$, where θ is arbitrary.

Q6A.2   (Requires knowledge of linear algebra.) Here is a different way to calculate eigenvectors. On quite basic theoretical grounds, one can associate each one of the spin observables S_x, S_y, and S_z with a 2 × 2 matrix

$$S_x \leftrightarrow \begin{bmatrix} 0 & s \\ s & 0 \end{bmatrix} \qquad S_y \leftrightarrow \begin{bmatrix} 0 & is \\ -is & 0 \end{bmatrix} \qquad S_z \leftrightarrow \begin{bmatrix} s & 0 \\ 0 & -s \end{bmatrix}$$

(Q6.33)

where s is a real number related to the intrinsic magnitude of the quanton's spin.

(a) Determine the eigenvalues and the corresponding eigenvectors (up to a possibly complex multiplicative constant) for each of these matrices. (The eigenvalues of one of these matrices represent the possible values that can result from an experiment to determine the observable corresponding to the matrix, and the corresponding eigenvector the eigenvector for that value.)

(b) Show that the eigenvectors listed in table Q6.1 are *consistent* with the constraints on the eigenvectors imposed by part (a) and the requirement that the vectors be normalized.

Comment   These vectors are not entirely *determined* by the results of part (a) and the normalization requirement because of the issue discussed in problem Q6S.5.

ANSWERS TO EXERCISES

Q6X.1   The calculations are as follows:

$$\langle +\theta \mid +\theta \rangle = \begin{bmatrix} \cos\frac{1}{2}\theta \\ i\sin\frac{1}{2}\theta \end{bmatrix}^* \cdot \begin{bmatrix} \cos\frac{1}{2}\theta \\ i\sin\frac{1}{2}\theta \end{bmatrix}$$

$$= \left(\cos\tfrac{1}{2}\theta\right)^* \left(\cos\tfrac{1}{2}\theta\right) + \left(i\sin\tfrac{1}{2}\theta\right)^* \left(i\sin\tfrac{1}{2}\theta\right)$$

$$= \cos^2\tfrac{1}{2}\theta + \left(-i\sin\tfrac{1}{2}\theta\right)\left(i\sin\tfrac{1}{2}\theta\right)$$

$$= \cos^2\tfrac{1}{2}\theta + \sin^2\tfrac{1}{2}\theta = 1 \qquad (Q6.34)$$

$$\langle -\theta \mid -\theta \rangle = \begin{bmatrix} i\sin\frac{1}{2}\theta \\ \cos\frac{1}{2}\theta \end{bmatrix}^* \cdot \begin{bmatrix} i\sin\frac{1}{2}\theta \\ \cos\frac{1}{2}\theta \end{bmatrix}$$

$$= \left(i\sin\tfrac{1}{2}\theta\right)^* \left(i\sin\tfrac{1}{2}\theta\right) + \left(\cos\tfrac{1}{2}\theta\right)^* \left(\cos\tfrac{1}{2}\theta\right)$$

$$= \left(-i\sin\tfrac{1}{2}\theta\right)\left(i\sin\tfrac{1}{2}\theta\right) + \cos^2\tfrac{1}{2}\theta$$

$$= \sin^2\tfrac{1}{2}\theta + \cos^2\tfrac{1}{2}\theta = 1 \qquad (Q6.35)$$

$$\langle +\theta \mid -\theta \rangle = \begin{bmatrix} \cos\frac{1}{2}\theta \\ i\sin\frac{1}{2}\theta \end{bmatrix}^* \cdot \begin{bmatrix} i\sin\frac{1}{2}\theta \\ \cos\frac{1}{2}\theta \end{bmatrix}$$

$$= \left(\cos\tfrac{1}{2}\theta\right)^* \left(i\sin\tfrac{1}{2}\theta\right) + \left(i\sin\tfrac{1}{2}\theta\right)^* \left(\cos\tfrac{1}{2}\theta\right)$$

$$= \left(\cos\tfrac{1}{2}\theta\right)\left(i\sin\tfrac{1}{2}\theta\right) + \left(-i\sin\tfrac{1}{2}\theta\right)\left(\cos\tfrac{1}{2}\theta\right)$$

$$= 0 \qquad (Q6.36)$$

Q6X.2   The calculation is as follows:

$$|\langle +z \mid -\theta \rangle|^2 = \left| \begin{bmatrix} 1 \\ 0 \end{bmatrix}^* \cdot \begin{bmatrix} i\sin\frac{1}{2}\theta \\ \cos\frac{1}{2}\theta \end{bmatrix} \right|^2$$

$$= \left| 1^* \left(i\sin\tfrac{1}{2}\theta\right) + 0\left(\cos\tfrac{1}{2}\theta\right) \right|^2$$

$$= \left| i\sin\tfrac{1}{2}\theta \right|^2 = \left(-i\sin\tfrac{1}{2}\theta\right)\left(i\sin\tfrac{1}{2}\theta\right)$$

$$= +\sin^2\tfrac{1}{2}\theta \qquad (Q6.37)$$

Q6X.3 $|\psi(0)\rangle = [0, 1]$ is already an energy eigenvector $|+z\rangle$, so we do not need to write it as a sum of energy eigenvectors. According to the time-evolution rule,

$$|\psi(t)\rangle = e^{-i(-E_0)t/\hbar} \begin{bmatrix} 0 \\ 1 \end{bmatrix} = \begin{bmatrix} 0 \\ e^{iE_0 t/\hbar} \end{bmatrix} \qquad (Q6.38)$$

Q6X.4   The area under the "curve" of the squared wavefunction in figure Q6.7b in the range $-1\text{ nm} \le x \le 0$ is half a square, while the area under the curve for $0 \le x \le +1$ nm is four half-squares, or two squares, so the total area in this range is 2.5 squares. The total area under the whole graph is 6 squares, so the probability is $\frac{2.5}{6}$, or $\frac{5}{12}$.

Q6X.5   The calculation looks like this:

$$\langle u \mid w \rangle = \sum_{\text{all } i} u_i^* w_i = \sum_{\text{all } i} [\overline{u}(x_i)]^* \overline{w}(x_i)$$

$$= \sum_{\text{all } x_i} [u(x_i)\sqrt{dx}]^* w(x_i)\sqrt{dx}$$

$$= \sum_{\text{all } x_i} [u(x_i)]^* w(x_i)\, dx = \int_{-\infty}^{\infty} [u(x)]^* w(x)\, dx$$

$$(Q6.39)$$

where the last step applies in the limit that $dx \to 0$.

Q6X.6   According to equation Q6.26, we have

$$|\psi(r_1, r_2, t)|^2 = |A(r)|^2 \left(e^{-i(p_0 r_1 - Et)/\hbar} + e^{-i(p_0 r_2 - Et)/\hbar} \right)$$
$$\times \left(e^{i(p_0 r_1 - Et)/\hbar} + e^{i(p_0 r_2 - Et)/\hbar} \right)$$

$$= |A(r)|^2 \left(e^{-i(p_0 r_1 - Et)/\hbar} e^{i(p_0 r_1 - Et)/\hbar} \right.$$
$$+ e^{-i(p_0 r_1 - Et)/\hbar} e^{i(p_0 r_2 - Et)/\hbar}$$
$$+ e^{-i(p_0 r_2 - Et)/\hbar} e^{i(p_0 r_1 - Et)/\hbar}$$
$$\left. + e^{-i(p_0 r_2 - Et)/\hbar} e^{i(p_0 r_2 - Et)/\hbar} \right)$$

$$= |A(r)|^2 \left(e^{i0} + e^{-i(p_0 r_1 - p_0 r_2)/\hbar} \right.$$
$$\left. + e^{i(p_0 r_1 - p_0 r_2)/\hbar} + e^{i0} \right) \qquad (Q6.40)$$

According to equation Q5.8a, $e^{i0} = 1$. According to equation Q5.9c, $e^{i\theta} + e^{-i\theta} = 2\cos\theta$. Plugging these results into equation Q6.40, we get

$$|\psi(r_1, r_2, t)|^2 = |A(r)|^2 \left\{ 1 + 2\cos[p_0(r_1 - r_2)/\hbar] + 1 \right\}$$
$$(Q6.41)$$

which is the same as equation Q6.27.

Q7 Bound Systems

Chapter Overview

Introduction

In this chapter, we will examine the energy eigenfunctions of bound systems, getting a first glimpse into why the possible values of the energy observable must be quantized. This chapter provides essential background to all the remaining chapters in the unit.

Section Q7.1:   An Introduction to Bound Systems

In newtonian mechanics (as discussed in chapter C11), a system consisting of a light-weight particle interacting with something essentially fixed will be bound if the light-weight particle is in a location corresponding to a valley in the interaction's potential energy function $V(x)$ and the system's energy E is lower than the valley walls. The particle is then trapped within the **classically allowed region** where $V(x) < E$, whose boundaries are the classical **turning points** where $E = V(x)$. Beyond those boundaries are **classically forbidden regions** where $V(x) > E$ and thus the particle's kinetic energy would have to be negative (which is absurd).

In this chapter we will consider bound systems from a quantum perspective.

Section Q7.2:   Energy Eigenfunctions

We are particularly interested in a bound system's energy eigenfunctions, because

1. We can say for sure whether a quanton is bound only if we know the value of E with certainty, which is true only if its wavefunction is an energy eigenfunction.
2. Energy is an important quantity that is often determined in experiments.
3. The time-evolution rule implies that we must know a quanton's energy eigenfunctions to determine how more general wavefunctions evolve in time.
4. Because the time-evolution rule also implies that energy eigenfunctions are **stationary** (i.e., have time-independent probability distributions), quantons tend to settle into energy eigenfunctions.

Section Q7.3:   A Quanton in a Box

A potential energy function such that $V(x) = 0$ for $0 \leq x \leq L$ and infinity outside that range is called a **box.** This potential energy function is a useful first approximation for many kinds of bound systems. Because any quanton in such a box is infinitely forbidden to exist where $V(x)$ is infinite, its wavefunction must be zero in these forbidden regions. The wavefunction of a quanton in a box will therefore be trapped between two perfectly reflecting barriers, like a standing wave on a string with two fixed ends. Combining this analogy with the de Broglie relation, one can argue that the energy eigenfunctions and associated energy values for a quanton in a box are

$$\psi_{E_n}(x) = \begin{cases} A \sin \dfrac{n\pi x}{L} & \text{if } 0 \leq x \leq L \\ 0 & \text{if } x < 0 \text{ or } x > L \end{cases} \qquad (Q7.9a)$$

$$E_n = \frac{h^2 n^2}{8mL^2} \qquad \text{where } n = 1, 2, 3, \dots \qquad (Q7.9b)$$

Purpose: These equations describe the possible energy eigenfunctions $\psi_{E_n}(x)$ and associated energy values E_n for a quanton in a box.

> **Symbols:** L is the box's length, A is a constant amplitude (which should be chosen so that the wavefunction is normalized), m is the quanton's mass, and h is Planck's constant.
> **Limitations:** These results apply only to the box potential energy function.
> **Note:** The requirement that n half-wavelengths of an eigenfunction fit between the boundaries requires that the energies be **quantized.**

Section Q7.4: The Simple Harmonic Oscillator

The simple harmonic oscillator is an even better model for many types of bound systems. Computing its energy eigenfunctions is presently beyond our grasp, but the allowed energies also turn out to be

> $$E_n = \frac{h\omega}{2\pi}\left(n + \tfrac{1}{2}\right) = \hbar\omega\left(n + \tfrac{1}{2}\right) \qquad \text{where } n = 0, 1, 2, 3, \ldots \qquad \text{(Q7.16)}$$
>
> **Purpose:** This equation specifies the possible energy values E_n for a simple one-dimensional harmonic oscillator.
> **Symbols:** ω is the oscillator's classical angular frequency $\equiv \sqrt{k_s/m}$, where k_s is the spring constant and m the oscillating object's mass, h is Planck's constant, and $\hbar \equiv h/2\pi$.
> **Limitations:** These results only apply to a quanton moving in one dimension and whose potential energy function is $V(x) = \tfrac{1}{2}k_s x^2 = \tfrac{1}{2}m\omega^2 x^2$.
> **Note:** This formula is conventionally written so that n starts at 0, not 1.

The corresponding energy eigenfunctions (see figure Q7.5) look much like those for the box, with an integer number of half-wavelengths trapped between the boundaries. However, the functions "leak" a little bit into the classically forbidden regions (we will discuss this more extensively in chapter Q10).

Section Q7.5: The Bohr Model of the Hydrogen Atom

A hydrogen atom is an intrinsically three-dimensional system. However, Niels Bohr devised a simple model, now called the **Bohr model,** that essentially makes the problem one-dimensional. An updated version of this model is based on the following (fairly wild) assumptions:

1. The electron is confined to a circular orbit of radius r.
2. We can calculate the electron's momentum p by using newtonian mechanics.
3. The de Broglie relationship applies.
4. The wavefunction must match itself as we go around the circular orbit.

As discussed in the section, this model predicts these possible energy values:

> $$E_n = -\frac{ke^2}{2a_0 n^2} \qquad \text{where } n = 1, 2, 3, \ldots \qquad \text{(Q7.24)}$$
>
> **Purpose:** This equation specifies the possible energies E_n for the electron in a hydrogen atom.
> **Symbols:** k is the Coulomb constant, e is the charge on a proton, and a_0 is the **Bohr radius** $= 0.053$ nm.
> **Limitations:** This is a *highly* oversimplified model: do not assume that its qualitative features are at all valid. The energy results, however, are correct.
> **Notes:** The numerical value of $ke^2/2a_0$ is 13.6 eV.

Q7.1   An Introduction to Bound Systems

Definition of what we mean by a *bound system*

A **bound system** is any system of interacting quantons where the nature of the interactions between the quantons keeps their relative separation limited: the quantons are thus *bonded* in the sense that we discussed in chapter C11. Understanding the quantum mechanics of bound systems is important because the basic microscopic building blocks of the universe (molecules, atoms, atomic nuclei, and even some subatomic particles) are bound systems whose structure and characteristics are shaped by quantum mechanics. We can even better understand the behavior of *macroscopic* systems such as crystalline solids or enclosed gases by considering them as bound systems shaped by quantum mechanics (as we will see in unit T).

Some simplifying assumptions

A full and careful analysis of a bound system involving many quantons can be very complicated. We will make this problem more tractable by focusing on the behavior of a *single* quanton that moves in what we imagine to be a *fixed* potential energy field $V(x)$ created by the other quantons in the system. For example, we might study a single electron moving in response to the electric field of an atomic nucleus, or a single atom moving in response to the electric field created by the other atoms in a crystal, or a proton moving in response to the strong-interaction field established by the other nucleons in a nucleus. In each of these cases, our single quanton of interest is so small in mass compared to the total mass of the other quantons that the field that the latter creates will not change (to a good degree of approximation) as the single quanton moves in response to it.

Review of classical physics of bound systems

Consider a newtonian particle moving in one dimension in response to an unchanging potential energy function $V(x)$. The total conserved energy associated with this particle's interaction with the rest of the system is $E = K + V(x)$ (we are assuming that the rest of the system is so massive that it is essentially at rest so that its kinetic energy is negligible). In what follows we will call E the *particle's* energy, even though it is technically the part of the *system's* energy associated with this particle's motion. In chapter C11, we saw that this particle is bound if it is in a **classically allowed** region (where $E > V$) enclosed by two **classically forbidden regions** or **barriers** (where $V > E$), as shown in figure Q7.1. Let's review why these regions are allowed and forbidden. Conservation of energy implies that $E = K + V(x)$, where K is the kinetic energy of the moving particle (we are considering the rest of the system to be so massive that its kinetic energy is negligible); thus $K = E - V(x)$. Since the particle's kinetic energy must be positive, the particle cannot be where $K = E - V(x) < 0$, so any region where $V > E$ is *classically forbidden*. The particle instead is constrained to move back and forth between the two **turning points** (points where $E = V$) that mark the boundaries of the classically allowed region. The particle can never escape this region as long as its energy E remains fixed.

Figure Q7.1

In classical mechanics, a particle is bound if its energy and position are such that it is in a *classically allowed* region (where $E > V$) bounded by two *classically forbidden* regions (where $V > E$). This diagram illustrates the definitions of these and other terms associated with such situations.

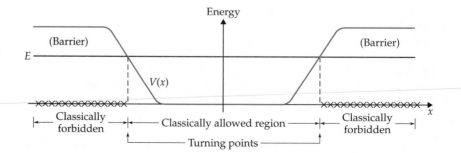

Note, however, that newtonian mechanics puts no particular constraints on the particle's energy: even if we insist that the particle be bound, it can have *any* energy E that is smaller than the heights of both barriers on either side.

Q7.2   Energy Eigenfunctions

In quantum mechanics, a quanton's energy E is (like the quanton's position) an observable and (like the position) is not necessarily well defined for a given wavefunction $\psi(x)$: the wavefunction may instead specify probabilities for a number of different energy values that we might obtain if we do an experiment to determine the quanton's energy. For a variety of reasons, it is useful when we consider bound systems to focus on the quanton's **energy eigenfunctions.**

As we discussed in chapter Q6, the collapse rule implies that a quanton's wavefunction just after we determine its energy is an energy eigenfunction. When a quanton's wavefunction is an energy eigenfunction, we know *exactly* what its energy is: any subsequent determination of the quanton's energy will yield that eigenfunction's corresponding energy value with certainty.

What is an energy eigenfunction?

Why focus our attention specifically on energy eigenfunctions? There are four important reasons for doing this.

Four reasons for studying energy eigenfunctions

The first is simply that if the quanton is in a state where its energy is some well-defined value E, this means that we can decide with certainty whether the quanton is bound or not. If the eigenfunction's corresponding energy value E is such that a classical particle would be bound by the potential energy function, then it turns out that its wavefunction will be trapped between the potential energy barriers on either side, making it quantum-mechanically bound also (chapter C11 explores this issue in greater depth). If the quanton's state is not an energy eigenfunction, then the quanton's energy is not well defined, and therefore whether it is bound or not might be ambiguous.

The second reason is simply that energy (like position) is an extremely important observable that experiments commonly determine. If we commonly determine a quanton's energy, it is advantageous not only to know what values of E we might obtain but also what the quanton's collapsed wavefunction after the experiment might look like.

The third reason is that the time-evolution rule singles out energy eigenfunctions as being special. If we want to know how a quanton's wavefunction evolves in time, we need to write that function as a weighted sum of energy eigenfunctions and then multiply the nth eigenstate in the sum by the factor $e^{-iE_n t/\hbar}$. To even begin this process, we have to know what a quanton's energy eigenfunctions (and associated energy values) are!

The fourth reason is that in many realistic applications of quantum mechanics, an initial wavefunction that is *not* an energy eigenfunction will fairly quickly evolve so that it *becomes* an energy eigenfunction. This is fundamentally so because the time-evolution rule implies that energy eigenfunctions have a special property that other wavefunctions do not: *the shape of the absolute square of an energy eigenfunction does not change in time* (for this reason, physicists often refer to energy eigenfunctions as **stationary** states). To see this, imagine that a quanton's wavefunction at time $t = 0$ is an energy eigenfunction $\psi_{E_n}(x)$. Since this already is expressed as a sum of energy eigenstates (the sum involves only one term in this case), the time-evolution rule implies that the quanton's wavefunction at time t is

Energy eigenfunctions are *stationary states*

$$\psi(x, t) = \psi_{E_n}(x)e^{-iE_n t/\hbar} \qquad (Q7.1)$$

The absolute square of this is

$$|\psi(x,t)|^2 = \left|\psi_{E_n}(x)e^{-iE_nt/\hbar}\right|^2 = \left|\psi_{E_n}(x)\right|^2 |e^{-iE_nt/\hbar}|^2 = \left|\psi_{E_n}(x)\right|^2 \quad (Q7.2)$$

since $|e^{i\theta}|^2 = 1$. You can see that $|\psi(x,t)|^2$ is indeed independent of time.

This is important, because a quanton's ability to interact with other things often depends on how its squared wavefunction changes with time. For example, it turns out that a charged quanton whose squared wavefunction *does* change shape with time will spontaneously radiate electromagnetic energy much more effectively than quantons whose wavefunction is stationary. A charged quanton whose initial wavefunction is *not* an energy eigenfunction will in fact radiate photons until its wavefunction *becomes* an energy eigenfunction. This process generally takes on the order of 10^{-8} s. Again, an exploration of this process is far beyond the level of this course, but this is another important reason for focusing on energy eigenfunctions.

The last rule means that we can, as a general rule, *assume* that the wavefunction of a bound quanton is an energy eigenfunction. Because of this, many texts blur the distinction between energy eigenfunctions and more general wavefunctions: it is simply *assumed* (usually without even explicitly saying so) that a quanton's wavefunction is an energy eigenfunction. In this text, however, I will use the notation $\psi_E(x)$ for an energy eigenfunction to distinguish it from a more general wavefunction $\psi(x)$.

Q7.3   A Quanton in a Box

The potential energy function for a box

The simplest of all models for bound systems in quantum mechanics is a quanton moving in one dimension in response to a potential energy function $V(x)$ that physicists somewhat whimsically call a **box.** The box potential energy function is zero within a certain range $0 \leq x \leq L$ (where L is the length of the box) and infinite outside of this range:

$$V(x) = \begin{cases} 0 & \text{if } 0 \leq x \leq L \\ \infty & \text{if } x < 0 \text{ or } x > L \end{cases} \quad (Q7.3)$$

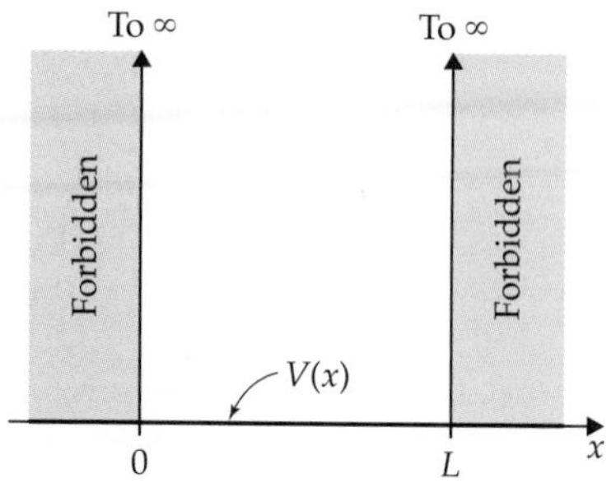

Figure Q7.2
The potential energy function $V(x)$ known as the *box*. The vertical axis in this graph corresponds to energy.

This potential energy function is illustrated in figure Q7.2. In classical mechanics, if $V(x)$ is a constant within a certain region, a particle is *free* (i.e., experiences zero external forces) in that region. On the other hand, the points $x = 0$ and $x = L$ (where the potential energy becomes infinite) are turning points for the particle no matter what its energy might be. Therefore, a classical particle moving in response to this potential energy function will behave as a free particle trapped between two impenetrable walls.

Finding the energy eigenfunctions in this case

What will an energy eigenfunction corresponding to a given energy E look like in this case? At the boundaries of the box at $x = 0$ and $x = L$, the quanton encounters infinite barriers. Since the quanton cannot possibly travel beyond these barriers, its wavefunction must be zero beyond these limits. In between the barriers, however, the quanton is completely free. We saw in chapters Q4 and Q6 that a free quanton's wavefunction has a definite wavelength λ given by the de Broglie relation. In this particular case, the quanton's wave will be trapped between two barriers where its wavefunction must go to zero, as if it were a string with two fixed ends. Therefore, by analogy to the string case, we would expect the quanton's wavefunction at a given instant to look something like a standing wave on a string with two fixed ends:[†]

$$\psi(x) = A \sin kx \quad (Q7.4)$$

[†]See problem Q7A.1 for a more mathematical argument for this result.

where $k \equiv 2\pi/\lambda$ and λ is such that an integer number of half-wavelengths fit between the box's boundaries. In this case, this means that the wavelength of the nth normal mode is

$$n\frac{\lambda_n}{2} = L \quad \Rightarrow \quad \lambda_n = \frac{2L}{n} \quad \text{where } n = 1, 2, 3, \ldots \quad (Q7.5)$$

Now, in the classically allowed region where $V(x) = 0$, the quanton's total energy is equal to its kinetic energy k:

$$E = K + V(x) = K + 0 = K = \frac{p^2}{2m} \quad (Q7.6)$$

Applying $p_n = h/\lambda_n$, we find that the quanton's energy when its wavefunction corresponds to the nth normal mode of waves in the box is

$$E_n = \frac{p_n^2}{2m} = \frac{1}{2m}\left(\frac{h}{\lambda_n}\right)^2 = \frac{h^2}{2m}\left(\frac{n}{2L}\right)^2 = \frac{h^2 n^2}{8mL^2} \quad (Q7.7)$$

We see that when the quanton's wavefunction has a definite wavelength λ_n, it has a well-defined and specific energy E_n, so these normal modes must be the quanton's energy eigenfunctions, with E_n being the associated energy eigenvalue for the nth eigenfunction. Equation Q7.5 implies that the wavenumber for the nth normal mode is

$$k_n = \frac{\pi n}{L} \quad (Q7.8)$$

Exercise Q7X.1

Verify equation Q7.8.

Therefore, the energy eigenfunctions and associated energy values for a quanton in a box are

$$\psi_{E_n}(x) = \begin{cases} A\sin\dfrac{n\pi x}{L} & \text{if } 0 \leq x \leq L \\ 0 & \text{if } x < 0 \text{ or } x > L \end{cases} \quad (Q7.9a)$$

$$E_n = \frac{h^2 n^2}{8mL^2} \quad \text{where } n = 1, 2, 3, \ldots \quad (Q7.9b)$$

The energy eigenfunctions and associated energy values for a quanton in a box

Purpose: These equations describe the possible energy eigenfunctions $\psi_{E_n}(x)$ and associated energy values E_n for a quanton in a box.
Symbols: L is the box's length, A is a constant amplitude (which should be chosen so that the wavefunction is normalized), m is the quanton's mass, and h is Planck's constant.
Limitations: These results apply only to the box potential energy function, where $V(x) = 0$ for $0 \leq x \leq L$ and infinite elsewhere.

These energy eigenfunctions are illustrated in figure Q7.3.
    We see that requiring the quanton's wavefunction to fit between the two boundaries means that the quanton's energy must be **quantized** (i.e., the energy can have only specific discrete values). Since the quanton's wavefunction must collapse to one of the energy eigenfunctions listed in equation Q7.9a when we do an experiment to determine its energy, the corresponding energy

The quanton's energies are quantized

Figure Q7.3
(a) The box potential energy function $V(x)$. (b) The energy eigenfunctions having the three lowest possible energies $E_1 = h^2/8mL^2$, $E_2 = 2^2 E_1$, and $E_3 = 3^2 E_1$ ($n = 1, 2,$ and 3, respectively). Note that as the wavelength decreases, the energy increases.

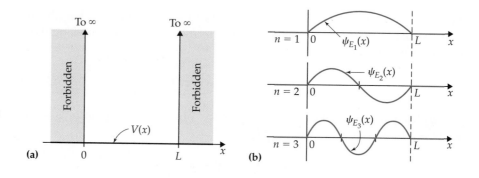

values listed in equation Q7.9b are the only possible results that we could get. This is very different from what classical physics would predict!

The box is a useful first approximation for many kinds of systems. We can apply it in almost any situation where the width of the classically allowed region does not change much as the quanton's energy increases. We will use this model extensively to model nuclei in a few chapters.

The box as a useful approximation

Example Q7.1   Electrons in a Nanostrip

Problem   It is becoming nearly possible to use advanced integrated-circuit technology to construct "nanostructures" that can confine electrons to a tiny strip of metal that we can model as a box (electrons have a much lower potential energy inside the metal than outside, as we saw in the chapter Q3 on the photoelectric effect). Imagine that we do an experiment to evaluate the energy of an electron in such a box that is 6.0 nm long. What possible results (in electronvolts) might we get?

Solution   It is easier to compute the energies if we multiply the top and bottom of equation Q7.6 by c^2, yielding

$$E_n = \frac{(hc)^2 n^2}{8mc^2 L^2} \qquad (Q7.10)$$

The rest energy mc^2 of the electron is 511,000 eV, and $hc = 1240$ eV $\cdot$ nm, so

$$E_n = \frac{(1240 \text{ eV} \cdot \text{nm})^2 n^2}{8(511,000 \text{ eV})(6.0 \text{ nm})^2} = 0.010 n^2 \frac{\text{eV}^2}{\text{eV}} = (0.010 \text{ eV}) n^2 \quad (Q7.11)$$

So the possible results are $E_1 = 0.010$ eV, $E_2 = 0.040$ eV, $E_3 = 0.090$ eV, $E_4 = 0.16$ eV, and so on.

Exercise Q7X.2

If the lowest possible electron energy that we get for such a box nanostructure is 0.09 eV, how long is it?

An example of how combinations of energy eigenfunctions evolve in time

Again, please remember that a quanton's wavefunction does not *have* to be an energy eigenfunction: it can have any arbitrary shape. But by the Fourier theorem (and as asserted by the time-evolution rule), we can write any arbitrary wavefunction as a weighted sum of energy eigenfunctions. Doing this allows us to determine how the wavefunctions evolve with time.

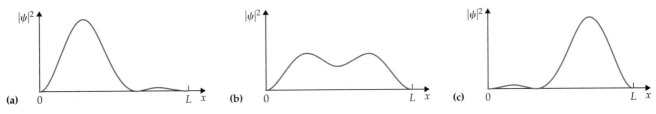

Figure Q7.4

(a) The absolute square of the wavefunction shown in equation Q7.13 (with $A = B$) at time $t = 0$. (b) The absolute square of the wavefunction at time $t = \frac{1}{2}\pi\hbar/(E_2 - E_1)$, which corresponds to one-quarter of a complete cycle of the shape of the waveform. (c) The same at time $t = \pi\hbar/(E_2 - E_1)$, halfway through the cycle.

As an example, imagine that a quanton in a box has a wavefunction at time $t = 0$ that is a superposition of the box's $n = 1$ energy eigenfunction $\psi_{E_1}(x)$ and its $n = 2$ eigenfunction $\psi_{E_2}(x)$:

$$\psi(x) = A\psi_{E_1}(x) + B\psi_{E_2}(x) \qquad (Q7.12)$$

where A and B are (for the sake of argument) real constants. The time-evolution rule implies that the quanton's wavefunction at time t will be

$$\psi(x, t) = Ae^{-iE_1t/\hbar}\psi_{E_1}(x) + Be^{-iE_2t/\hbar}\psi_{E_2}(x) \qquad (Q7.13)$$

One can show that the absolute square of this wavefunction is *not* time-independent (in contrast to equation Q7.2), but has a shape that oscillates at a frequency that is proportional to the energy difference $E_2 - E_1$ (see problem Q7S.9). Figure Q7.4 shows the shape of the absolute square of this wavefunction at $t = 0$ and one-quarter and one-half cycle later. We see that this wavefunction essentially shows the quanton's probability peak bouncing back and forth within the box at this frequency, behavior that is reminiscent of what a classical particle trapped in a box would do.

Q7.4   The Simple Harmonic Oscillator

We have discussed earlier in units C and N how useful the harmonic oscillator is as a model for many physical systems. It would therefore be nice to know something about the possible energies and energy eigenfunctions for a quanton responding to a harmonic oscillator potential energy function

$$V(x) = \frac{1}{2}k_s x^2 = \frac{1}{2}m\left(\frac{k_s}{m}\right)x^2 = \frac{1}{2}m\omega^2 x^2 \qquad (Q7.14)$$

where k_s here is the effective spring constant of the oscillator and $\omega \equiv \sqrt{k_s/m}$ is its classical angular frequency of oscillation.

This problem is, however, much more complicated than the box problem. First, as the quanton's energy increases, the distance between the classical turning points does not remain fixed (as it does in the case of the box). Since the turning points are positions x_{tp} where $V(x_{tp}) = E$, we have in this case that

Really solving the harmonic oscillator would be hard

$$\frac{1}{2}m\omega^2 x_{tp}^2 = E \quad \Rightarrow \quad x_{tp} = \pm\sqrt{\frac{2E}{m\omega^2}} \qquad (Q7.15)$$

Thus the distance $L = 2|x_{tp}|$ between the turning points increases as the square root of E (this is illustrated in figure Q7.5).

To compound the problem, $V(x)$ is not zero (or even constant) within the classically allowed region, so the eigenfunction corresponding to a definite energy $E = p^2/2m + V(x)$ will not necessarily be anything like that for a free

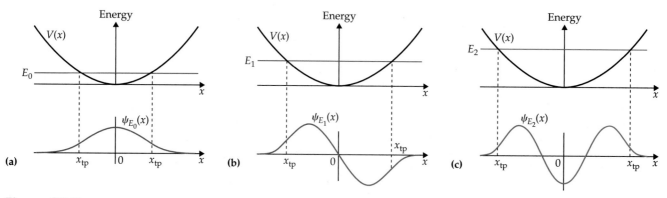

Figure Q7.5
The three energy eigenfunctions with the lowest energies for the one-dimensional quantum harmonic oscillator. The top graph for each energy shows the oscillator potential energy function $V(x) = \frac{1}{2}m\omega^2 x^2$, and the bottom graph shows the corresponding energy eigenfunction.

quanton with energy E. We have not yet developed in this text any method for calculating the shape of an energy eigenfunction if $V(x)$ varies with x.

We will develop such a method in chapter Q10. At present, let me simply quote the results that we get when we do apply this method. The possible energies for a quanton in this situation turn out to be

$$E_n = \frac{h\omega}{2\pi}\left(n + \tfrac{1}{2}\right) = \hbar\omega\left(n + \tfrac{1}{2}\right) \qquad \text{where } n = 0, 1, 2, 3, \ldots$$
$$(\text{Q7.16})$$

Purpose: This equation specifies the possible energy values E_n for a simple one-dimensional harmonic oscillator.
Symbols: ω is the oscillator's classical angular frequency $\equiv \sqrt{k_s/m}$, where k_s is the spring constant and m the oscillating object's mass, h is Planck's constant, and $\hbar \equiv h/2\pi$.
Limitations: This only applies to a one-dimensional simple harmonic oscillator.
Note: This formula is conventionally written so that n starts at 0, not 1. Be aware of this! The corresponding energy eigenfunctions are drawn in figure Q7.5.

If you compare the eigenfunctions shown in figure Q7.5 with those shown in figure Q7.3 for the box, you will note that the shapes of these functions are *very* similar: the $n = 0$ eigenfunction corresponds to essentially a half-wavelength trapped within the allowed region, the $n = 1$ eigenfunction has essentially a full wavelength trapped in that region, and so on. We can see that the quantization of energy in both cases arises essentially because the wavefunction has to fit an integer number of half-wavelengths within the allowed region, and thus only certain wavelengths (and thus certain energies) work.

There are some interesting differences, though. Note how the wavefunctions get wider as energy increases because the distance between turning points increases. Also, because the potential energy function in the harmonic oscillator case does not go instantly to infinity at the turning points, it turns out that the energy eigenfunctions "leak" into the classically forbidden regions a little bit (we will see why in chapter Q10). This is related to a quantum-mechanical phenomenon called *tunneling*, which we will discuss more thoroughly in chapter Q11.

Q7.5   The Bohr Model of the Hydrogen Atom

Another physical system of great interest is the atom. In this section, we will consider a simplified model of the *hydrogen atom*, the simplest of all atoms, which consists of a single electron bound to a single proton by their mutual electrostatic attraction.

This system presents even more challenges than the harmonic oscillator. Even a simple hydrogen atom is an intrinsically *three*-dimensional system, and we do not have any reason to believe that its three-dimensional energy eigenfunctions would have much in common with the one-dimensional eigenfunctions we have considered to this point. Even so, we will see that if we chip away at the complexities of this problem, we can create a simplified *one*-dimensional model of the hydrogen atom that, even though it is obviously too simple to be fully correct, does deliver some useful physical insight.

The *classical* model of the hydrogen atom imagines the electron to orbit the much more massive proton in much the same way as a planet orbits the more massive sun. We saw in chapter N12 that conservation of angular momentum implies that such an orbit must lie in a *plane*. Let's assume that something similar applies in the case of quantum mechanics: let's assume that we can pretend that the electron is confined to a certain plane in space. This reduces the three-dimensional problem to a two-dimensional problem.

There are also constraints on an electron's distance from the proton. We saw in chapter N13 that a planet moves around the sun in an elliptical orbit. Since the electrostatic force is an inverse-square force just as the gravitational force is, a classical electron orbiting a proton will behave in exactly the same way, according to newtonian mechanics. Now, the orbits of most of the planets happen to be nearly circular. Let's assume that this is true of the electron as well: let's assume that we can learn the most important things about the electron's wavefunction if we pretend that it is essentially confined to move in an approximately *circular* orbit of radius r. This reduces the problem to a basically *one*-dimensional problem: we are most interested in the characteristics of the electron's wavefunction in a circular band of radius r around the proton.

Now a *classical* electron in a circular orbit would have a constant speed and thus a constant momentum magnitude p. The de Broglie relation thus *suggests* that the electron's wavefunction as we go around this band should have a well-defined and constant wavelength $\lambda = h/p$.

The wave nature of the electron's wavefunction puts an additional constraint on the problem that has no classical counterpart. If we look at the electron's wavefunction at a particular instant in time, it must have a single, well-defined value at all points in space. This means that if we start at some point on our circular band of radius r and go exactly once around the band, the electron's wavefunction must return to whatever value it had when we started.

This in turn means that if the wave oscillates with a definite wavelength λ, it must go through an integer number of complete oscillations as we go around the band exactly once (see figure Q7.6). Since the distance around the

A full analysis of the hydrogen atom would be difficult

The first assumption: the electron is confined to a plane

The second assumption: the electron is confined to a circle

The third assumption: the wave must match itself as we go once around the circle

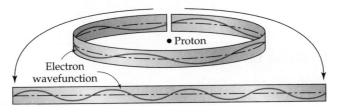

Figure Q7.6
The wavefunction of an orbiting electron plotted on a band encircling the proton at a constant radius r. If we unwrap the band and lay it flat, we see a sinusoidal wave that must go through an integer number of cycles (in this case, four) so that it matches itself as we go around the circle exactly once.

band is $2\pi r$, this condition implies that

$$2\pi r = n\lambda \qquad \text{where } n = 1, 2, 3, \ldots \qquad (Q7.17)$$

The fourth assumption: we can find p (and thus λ) by using classical mechanics

Now we are ready to put the model together. Newton's second law and Coulomb's law imply that for a *classical* electron in a circular orbit of radius r, we have

$$\frac{mv^2}{r} = ma = F_{\text{net}} = \frac{ke^2}{r^2} \qquad (Q7.18)$$

since an object's acceleration in a circular orbit is v^2/r and both the electron and the proton have the same magnitude of charge e (in this equation k is the Coulomb constant). Multiplying both sides of this equation by mr, we get

$$p^2 = (mv)^2 = \frac{mke^2}{r} \qquad (Q7.19)$$

Using this equation, the de Broglie relation $p = h/\lambda$, and equation Q7.17, we can eliminate p in favor of r and solve for r to get

Equation for possible (quantized) orbital radii

$$r_n = \frac{n^2 h^2}{4\pi^2 mke^2} = \frac{n^2 \hbar^2}{mke^2} \qquad \text{where } n = 1, 2, 3, \ldots \qquad (Q7.20)$$

The subscript on the r emphasizes that only orbits with certain *discrete* radii are consistent with the constraints of classical mechanics, the de Broglie relation, and the wave matching itself as we go once around the atom.

Exercise Q7X.3

Verify equation Q7.20. (This is important!)

We define the **Bohr radius** a_0 to be the value of this radius when $n = 1$:

The Bohr radius

$$a_0 \equiv \frac{h^2}{4\pi^2 mke^2} = \frac{\hbar^2}{mke^2} = 0.053 \text{ nm} \qquad r_n = n^2 a_0 \qquad (Q7.21)$$

Exercise Q7X.4

Verify the numerical value of the Bohr radius. (*Hint:* This is easier if you multiply the top and bottom by c^2 and use $mc^2 = 511,000$ eV and $hc = 1240$ eV · nm. It also helps if you first show that $ke^2 = 1.44$ eV · nm.)

The classical *energy* of the orbiting electron in this case is

$$E = \frac{1}{2}mv^2 + V(r) = \frac{1}{2}mv^2 - \frac{ke^2}{r} \qquad (Q7.22)$$

Comparing this with equation Q7.18, we can show that

$$E = -\frac{ke^2}{2r} \qquad (Q7.23)$$

Exercise Q7X.5

Verify equation Q7.23.

Finally, plugging $r_n = n^2 a_0$ (equation Q7.21) into $E = -ke^2/2r$ (equation Q7.23), we find that the quantized energy values for the electron in the hydrogen atom are

$$E_n = -\frac{ke^2}{2a_0 n^2} \qquad \text{where } n = 1, 2, 3, \ldots \qquad \text{(Q7.24)}$$

Purpose: This equation specifies the Bohr model's prediction for the possible energies E_n for the electron in a hydrogen atom.
Symbols: k is the Coulomb constant, e is the charge on a proton, and a_0 is the *Bohr radius* = 0.053 nm.
Limitations: This is a *highly* oversimplified model: do not assume that its qualitative features are at all correct. The energy results, however, are astonishingly good for such a crude model.
Notes: The numerical value of $ke^2/2a_0$ is 13.6 eV.

Possible electron energies for hydrogen atom

Therefore, if we do an experiment to evaluate the total energy of the orbiting electron in a hydrogen atom, the possible results are $E_1 = -13.6$ eV, $E_2 = -(13.6 \text{ eV})/2^2 = -3.4$ eV, $E_3 = -(13.6 \text{ eV})/3^2 = -1.5$ eV, and so on. An electron at rest at $r = \infty$ has an energy of zero according to the definition of the electrostatic potential energy; so the fact that the energies here are *negative* simply means that an electron bound into a hydrogen atom has *less* energy than it would at rest at infinity. (Even though these energy values are negative, note that they increase as n increases.)

What is amazing (considering the wild approximations and assumptions that we made throughout the construction of this model) is that these quantized energy values turn out to be *almost exactly right!* These results agree extremely well with both experimental results and a more sophisticated, fully three-dimensional analysis. We call this model the **Bohr model** of the hydrogen atom because it is an updated version of a model proposed by the Danish physicist Niels Bohr in 1913. Bohr cleverly found exactly how to zero in on the crucial physics of this situation and capture its essence in a nice, simple, and memorable model.

Even though the model is simplified, the results are nearly exact!

Bohr's efforts in this regard (which predated the full theory of quantum mechanics by about 13 years) were partly driven by serious problems with the classical model of the atom. In the classical model, the electron constantly accelerates as it moves in its orbit around the proton. We saw at the end of unit E that any accelerating charged object must emit electromagnetic radiation. The electromagnetic waves that the electron must radiate drain away its kinetic energy, causing it to spiral into the proton in a tiny fraction of a second (see problem E16S.11). Hydrogen atoms are, needless to say, *not* observed to collapse in this way, so there must be something very wrong with this picture.

How quantum mechanics explains the atom's stability

Bohr simply *postulated* that the special circular orbits of radius $r_n = n^2 a_0$ did not radiate. The full theory of quantum mechanics explains *why*. The squared energy eigenfunctions of the electron (like standing waves on a string) do not change shape as time passes, while other squared wavefunctions *do* change shape. The net electric field produced by an electron (according to quantum mechanics) turns out to be determined by the shape of the absolute square of its wavefunction (which is analogous to the time average of a standing wave). If the shape of the squared wavefunction changes with time, the electric field will change with time and the electron will radiate electromagnetic waves. But if the shape of the squared wavefunction does *not* change in time, the electric field, like the squared wavefunction, is stationary and thus no electromagnetic waves are radiated. This explains how it is possible for an electron whose wavefunction is an energy eigenfunction to remain bound to a nucleus without catastrophically spiraling into it.

Why electron energies are quantized in this situation

It is worth reviewing why the possible energies of a hydrogen atom in this model are quantized. The condition that the wavefunction connect with itself as we go once around the atom means that we have to fit an integer number of wavelengths within the circumference of the orbit. This is exactly analogous to the constraint that an integer number of half-wavelengths have to fit along the length of the box. In both cases, the wavelike nature of the quanton's wavefunction is what makes these conditions necessary.

You should strive in the long run to be able to reproduce (on your own) the key derivations in this chapter. The derivations of the allowed energies in the box model and the Bohr model in particular are important for every well-educated person to know, because these models have many applications and make using and understanding quantum physics much easier.

The limitations of the Bohr model

However, you should also recognize that in spite of its pedagogical and historical importance, the Bohr model is, from a modern perspective, a poor and even misleading picture of what is *really* going on, and so should not be taken literally at all. After Bohr's great success with the hydrogen atom, many physicists tried with little success to extend Bohr's model to other atoms and other systems. In other words, Bohr's model proved ultimately to be a conceptual dead end that just happened to give the right answers in one specific case.

The full theory of quantum mechanics, in contrast, provides a coherent approach that correctly handles not only the hydrogen atom but also a host of other systems. One sign of a really good model is the breadth of its ability to explain, and in this regard, the full theory of quantum mechanics (which we will get just a taste of in chapters Q9 and Q11) is far superior to the Bohr model.

On the other hand, another feature of a good model is that it is simple. In *this* regard, the Bohr model is way ahead of the full theory of quantum mechanics, whose treatment of even the hydrogen atom is very difficult. This is why the Bohr model, limited and misleading as it is, is still remembered and taught.

Exercise Q7X.6

(Important!) Show that the magnitude L of the electron's angular momentum in the Bohr model is also quantized, with $L_n = nh/2\pi = n\hbar$. (Since Bohr's work predated the de Broglie relation by more than a decade, Bohr actually *postulated* that $L_n = n\hbar$ rather than the third assumption we made.)

TWO-MINUTE PROBLEMS

Q7T.1 A quanton moving in a system whose $V(x)$ is given by the box potential energy function can never be unbound, no matter what its total energy might be, true (T) or false (F)?

Q7T.2 An electron moving near a proton can never be unbound, no matter what its energy might be, T or F?

Q7T.3 A friend claims that the value for n for a given energy eigenfunction is always equal to the number of

"bumps" (i.e., crests and troughs) in the wavefunction. This assertion is
A. True for all the models in this chapter
B. True for the one-dimensional models but not the Bohr model
C. True for the Bohr model but not the one-dimensional models
D. Not true for any of the models

Q7T.4 Consider a quanton in a box. How does the wave-length λ of an energy eigenfunction in this case depend on the energy E corresponding to that eigenfunction?
A. $\lambda \propto E$
B. $\lambda \propto \sqrt{E}$
C. λ is independent of E
D. $\lambda \propto 1/E$
E. $\lambda \propto 1/\sqrt{E}$
F. Other (specify)

Q7T.5 Consider a quanton in a hydrogen atom. How does the wavelength λ of an energy eigenfunction (as measured around the circular band) depend on the energy E associated with that eigenfunction?
A. $\lambda \propto |E|$
B. $\lambda \propto \sqrt{|E|}$
C. λ is independent of E

D. $\lambda \propto 1/|E|$
E. $\lambda \propto 1/\sqrt{|E|}$
F. Other (specify)

Q7T.6 Imagine that we have a box containing an electron with the lowest possible energy E_1. Imagine that we slowly squeeze the box, making its length L smaller. What will happen to the energy of the electron? (Try to answer this *without* looking at any formulas in the book.) The electron's energy will
A. Increase
B. Decrease
C. Remain the same

Q7T.7 (Do problem Q7T.6 first.) The electron described in problem Q7T.6 will *resist* being compressed in the manner described by exerting a resisting pressure on the walls of the box, T or F?

HOMEWORK PROBLEMS

Basic Skills

Q7B.1 Imagine an electron is confined to a box that is roughly the diameter of an atom (≈ 0.2 nm). If we perform an experiment to evaluate the electron's energy, what are the three lowest values that we could get?

Q7B.2 Imagine that an electron is confined to a very thin wire 5.0 nm long. We do an experiment to evaluate the energy of this electron, and we find it to be 0.376 eV. What is the value of n for the electron's wavefunction after the experiment is over?

Q7B.3 Imagine that we can model an atomic nucleus as a box for protons and neutrons whose length is equal to the diameter of the nucleus (≈ 8 fm $= 8 \times 10^{-15}$ m). If we do an experiment to measure the energy of a neutron in such a nucleus, roughly what is the smallest possible result? (*Hint:* mc^2 for a proton is about 938 MeV.)

Q7B.4 Imagine that we do an experiment to measure the energy of atoms vibrating in a certain kind of crystal, and we find that the lowest energy we get is 0.31 eV. What would the classical phase rate ω of the oscillation of one of the crystal's atoms be in this context? (*Hint:* treat the atom as a harmonic oscillator and use equation Q7.12.)

Q7B.5 Imagine that the wavefunction for an electron in a hydrogen atom is equal to the $n = 7$ energy eigenfunction for that atom. What is the electron's energy? What is the approximate radius of its effective "orbit"?

Q7B.6 Physicists have been able to create hydrogenlike atoms in which the electron has been replaced by a *muon*, an elementary particle that has the same basic properties as the electron except its mass is 210 times larger and its lifetime is a few microseconds. Assuming that the Bohr model accurately describes such an atom, what is the lowest energy that the muon in this atom can have? What is the radius of its effective "orbit" when it has this energy?

Synthetic

Q7S.1 Consider a *photon* in a perfectly mirrored box of length L. This is a quanton-in-a-box problem, but the results that we found in section Q7.3 do not apply because our derivation there assumed that the quanton was nonrelativistic. The main difference between a photon and a nonrelativistic quanton is the relationship between energy and momentum: for a photon, $E = pc$, but for a nonrelativistic quanton in a box, $E = p^2/2m$. Use this to determine the possible energies of a photon in a perfectly mirrored box.

Q7S.2 Here is a very crude approach to the modeling of the simple harmonic oscillator. Let's assume that within the classically allowed region, the quanton's wavefunction is sinusoidal with a wavelength $\lambda = h/p$ computed *as if* $V(x)$ were zero in this region. The energy eigenfunctions then are the same as for the "box" problem, except that L is now the distance between the oscillator turning points, which varies with energy E. Use equation Q7.15 to show that the effective $L = (8E/m\omega^2)^{1/2}$, and plug this into equation Q7.9 to find the possible energies for the harmonic oscillator in this model. Compare and contrast your results with the exact results given by equation Q7.16.

Q7S.3 Consider a pendulum bob of mass 0.01 g suspended by a massless string of length $L = 12$ cm. The bob in this case (for small oscillations) will behave as if it were a simple harmonic oscillator with $\omega = \sqrt{g/L}$ (see section N11.6). According to equation Q7.16, the lowest possible energy E_0 of such a harmonic oscillator is *still* nonzero. What is the pendulum's classical amplitude of oscillation when it has the lowest possible quantum energy? Do you think that it would be easy to detect these oscillations?

Q7S.4 Imagine that the proton and electron in a hydrogen atom were *not* charged. They could still form an atom based on the gravitational (instead of electrostatic) attraction between the two quantons. What would be the minimum energy of the electron in this gravitational atom? What would be its minimum size?

Q7S.5 Imagine that the potential energy associated with a light atom of mass m interacting with a massive atom is given by the three-dimensional harmonic oscillator potential energy function $V(r) = \frac{1}{2}m\omega^2 r^2$. Using the same basic approach that we used for the Bohr model, find a formula for the possible energies of this system. How does your formula compare to the result (given by equation Q7.16) for the *one*-dimensional harmonic oscillator? (*Hint:* The magnitude of the springlike force exerted on the light atom is $F = |dV/dr| = m\omega^2 r = k_s r$.)

Q7S.6 The potential energy of the strong interaction between two quarks has the approximate form $V(r) = br$, where b is some constant. Assume that we have a light quark of mass m interacting with a very massive quark. Using the same basic approach that we used for the Bohr model, find the possible energies and effective radii of this system. If the light quark's rest energy $mc^2 = 310$ MeV for the light quark and $b \approx 15$ tons $\approx 150{,}000$ N (these approximate actual values for this system), estimate the minimum radius of the lighter quark's effective "orbit." (*Hints:* The magnitude of the *force* exerted on the little quark in its orbit is given by $F = |dV/dr| = b$. You should find that $E_n \propto n^{2/3}$.)

Q7S.7 Imagine a universe in which there was no electrostatic interaction and electrons in atoms were bound to protons via the gravitational interaction. Use the Bohr model to (a) estimate the *size* of such a gravitational hydrogen atom when the electron has its lowest possible total energy and (b) how much energy would have to be added to such a gravitational atom to make the electron free.

Q7S.8 Consider a nitrogen molecule bouncing around a cubical box 10 cm on a side. Pretending for the moment that the molecule can only move in the x direction, what is the approximate value of n for the energy eigenfunction of this molecule if it has energy $E \approx 0.025$ eV that one would expect from

random thermal motion? Estimate the ratio $\Delta E/E$ for this value of n, where ΔE is the energy difference between adjacent possible energies in the neighborhood of E. Would the energies available to this molecule *seem* continuous in this case, as classical mechanics would predict? (*Hint:* 1 mol of 6.02×10^{23} nitrogen molecules has a mass of 28 g.)

Q7S.9 Imagine a quanton in a box whose wavefunction at time $t = 0$ is a superposition of the box's nth energy eigenfunction $\psi_{E_n}(x)$ and its mth eigenfunction $\psi_{E_m}(x)$:

$$\psi(x) = A\psi_{E_m}(x) + B\psi_{E_n}(x) \qquad (Q7.25)$$

where m and n are integers such that $n > m$, and A and B are (for the sake of simplicity) real constants. The time-evolution rule implies that the quanton's wavefunction at time t will be

$$\psi(x, t) = Ae^{-iE_m t/\hbar}\psi_{E_m}(x) + Be^{-iE_n t/\hbar}\psi_{E_n}(x)$$

$$(Q7.26)$$

Show that the absolute square of this wavefunction is *not* time-independent (in contrast to equation Q7.2) but instead has the form $f(x) + g(x)\cos\omega t$, where the angular frequency of oscillation $\omega = (E_n - E_m)/\hbar$.

Rich-Context

Q7R.1 Early studies of radioactivity turned up some examples of radioactive atoms that decay by emitting an electron. A straightforward model for this would be that the nucleus really contains only protons and electrons, with enough electrons inside the nucleus to cancel the charge of about one-half of the protons, making them appear to be neutral. This model would then explain why sometimes nuclei appear to emit electrons: one of the electrons in nucleus has simply escaped somehow.

This model, however, *cannot* be correct. Let's treat the nucleus essentially as if it were a one-dimensional box of length L equal to the diameter of the nucleus $\approx 8 \times 10^{-15}$ m (this will not be an accurate model, but it should give a good order-of-magnitude estimate). Show that an electron's kinetic energy in this nucleus will *far* exceed the attractive electrostatic potential energy of its interaction with all the (50 or so) protons in a nucleus of that size (even ignoring the repulsive effects of the other electrons). Argue that this means that the electron will inevitably escape the nucleus. You will need to make appropriate estimates.

(This model was rendered obsolete in 1932 when the neutron was discovered. We will study a better model of this process in chapters Q13 and Q14.)

Advanced

Q7A.1 Here is a more detailed argument about what the wavefunction of a quanton in a box should be. As we discussed in chapter Q6, a free quanton with a definite energy that is traveling in the $+x$ direction will have a time-dependent wavefunction that looks like this:

$$\psi_p(x, t) = Be^{i(px-Et)/\hbar} \qquad (Q7.27)$$

where B is some (possibly complex) constant. However, the box boundaries will completely reflect the traveling wave, so the quanton's total wavefunction in the box will be a superposition of an upright right-going wave and an inverted left-going wave

$$\psi_E(x, t) = Be^{i(px-Et)/\hbar} - Be^{i(-px-Et)/\hbar} \qquad (Q7.28)$$

where p is the magnitude of the momentum $= \sqrt{2mE}$.

(a) Explain why we must put a minus sign in front of px to make the second term a left-going wave. [*Hint:* The combined wave is an energy eigen-state, so it must have the form $f(x)e^{-iEt/\hbar}$.] Also argue that these two waves are really inverted with respect to each other at the boundary $x = 0$.

(b) Show that we can rewrite this function in the form

$$\psi_E(x, t) = 2Bie^{-iEt/\hbar} \sin\frac{px}{\hbar} \qquad (Q7.29)$$

This is the quantum analogy to the standing wave formula given by equation Q1.8: it consists of a stationary sinusoidal position-dependent wave with an amplitude that oscillates in time.

(c) Show that

$$|\psi_E(x, t)|^2 = |2B|^2 \left(\sin\frac{px}{\hbar}\right)^2 \qquad (Q7.30)$$

Therefore, the aspect of the wavefunction that determines the spatial shape of the energy eigenfunction is just

$$\psi_E(x) = \sqrt{|2B|^2} \sin\frac{px}{\hbar} \qquad (Q7.31)$$

If we define $A = \sqrt{|2B|^2}$ and $k = p/\hbar$, we have equation Q7.4.

ANSWERS TO EXERCISES

Q7X.1 If we plug $\lambda = 2L/n$ into $k = 2\pi/L$, we get

$$k = \frac{2\pi}{\lambda} = \frac{2\pi}{2L/n} = \frac{n\pi}{L} \qquad (Q7.32)$$

Q7X.2 An energy E_1 of 0.09 eV is 9 times the value of E_1 for the box in example Q7.1. Since the lowest energy of the box $\propto 1/L^2$, if E_1 goes up by a factor of 9, then L must go down by a factor of 3, so $L = (6.0 \text{ nm})/3 = 2.0$ nm.

Q7X.3 Equation Q7.17 implies that $\lambda = 2\pi r/n$. If we plug $p = h/\lambda$ and $\lambda = 2\pi r/n$ into equation Q7.19, we get

$$\frac{mke^2}{r} = p^2 = \frac{h^2}{\lambda^2} = \frac{h^2}{(2\pi r/n)^2} = \frac{h^2 n^2}{4\pi^2 r^2} \qquad (Q7.33)$$

Multiplying both sides by r^2/mke^2 (and attaching the subscript to the r to remind ourselves that the possible discrete values of r depend on n), we get equation Q7.20.

Q7X.4 Let's check the given value of ke^2:

$$ke^2 = \left(8.99 \times 10^9 \frac{J \cdot m}{C^2}\right)(1.60 \times 10^{-19} C)^2$$
$$\times \left(\frac{1 \text{ eV}}{1.6 \times 10^{-19} J}\right)$$
$$= 1.44 \times 10^{-9} \text{ eV} \cdot m \left(\frac{1 \text{ nm}}{10^{-9} m}\right) = 1.44 \text{ eV} \cdot \text{nm}$$
$$(Q7.34)$$

Now, taking the advice offered in the hint, we have

$$a_0 = \frac{h^2}{4\pi^2 mke^2} = \frac{(hc)^2}{4\pi^2 (mc^2)(ke^2)}$$
$$= \frac{(1240 \text{ eV} \cdot \text{nm})^2}{4\pi^2(511,000 \text{ eV})(1.44 \text{ eV} \cdot \text{nm})} = 0.053 \text{ nm}$$
$$(Q7.35)$$

Q7X.5 Multiplying both sides of equation Q7.18 by r, we get

$$mv^2 = \frac{ke^2}{r} \qquad (Q7.36)$$

Using this to eliminate mv^2 in equation Q7.22, we get

$$E = \frac{1}{2}mv^2 - \frac{ke^2}{r} = \frac{ke^2}{2r} - \frac{ke^2}{r} = -\frac{ke^2}{2r} \qquad (Q7.37)$$

Q7X.6 The magnitude L of the angular momentum of an object in a circular orbit is $L = rmv = rp$. According to the de Broglie relation, $p = h/\lambda$, and according to equation Q7.17, $\lambda = 2\pi r/n$. Combining these relationships, we get

$$L_n = rp = r\frac{h}{\lambda} = r\frac{h}{2\pi r/n} = \frac{h}{2\pi}n = n\hbar \qquad (Q7.38)$$

So we see that the electron's orbital angular momentum is quantized in units of $\hbar$. This turns out to be a quite general statement that applies to many kinds of systems.

Q8

Spectra

Chapter Overview

Introduction

In this chapter, we look at the implications of quantized energy levels regarding the emission and absorption of light. We will also look in some detail at the physics of spin, which is important preparation for chapter Q9 and chapters Q12 through Q15.

Section Q8.1: Energy-Level Diagrams

An energy-level diagram displays energy levels as short horizontal lines drawn to the right of a vertical energy axis. In the simplest form of the diagram, the horizontal axis means nothing. Every kind of system has a unique energy-level diagram.

Section Q8.2: The Spontaneous Emission of Photons

A charged quanton's electric field is as if its charge were distributed with a density proportional to $|\psi|^2$. The electric field of a quanton in an **energy eigenstate** (i.e., whose wavefunction is an energy eigenfunction ψ_E) will be static because $|\psi_E|^2$ is static, but the electric field of any other kind of wavefunction is dynamic, implying that the quanton can generate photons.

Spontaneous emission of a photon is possible even when the quanton *is* in an energy eigenstate because quantum field theory predicts that random fluctuations in the background electromagnetic field keep the quanton from ever being *exactly* in an energy eigenstate, enabling it to undergo a **transition** to a lower energy level. In such circumstances, the photon will carry away the energy *difference* between the levels.

Section Q8.3: Spectral Lines

When some physical process boosts quantum systems to **excited states** whose energy levels are above the lowest-energy state (the **ground state**), the systems will emit photons as they decay back to lower energy levels. If that light is dispersed by a prism or grating, we see bright **spectral lines** (or **emission lines**) at specific wavelengths. One can use the pattern of these lines to determine characteristics of the quantum system.

Section Q8.4: Absorption Lines

An incoming photon can kick a quanton from a lower energy level to a higher energy level if the photon's energy is exactly equal to the energy difference between the two levels. This is analogous to *resonance:* a photon with the right frequency essentially resonates with the quantum system and thus effectively transfers energy to it. This means that quantum systems will preferentially absorb photons from white light that are equal to the energy difference between the two energy levels. This creates dark **absorption lines** against the continuous white background. These lines have the same wavelengths as the emission lines, so **absorption spectroscopy** yields the same information about a quantum system that studying its emission spectrum does.

Section Q8.5: The Physics of Spin

Cold interstellar hydrogen emits photons with a wavelength of 21 cm. The Bohr model cannot explain this emission line, which arises ultimately from spin phenomena.

Both the electron and the proton in a hydrogen atom behave in many circumstances as if they were spinning balls of charge. A rotating charge is a circulating current, and thus it should behave as an electromagnet. Stern and Gerlach directly demonstrated this magnetic aspect by sending a beam of electrons through vertical but vertically varying magnetic field. Such a field exerts a vertical force on each electron that is directly proportional to the projection of the electron's spin axis on the magnetic field direction.

One would expect electrons to hit the final display screen in a variety of vertical positions reflecting their initially random spin orientations. Stern and Gerlach found instead that the electrons hit in only two positions, implying that their spin projection on the field direction is *quantized*. This is ultimately so because, just as an orbiting electron's wave function must match itself as the electron goes once around its orbit, so the electron's wavefunction has to match itself in a certain way during a complete rotation.

The ultimate rules for describing the possible values for the projection of either orbital or spin angular momentum on a given axis are as follows:

$$L_z = \ell\hbar, (\ell - 1)\hbar, \ldots, -\ell\hbar \qquad \text{where } \ell = 0, 1, 2, \ldots \qquad \text{(Q8.8a)}$$

$$S_z = s\hbar, (s - 1)\hbar, \ldots, -s\hbar \qquad \text{where } s = 0, \tfrac{1}{2}, 1, \tfrac{3}{2}, \ldots \qquad \text{(Q8.8b)}$$

Purpose: These equations specify the possible values for the projections L_z and S_z of a quanton's orbital and spin angular momenta on the z axis, respectively.
Symbols: $\hbar \equiv h/2\pi$, and h is Planck's constant.
Limitations: These equations have no known limitations. Indeed, though the results are stated for projections along the z axis, they apply equally well to any arbitrary axis.

Both the electron and the proton have $s = \tfrac{1}{2}$ and so have two possible projections ($+\tfrac{1}{2}\hbar$ and $-\tfrac{1}{2}\hbar$) on a given axis.

The 21-cm hydrogen line results because the magnetic interaction between the electron and proton splits the hydrogen atom's ground state into two slightly different energy levels depending on the relative alignment of the electron and proton spins. An electron in the higher level can undergo a spin flip to emit the 21-cm photon.

Section Q8.6: The Pauli Exclusion Principle

A **fermion** is a quanton with s such that $s = n + \tfrac{1}{2}$, where $n = 0, 1, 2, \ldots$, whereas a **boson** has integer spin. The **Pauli exclusion principle** states that two fermions cannot have exactly the same quantum state. Since electrons, protons, and neutrons all have $s = \tfrac{1}{2}$, this means that only two such quantons (with opposite spins) can occupy a given energy level. This principle explains many experimental observations including missing spectral lines in helium and correctly predicts the lowest absorption line of cyanine.

Q8.1   Energy-Level Diagrams

How to draw an energy-level
diagram

We can conveniently depict the set of possible energy values E_n (often called
energy levels) for a bound quanton by using an **energy-level diagram.** Such
a diagram simply represents each possible energy as a horizontal line drawn
next to a vertical energy scale. In the simplest form of such a diagram, the
horizontal axis is meaningless.

Let me illustrate with a specific example. Imagine that we have an electron trapped in a box of width $L = 0.50$ nm. In this case, since the rest energy
of an electron is $mc^2 = 511{,}000$ eV, the electron's possible energies are given by

$$E_n = \frac{h^2 n^2}{8mL^2} = \frac{(hc)^2 n^2}{8mc^2 L^2} = \frac{(1240 \text{ eV} \cdot \text{nm})^2 n^2}{8(511{,}000 \text{ eV})(0.50 \text{ nm})^2} = (1.5 \text{ eV})n^2 \quad \text{(Q8.1)}$$

where $n = 1, 2, 3 \ldots$. The energy levels in this case are thus $E_1 = 1.5$ eV,
$E_2 = (1.5 \text{ eV})(4) = 6.0$ eV, $E_3 = (1.5 \text{ eV})(9) = 13.5$ eV, and so on. Figure
Q8.1a shows an energy-level diagram for this situation.

For the sake of comparison, figure Q8.1b shows the energy levels for the
quanton moving in response to a harmonic oscillator potential energy function whose effective spring constant is such that $(h/2\pi)\omega = \hbar\omega = 4.0$ eV. As
we saw in chapter Q7, the possible energies for this system are given by

$$E_n = \frac{\hbar\omega}{2\pi}\left(n + \tfrac{1}{2}\right) = \hbar\omega\left(n + \tfrac{1}{2}\right) \qquad \text{where } n = 0, 1, 2, \ldots \qquad \text{(Q8.2)}$$

so the possible energies are thus $E_0 = \tfrac{1}{2}\hbar\omega = 2.0$ eV, $E_1 = \left(\tfrac{3}{2}\right)\hbar\omega = 6.0$ eV,
$E_2 = \left(\tfrac{5}{2}\right)\hbar\omega = 10$ eV, and so on.

Figure Q8.1c shows an energy-level diagram for an electron in a hydrogen atom, which we found in chapter Q7 to have energies

$$E_n = -\frac{ke^2}{2a_0 n^2} = -\frac{13.6 \text{ eV}}{n^2} \qquad \text{where } n = 1, 2, 3, \ldots \qquad \text{(Q8.3)}$$

The electron's possible energy levels are thus $E_1 = -13.6$ eV, $E_2 =
-(13.6 \text{ eV})/4 = -3.4$ eV, $E_3 = -(13.4 \text{ eV})/9 = -1.5$ eV, and so on. Note that
there are an infinite number of energy levels between the one for $n = 4$ (the
uppermost shown on the diagram) and $E = 0$ (which is the maximum possible energy that a *bound* electron in the hydrogen atom can have).

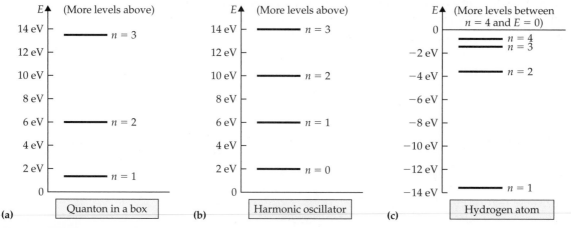

(a) Quanton in a box   **(b)** Harmonic oscillator   **(c)** Hydrogen atom

Figure Q8.1
Energy-level diagrams for various systems.

Note that the spacing between energy levels *increases* as *n* increases in the case of the box, *remains constant* in the case of the harmonic oscillator, and *decreases* as *n* increases in the case of the hydrogen atom. The point is that each kind of system has a characteristic energy-level diagram.

Each kind of system has a characteristic diagram

Q8.2   The Spontaneous Emission of Photons

As we saw in chapter Q7, the absolute square of an energy eigenfunction is a completely *stationary* wave without *any* time dependence (not even the up-and-down motion of a standing wave), whereas wavefunctions that are *not* energy eigenfunctions do have absolute squares that *do* vary in time. It turns out that the electric field of a charged quanton at any instant of time is determined by the absolute square of the quanton's wavefunction, as if the quanton's charge were smeared throughout space with a density proportional to the absolute square of the wavefunction. This means that the electric field produced by a quanton in an **energy eigenstate** (i.e., whose wavefunction is an energy eigenfunction) is *static*. On the other hand, the electric field of the quanton having any *other* kind of wavefunction is dynamic. A static electromagnetic field cannot create an electromagnetic wave, which is dynamic by its very nature. This implies that when a quanton is in an energy eigenstate, it *cannot* radiate electromagnetic waves (photons); but it can and will if it is *not* in an energy eigenstate. This (as I suggested in chapter Q7) explains both why systems tend to settle down into energy eigenstates and why atoms in such states are stable against the collapse that would inevitably result if their electrons could radiate away their orbital energy.

Why a quanton in an energy eigenstate should not radiate photons

It turns out, however, that even quantons in upper-level energy eigenstates have a small probability per unit time of decaying to a *lower* energy level by emitting a photon that carries away the difference in energy between the two states. This **spontaneous emission** of photons is possible (according to relativistic quantum field theory) because the quanton is continually buffeted by small quantum fluctuations of the background electromagnetic field that keep the quanton's wavefunction from being an *exact* energy eigenfunction. Only when the quanton is in its **ground state** (the energy eigenstate with the lowest possible energy value) is it *completely* stable (and then, only because there is no *lower*-energy eigenstate to which it can decay!).

Why a quanton in an *upper-level* eigenstate will radiate anyway

A quanton in any **excited state** (i.e., an energy eigenstate with an energy *above* the ground state) will thus *eventually* radiate a photon and move to a lower energy level, a process that is generally called a **transition** and sometimes (colloquially) a **quantum jump.** The average time that a quanton remains in an excited state depends on a variety of factors and ranges from about 100 ns to many seconds in atoms.

Q8.3   Spectral Lines

Since the photon carries away the difference between the initial and final energy levels, and since these levels are quantized in a bound system, it follows that the photons radiated by a bound quanton will have only certain discrete energies (and thus wavelengths) that correspond to the possible energy *differences* between that quanton's energy levels. When the light emitted by a large number of such quantons is dispersed by a prism according to color, one sees bright **spectral lines** (also known as **emission lines**) at the specific colors corresponding to these wavelengths.

For example, if an electron in a box whose length is 0.50 nm decays from the $n = 2$ to the $n = 1$ energy level (see equation Q8.1), the emitted photon carries away an energy of $E_{ph} = E_2 - E_1 = 6.0\ eV - 1.5\ eV = 4.5\ eV$. The wavelength of such a photon is

$$\lambda = \frac{hc}{E_{ph}} = \frac{1240\ eV \cdot nm}{4.5\ \cancel{eV}} = 280\ nm \qquad (Q8.4)$$

which is in the ultraviolet range. Similarly, you can show that the photon that would be produced by an $n = 3$ to $n = 2$ transition has a wavelength of 165 nm; the photon that would be produced by an $n = 3$ to $n = 1$ transition has a wavelength of 103 nm; and so on (both of these photons are in the deep ultraviolet).

Exercise Q8X.1

Verify that the stated wavelengths of photons for the $n = 3$ to $n = 2$ and $n = 3$ to $n = 1$ transitions are correct.

Formulas for the wavelengths of spectral lines

We can find a general formula for predicting the wavelengths of the photons emitted by a quanton in a box as follows:

$$\frac{hc}{\lambda} = E_{ph} = E_{init} - E_{final} = \frac{(hc)^2}{8mc^2L^2}\left(n_{init}^2 - n_{final}^2\right) \qquad (Q8.5a)$$

$$\Rightarrow \quad \lambda = \frac{hc}{E_{ph}} = \frac{8mc^2L^2}{hc\left(n_{init}^2 - n_{final}^2\right)} \qquad (Q8.5b)$$

Similarly, the possible wavelengths of light emitted by a harmonic oscillator are

$$\lambda = \frac{hc/\hbar\omega}{n_{init} - n_{final}} = \frac{2\pi c/\omega}{n_{init} - n_{final}} \qquad (Q8.6)$$

Exercise Q8X.2

Verify equation Q8.6.

The possible wavelengths of light that might be emitted by a hydrogen atom are

$$\lambda = \frac{2(hc)a_0}{ke^2}\left(\frac{n_{init}^2 n_{final}^2}{n_{init}^2 - n_{final}^2}\right) \qquad (Q8.7)$$

Exercise Q8X.3

Show that the constant $2(hc)a_0/ke^2 = 91.3$ nm.

Predicting visible spectra

If we somehow excited a number of quantons to random energy levels and then dispersed the light that they emit into a spectrum by using a prism or something similar, we would (as already mentioned) see bright lines at certain specific wavelengths. Figure Q8.2 shows an energy-level diagram for

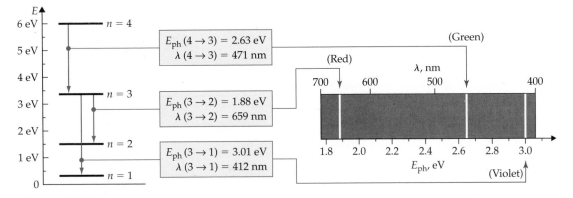

Figure Q8.2
The transitions of an electron in a box 1.0 nm wide that give rise to visible photons. The spectrum shown on the right is what you would see if the light from a number of excited electrons were dispersed by a prism.

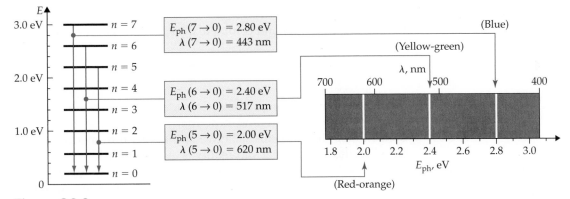

Figure Q8.3
Energy-level diagram and hypothetical spectral lines for a harmonic oscillator with $\hbar\omega = 0.40$ eV. Note that all transitions with the same Δn will produce photons of the same wavelength: so the transitions $8 \rightarrow 1$, $9 \rightarrow 2$, and so on therefore emit photons of the same wavelength as those emitted by the $7 \rightarrow 0$ transition.

an electron in a box of width $L = 1.0$ nm [whose energy levels are therefore $E_n = (0.376 \text{ eV})n^2$]. The three transitions that give rise to *visible* photons (700 nm $\leq \lambda \leq$ 400 nm) are linked to the spectrum that appears on the right side of the diagram (many more transitions are *possible*, of course, but no other transitions in this case produce *visible* photons).

Figure Q8.3 shows a similar diagram for a harmonic oscillator with $\hbar\omega = 0.40$ eV: again, the three transitions $\Delta n = 5$, $\Delta n = 6$, and $\Delta n = 7$ are the only ones that produce visible photons in this case.

Figure Q8.4 shows an energy-level diagram for the hydrogen atom, with *one* of the four transitions that produce visible photons shown. *You* can easily predict the spectrum of hydrogen by completing the following exercises.

Exercise Q8X.4

Argue that *all* transitions to the $n = 1$ energy level of hydrogen produce photons that are *not* in visible range.

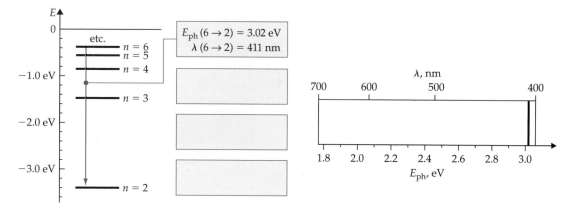

Figure Q8.4

Energy-level diagram and spectral chart for the hydrogen atom, with one visible transition and spectrum line illustrated.

Exercise Q8X.5

Fill in the blank boxes and complete the spectral chart for hydrogen shown in figure Q8.4.

The spectrum of real hydrogen atoms turns out to be essentially *identical* to what the Bohr model predicts. Considering how strange the postulates of quantum mechanics are and how many assumptions we made to get the Bohr model, this agreement between the quantum-mechanical predictions based on the Bohr model and reality is almost incredible!

How to get hydrogen atoms to emit light

To get such quantum systems to emit light in the first place, it is necessary to make sure that at all times there are at least some systems in energy levels above the ground state. In the case of hydrogen, this is usually accomplished by putting rarefied hydrogen gas in a tube and then running an electric current through it. The moving electrons in the electric current bump into the hydrogen atoms, sometimes knocking their electrons into excited states. The action of the current thus ensures that there are always *some* hydrogen atoms that are poised to decay and thus emit photons.

The main thing that I'd like you to notice is that in general, every kind of system produces a distinct spectrum. One can tell simply by looking at the light emitted from a set of identical systems whether it was produced by quantons in a box, harmonic oscillators, or hydrogen atoms. The spectra of multielectron atoms are more complex and hard to predict than the spectrum of hydrogen (mostly because interactions between the electrons have a variety of complicated effects), but the spectrum of each atom is still distinct from all other atoms.

One can also learn about the structure of a quantum system from its spectrum

Conversely, by closely studying the spectrum of light emitted by a quantum system, one can often infer things about the structure and potential energy function for that system. Studying atomic spectra, for example, has enabled physicists to develop useful models for understanding atomic structure and even uncover new fundamental principles of physics.

Q8.4   Absorption Lines

What are absorption lines?

When we illuminate a transparent gas with white light (light that is a mixture of many wavelengths), we often see that the light that comes through the

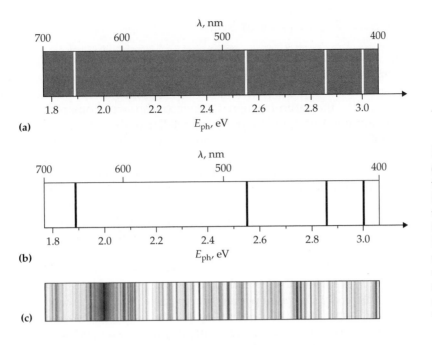

Figure Q8.5

(a) A spectrum chart showing the emission lines for atomic hydrogen. (b) A chart showing the absorption lines of atomic hydrogen. Every quantum system has its own unique pattern of spectral lines. (c) A photograph of a tiny portion of the actual absorption spectrum from the sun displaying many absorption lines from a variety of atoms and ions.

mixture has certain specific wavelengths removed, the same wavelengths that it would *emit* if the gas atoms (or molecules) were excited (see figure Q8.5). We call the resulting dark lines in the spectrum of the transmitted light **absorption lines.** What is happening here is that the gas is selectively absorbing from the white light the particular photons that have exactly the right energy to kick an electron from one energy level to another higher energy level. Since energy must be conserved, the photon has to bring to the atom an energy equal to the difference between the two levels, which is the same energy that it would release if it were to undergo the reverse spontaneous transition.

Why do the gas atoms or molecules *selectively* absorb light this way? Conservation of energy is not the whole story here. For example, an electron in an atom or molecule *could* accept a more energetic photon than is strictly needed to cause the transition and emit the leftover energy in the form of another photon, but this does not happen very often.

The physics involved here is similar to the physics of *resonance* (see chapter Q1). Just as a disturbance most effectively transfers energy to standing waves in a classical system when it has just the right frequency, so a photon (which is an electromagnetic disturbance) can most effectively transfer its energy to the electron if it has the right frequency. When the photon has an energy exactly equal to the difference between two energy levels, it "resonates" with the transition and thus most effectively transfers its energy to the electron. Photons with energies that are nearly but not quite right do *not* resonate with the transition, so their absorption is far less probable.

Since absorption spectra (like emission spectra) are different for different kinds of substances, absorption spectra can also be used to identify substances. Astronomers commonly use **absorption spectroscopy** to identify atoms and molecules in stellar atmospheres, since the essentially white light produced at a star's surface passes through the stellar atmosphere on its way to earth. Absorption lines can also identify substances in gas clouds that might exist between a star and the earth (subtle characteristics of the absorption lines allow astronomers to distinguish what is in the stellar atmosphere and what is not). Chemists, geologists, and biologists often use absorption spectroscopy as well.

Why absorption lines exist

Q8.5   The Physics of Spin

The 21-cm line of hydrogen

The spectrum of hydrogen contains one observed emission line that is impossible to explain by using the Bohr model. Cold hydrogen gas in interstellar space is observed to produce an emission line with a wavelength of 21 cm, which is in the radio-frequency range (radio astronomers in fact use this emission line to locate clouds of interstellar hydrogen, as illustrated in figure Q8.6). This line must come from energy levels separated by only about 5.9×10^{-6} eV.

According to the simple Bohr model, such a spectral line would have to come from a transition between two energy levels with extremely high n. This is not really a plausible explanation, though, because this interstellar hydrogen is very *cold*, meaning that collisions between atoms will not have nearly the energy required to kick a hydrogen atom to a high enough energy level to be involved in a transition. Moreover, if the hydrogen atoms were excited to such high energy levels, we would see emission lines at a variety of nearby wavelengths instead of just a single isolated 21-cm line.

It turns out that this spectral line arises from the interaction between the electron's spin and the proton's spin. We discussed how spins behave in chapters Q5 and Q6, but this spectral line provides a good occasion to discuss the physics of spin in greater detail.

A newtonian model of spin

The model of electron spin I am about to describe is a semiclassical model that is not fully correct but nonetheless delivers some useful physical insight. Imagine that both the electron and the proton are like tiny balls that spin around some axis with some angular momentum $\vec{S}$. Since both are charged, this means that their rotation will create an endlessly circulating electric current, which (in effect) makes them both tiny electromagnets. In the case of an electron (which is negatively charged), the direction of an arrow pointing from the south to north pole of the equivalent magnet points *opposite* to the direction of $\vec{S}$ (see figure Q8.7a); in the case of a proton, such an arrow is parallel to the direction of $\vec{S}$ (see figure Q8.7b).

How a Stern-Gerlach device works

In the first decades of the 20th century, physicists studying the light emitted by excited atoms in magnetic fields noticed effects that suggested that atomic electrons were behaving as if they were tiny magnets. In 1921,

Figure Q8.6
A map of the 21-cm radio emissions from the spiral galaxy M51.

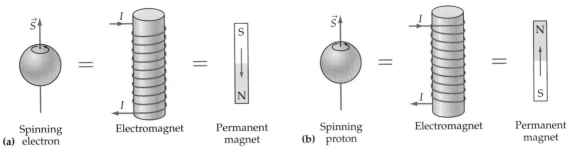

Spinning
(a) electron   Electromagnet   Permanent
                                          magnet
Spinning
(b) proton   Electromagnet   Permanent
                                          magnet

Figure Q8.7
(a) An electron spinning with angular momentum $\vec{S}$ represents a circulating current and thus should behave as an electromagnet. Because the electron's charge is negative, the south-to-north direction of its equivalent electromagnet points opposite to the electron's spin $\vec{S}$. (b) The south-to-north direction of the equivalent electromagnet for a proton is parallel to the proton's spin $\vec{S}$.

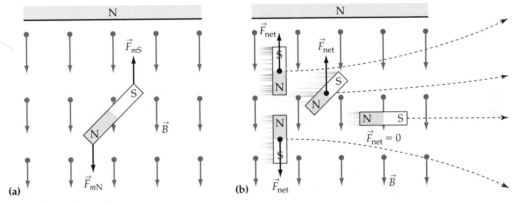

Figure Q8.8
(a) In a vertically downward magnetic field whose strength increases vertically, the upper pole of a magnet experiences a stronger force than the lower pole. (b) This means that small magnets moving horizontally in such a field will experience a net force that depends on their orientation relative to the field.

Otto Stern and Walter Gerlach set out to perform a now-famous experiment to directly demonstrate the magnetic nature of the electron. Imagine that we were to send a beam of electrons through a region where the magnetic field points in the $-z$ direction but whose strength increases with z. The magnetic field at points near the north pole of an external magnet oriented so that its north pole faces downward is qualitatively like this (see problem Q8S.11 for a discussion of how Stern and Gerlach actually created a stronger and more uniform field fitting this description). Such a magnetic field would tug upward on the south pole but downward on the north pole of the electron's equivalent magnet, as shown in figure Q8.8a. The *net* force on a given electron will depend on its orientation: if the electron's south pole is higher in the field, the net force on the electron will be upward; but if its north pole is higher, the net force will be downward. Therefore, an electron passing through this region should experience an upward or downward deflecting force that is directly proportional to S_z, the projection of its spin angular momentum on the z axis, as illustrated in figure Q8.8b.

Therefore, if we send an electron beam through a region containing a magnetic field of this description, these upward or downward forces on individual electrons will cause the beam to fan out vertically, as shown in figure Q8.9a. The vertical position of a given electron on a final display screen

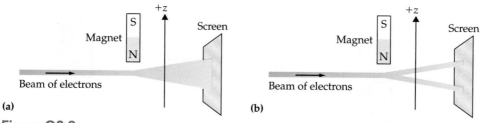

Figure Q8.9

(a) If the randomly oriented electrons in a beam act as spinning electromagnets, we might expect them to experience a range of deflections in response to the field of the large magnet. (b) The actual results shows only two possible deflections, showing that an electron's spin orientation can only have two possible values.

will be a direct measure of the value of that electron's value of S_z. By reorienting the external magnetic field along the x or y axes, one can in principle similarly measure S_x and S_y or indeed the projection of $\vec{S}$ along any axis.

Exercise Q8X.6

One might think in figure Q8.8 that the electron magnets would simply flip over like compass needles to be aligned with the field, and therefore would all get deflected in the same direction. Use the newtonian spin model to argue that an electron cannot in fact just "flip over" but its spin will instead *precess* around the z axis. Also argue that this means that any given electron's value of S_z will *not change* as it moves through the magnetic field.

The quantization of the spin projection

Since the electrons in the incoming beam should be randomly oriented, one would expect the electrons to end up at the screen with a continuous range of vertical positions. When one actually performs the experiment, though, one sees that the electrons arrive at only *two* vertical positions (see figure Q8.9b). This directly implies that the value of S_z for a given electron can have only one of *two* possible values, as stated in chapter Q5: the values of this spin projection are *quantized.*

Our discussion of the Bohr model in chapter Q7 can help us understand why this must be so. We argued there that for a quanton's wavefunction to be self-consistent, it must match itself as it goes once around its orbit, and that this implies that the magnitude L of the electron's orbital angular momentum must be an integer multiple of $\hbar$. The same kind of reasoning applies to *rotating* the quanton's wavefunction around a given axis.

Technically, the *real* requirement for spinning quantons (but not for orbiting quantons for reasons beyond us here) is only that the wavefunction's *absolute square* match itself after a complete rotation. This makes sense, actually, because it is the absolute square that is connected to physical predictions. If this is true, then a quanton's wavefunction need only undergo *one-half* of a complete oscillation during a complete rotation around an axis, because the square of a positive half of an oscillation is the same as the square of the negative half. Therefore the magnitude S of a quanton's spin angular momentum should be quantized in steps of $\frac{1}{2}\hbar$ instead of $\hbar$.

Since the electron could be spinning either clockwise or counterclockwise around the axis, it makes sense that since an electron has the smallest nonzero spin magnitude of $\frac{1}{2}\hbar$, the projection of its spin angular momentum $\vec{S}$ on any axis can be either $+\frac{1}{2}\hbar$ or $-\frac{1}{2}\hbar$. This qualitatively explains the Stern-Gerlach results.

The three-dimensional nature of the wavefunction makes things a bit more complicated than this simple model would suggest. A full analysis of the problem for general quantons implies that the projection of either a quanton's orbital angular momentum $\vec{L}$ or its spin angular momentum $\vec{S}$ on a given axis has possible values that range from a maximum positive value to a negative value of the same magnitude in integer steps of $\hbar$:

$$L_z = \ell\hbar, (\ell - 1)\hbar, \ldots, -\ell\hbar \qquad \text{where } \ell = 0, 1, 2, \ldots \qquad \text{(Q8.8a)}$$

$$S_z = s\hbar, (s - 1)\hbar, \ldots, -s\hbar \qquad \text{where } s = 0, \tfrac{1}{2}, 1, \tfrac{3}{2}, \ldots \quad \text{(Q8.8b)}$$

Purpose: These equations specify the possible values for the projections L_z and S_z of a quanton's orbital and spin angular momenta on the z axis, respectively.
Symbols: $\hbar \equiv h/2\pi$, and h is Planck's constant.
Limitations: These equations have no known limitations. Indeed, although the results are stated for projections along the z axis, they apply equally well to any arbitrary axis.

Possible angular momentum projections on an axis

Electrons, protons, and neutrons (and indeed most elementary quantons) have intrinsic rotation rates such that $s = \tfrac{1}{2}$. For such quantons, equation Q8.8b implies that there are only two possible spin projections on a given axis, $+\tfrac{1}{2}\hbar$ and $-\tfrac{1}{2}\hbar$, consistent with the Stern-Gerlach result.

Exercise Q8X.7

A proton has a mass m_p that is nearly 2000 times the electron's mass m_e. Assume a model where the electron and proton are spinning balls with roughly the same radius. Argue that this implies that the proton must be spinning 2000 times more slowly than the electron, and thus creates a magnetic field that is substantially weaker than the electron's magnetic field.

Like the Bohr model, this simple model of quanton spin is oversimplified and should not be pressed too far. For example, if the electron is really a spinning ball, then collisions with other electrons should be able to give it different magnitudes of spin angular momentum, right? If, for example, collisions can give some electrons spins corresponding to values of s higher than $\tfrac{1}{2}$, then a Stern-Gerlach experiment would split a beam of such electrons into *more* than two paths. This is *never* observed. In fact, no one in any context has ever observed an electron to have a spin angular momentum corresponding to anything *but* $s = \tfrac{1}{2}$. Thus, *unlike* a spinning ball, an electron's spin seems to be one of its intrinsic characteristics, like its mass or its charge. This should caution us about taking the spinning ball model too literally (see problem Q8S.10 for another serious problem with the naive interpretation).

Due notice about simplifications

I have also greatly simplified the description of a Stern-Gerlach experiment. For example, the electrostatic charge of the electrons in any realistic experiment would give rise to forces that overwhelm the small magnetic forces we are interested in (see problem Q8S.12), so we cannot actually use a beam of bare electrons. Stern and Gerlach instead used a beam of neutral silver atoms. The current loops represented by all the spinning and orbiting electrons in a silver atom happen to cancel each other out *except* for the spin of a

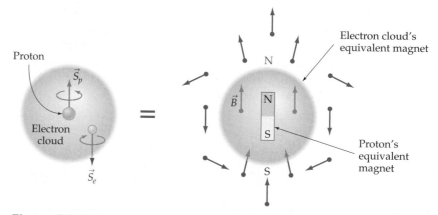

Figure Q8.10

The electron in a hydrogen atom, instead of behaving as a particle with a definite location, behaves as if it were a spherical cloud of charge surrounding the proton. This means that the magnetic effect of the spinning electron is smeared out throughout the cloud, as if the cloud were a large, insubstantial magnet surrounding the proton. The magnetic field *inside* a magnet points from its south pole to its north pole. Therefore, if the electron and proton spins are antialigned, the proton's equivalent magnet will be aligned with the cloud's internal magnetic field. This will be the proton and electron's lowest-energy alignment state.

single electron, so measuring the deflection of a silver atom still effectively determines S_z for a single electron. T. E. Phipps and J. B. Taylor repeated the experiment in 1927 with neutral hydrogen atoms (*Phys. Rev.* **29,** p. 309). Since exercise Q8X.8 implies that the magnetic effect of the proton is negligible compared to that of the electron, this even more directly determines S_z for a single electron.

I also did not mention that the experiment must be performed in a high vacuum (so that gas molecules do not deflect the atoms) and did not describe how one might create the described magnetic field without violating Gauss's law or how one might detect the deflected atoms. These are just some of the details about the experiment that I have ignored or brushed over quickly. However, the meaning of this experiment and its implications are not at all changed by any of these complications.[†]

Back to the 21-cm line (finally)

We can now understand fully the origin of the 21-cm spectral line of hydrogen. Because of spin quantization, the electron's spin can only be either completely aligned with or completely antialigned with the axis defined by the proton's spin. This means the equivalent magnets for the proton and electron are either antialigned (if the spins are aligned) or aligned (if the spins are antialigned). As figure Q8.10 shows, when the electron spin is downward, the electron cloud creates an upward magnetic field in the vicinity of the proton. The system has its lowest energy when the proton's effective magnet is aligned with this field (i.e., when the electron and proton spins are antialigned). Therefore, the lowest *orbital* energy level of the hydrogen atom is actually split into two slightly different energy levels, depending on the relative alignment of the electron and the proton spins. The electron can therefore emit a photon by flipping its spin from the aligned to the antialigned state. The difference in energy is small because magnetic effects are weak in

---

[†]For an excellent discussion of the issues involved in a realistic Stern-Gerlach experiment, see A. P. French and E. F. Taylor, *Introduction to Quantum Physics*, Norton, 1978, pp. 428–438.

general and because the proton's equivalent magnet is very weak compared to the electron, but it is enough (5.9 μeV) to allow the electron to emit a photon with a 21-cm wavelength during the flip. Note that this energy difference is so small that random thermal collisions can boost an appreciable number of even very cold hydrogen atoms into the excited state.

Q8.6   The Pauli Exclusion Principle

Atomic spectra also provided the first evidence for a very important principle known as the **Pauli exclusion principle.** In its earliest form, this principle stated that *no two electrons can have exactly the same quantum state.* This principle was proposed by W. F. Pauli in 1925 to explain certain puzzling features of atomic spectra. In particular, there are spectral lines that might be expected to exist in multielectron atoms that do not in fact appear. Pauli noted that the missing spectral lines corresponded to transitions that would require two electrons to have exactly the same state either initially or finally.

The Pauli exclusion principle for electrons

For example, consider the spectrum of helium, which is the simplest multielectron atom. Since the component of an electron's spin angular momentum in any given direction has only two possible quantized values, the spins of the two electrons in a helium atom can be either completely aligned with each other or completely antialigned; no other possibilities exist.

Evidence for this principle in the spectrum of helium

Consider first a helium atom where one electron is in the ground state and another is in an excited state and is antialigned with the first (transitions that involve spin flips turn out to be quite a bit less probable than other transitions, so once we set up the electrons with their spins in a certain relative orientation, they tend to stay that way a while). The spectral lines from a set of such helium atoms display transitions that correspond to a slightly more complicated version of the hydrogen spectrum (due to interelectron interactions). In particular, there is a family of lines whose wavelengths are *much* shorter than the others (the longest wavelength in the family corresponds to 58.4 nm) that pretty clearly reflect transitions to the very low $n = 1$ level.

On the other hand, when we look at the spectrum of helium atoms whose electrons are aligned, this family of lines disappears! The Pauli exclusion principle explains why. The excited electron cannot decay to the $n = 1$ level if it has the *same* spin orientation as the electron already at that level, because then they would have exactly the same quantum state (both their basic spatial wavefunction and their spin orientation would be the same). On the other hand, if the electrons' spins are antialigned, then the excited electron can decay to the $n = 1$ level; even though their wavefunctions are the same $n = 1$ energy eigenfunction, they have different spin states and so have a different total quantum state.

The physicists who developed relativistic quantum field theory in the 1950s were able to show that the Pauli exclusion principle applies to *any* quanton whose spin magnitude number s is $\frac{1}{2}, \frac{3}{2}, \frac{5}{2},$ or any odd integer divided by 2. Physicists call such quantons **fermions** (after Enrico Fermi, who contributed to the study of such quantons). The generalized Pauli exclusion principle states that

The most general statement of the exclusion principle

> *No two identical fermions can have exactly the same quantum state.*

Electrons, protons, neutrons, quarks, and indeed most subatomic particles are spin-$\frac{1}{2}$ fermions. Since spin-$\frac{1}{2}$ quantons have exactly two possible spin orientations, at most two electrons or two neutrons or two protons can have the same energy eigenfunction and thus occupy the same energy level. This is a very important principle for understanding both atomic structure (as we will see in chapter Q9) and nuclear struture (as we will see in chapters Q12 through Q15).

The interesting thing is that the Pauli exclusion principle does *not* apply to any quanton whose spin magnitude number s is a simple integer. Such quantons are called **bosons** (after Satyrendra Bose, a physicist who studied this class of quantons). Identical bosons can happily occupy the same quantum state (indeed, there is an enhanced probability that they will do so). Photons are bosons, as is any system of quantons whose angular momenta cancel each other out (such as a helium atom).

This curious link between the Pauli exclusion principle and quanton spin is a subtle consequence of relativistic quantum field theory. Richard Feynman, one of the most brilliant theoretical physicists of the 20th century, once asserted that anything that physicists *really* understood could be explained to first-year students. David Goodstein, a colleague of Feynman's at Cal Tech, challenged him to explain this link. After several days, Goodstein reported that Feynman admitted defeat, and said, "That means we really don't understand it."

At the level of this class, then, we will just take the Pauli exclusion principle (as Pauli himself did) as a basic hypothesis about how the universe works.

Example Q8.1   Cyanine Dye Molecule

Problem   A simple cyanine dye molecule has the following chemical structure:

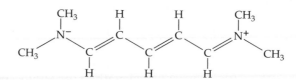

In this molecule there is a long central carbon chain, with a total of six bonds between the two nitrogen atoms. One electron per bond is essentially free to move up and down this chain. The effective potential energy for such an electron is as follows:

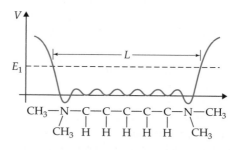

The distance between atoms in the C—C bond is about 0.14 nm, and the length of the C—N bond is about the same. Predict the wavelength of the lowest-energy photon that the electrons in the central chain of this dye molecule can absorb.

Model   Let us model this molecule as a one-dimensional box. This model will not be perfect because the walls of the potential energy function are not vertical and infinite, so it is a little bit difficult to determine exactly where the box's ends are. If we define $b = 0.14$ nm to be the length of a bond, and if we recognize that potential rises sharply somewhat beyond the nitrogen molecule at each end of the chain, then we might estimate the length of the box to be very roughly $L \approx 6.6b$. Now this box contains six electrons. According to the Pauli principle, only two electrons can occupy a given energy level (and even then only if they have opposite spins). Therefore, when all the electrons are in their lowest possible states, the $n = 1, 2$, and 3 levels will be filled with two electrons each. Therefore, the lowest-energy photon this molecule can absorb will kick an electron in the $n = 3$ level to the $n = 4$ level (the lowest open energy level). Knowing the box length, we can compute the energy difference between these levels and from that find the photon wavelength.

Solution   The energy difference between the $n = 4$ and $n = 3$ levels is

$$\Delta E = \frac{(hc)^2}{8mc^2 L^2}\left(n_i^2 - n_f^2\right) \approx \frac{(1240 \text{ eV} \cdot \text{nm})^2 (4^2 - 3^2)}{8(511{,}000 \text{ eV})[6.6(0.14 \text{ nm})]^2} = 3.08 \text{ eV} \quad (\text{Q8.9})$$

The corresponding photon wavelength is

$$\lambda = \frac{hc}{E_{\text{ph}}} = \frac{hc}{\Delta E} = \frac{1240 \text{ eV} \cdot \text{nm}}{3.08 \text{ eV}} = 400 \text{ nm} \qquad (\text{Q8.10})$$

Evaluation   The lowest observed absorption line for this particular dye molecule is broad (due to a variety of effects), but its center is roughly at 410 nm.

TWO-MINUTE PROBLEMS

Q8T.1   Imagine that the energy levels of a certain system are proportional to $\sqrt{n}$, where $n = 1, 2, 3, \ldots$. Imagine also that for some reason only transitions such that $\Delta n = 1$ are physically possible. In a spectrum chart (like the ones shown in figures Q8.2 through Q8.4), the emission lines produced by this system
A.   Get closer together as their energy increases
B.   Are evenly spaced
C.   Get farther apart as their energy increases

Q8T.2   Imagine that the energy of the longest-wavelength photon emitted by an electron in a box of length L is E_{ph}. What is the energy of the photon with the fourth-longest wavelength emitted by an electron in this box?
A. $3E_{\text{ph}}$
B. $4E_{\text{ph}}$
C. $5E_{\text{ph}}$
D. $7E_{\text{ph}}$
E. $8E_{\text{ph}}$
F. $9E_{\text{ph}}$
T.   Some other energy (specify)

Q8T.3   Two electrons are placed into separate (very tiny) one-dimensional boxes A and B. When the electrons are excited to higher energy levels and then decay to lower levels, they emit photons. The longest-wavelength spectral line from the electron in box A is observed to have the same wavelength as the third-longest spectral line from the electron in box B. If the lengths of the boxes are L_A and L_B, respectively, then this observation implies that L_A/L_B is equal to
A.   3/7
B.   7/3
C. $\sqrt{3/7}$
D. $\sqrt{7/3}$
E.   9/49
F.   49/9
T.   Some other ratio (specify)

Q8T.4   Imagine that we measure the wavelengths of the spectral lines of a vibrating diatomic molecule involving atoms whose masses we know. We can deduce the spring constant k_s of the effective spring holding the atoms together from these data, true (T) or false (F)?

Q8T.5 Imagine that light is emitted by electrons that could be either in a "box" of a certain unknown length L or in a harmonic-oscillator potential energy function with a certain unknown effective spring constant k_s. Imagine that the electrons emit only a single spectrum line in the visible range. Measuring the wavelength of this line is sufficient to determine which kind of potential energy function is involved, T or F?

Q8T.6 Imagine that we have a beam of quantons that have spin $s = \frac{5}{2}$. If we evaluate the component of the quantons' spin along a certain axis, how many results might we get?
A. No limit
B. 11
C. 8
D. 6
E. 5
F. 1

Q8T.7 Imagine that we put 20 noninteracting quantons into a one-dimensional box of length L. The quan-

tons quickly settle into the lowest possible energy states they can get into. Roughly how many times larger will the quantons' total energy be if they are fermions as opposed to bosons? (Choose the closest answer.)
A. About 400 times
B. About 40 times
C. About 10 times
D. About 4 times
E. About the same

Q8T.8 We can model cyanine dye molecules as if they contained N electrons trapped in a "box," like the potential energy function, where N is some even integer. The energy of a single quanton in a box is given by $E_n = E_1 n^2$, where n is an integer and E_1 is the energy of the ground ($n = 1$) energy level. If for a given molecule there are four electrons trapped in the molecule's "box," what is the lowest possible *total* energy of these electrons?
A. $E_{tot} = E_1$
B. $E_{tot} = 4E_1$
C. $E_{tot} = 8E_1$
D. $E_{tot} = 10E_1$
E. $E_{tot} = 30E_1$

HOMEWORK PROBLEMS

Basic Skills

Q8B.1 Imagine that an electron is trapped in a box whose width is 0.85 nm. What is the longest wavelength of light that could be emitted by this system? Is light of that wavelength visible?

Q8B.2 Imagine that an electron is trapped in a box whose width is 0.32 nm. What is the longest wavelength of light that could be emitted by this system? Is light of that wavelength visible?

Q8B.3 For the diatomic CO molecule, the value of $\hbar\omega$ is measured to be 0.269 eV. What is the wavelength of the photon emitted by a $3 \to 2$ vibrational transition of this molecule?

Q8B.4 The longest wavelength associated with vibrational transitions of the diatomic H_2 molecule is 4540 nm. What is the value of $\hbar\omega$ for this molecule?

Q8B.5 Argue that a transition from *any* level in the hydrogen atom to the $n = 3$ level will produce a photon whose wavelength is too long to be visible.

Q8B.6 Find the wavelength of the photon emitted during a $4 \to 3$ transition in a hydrogen atom.

Q8B.7 Find the wavelength of the photon emitted during a $5 \to 4$ transition in a hydrogen atom.

Q8B.8 Imagine that we do an experiment to evaluate the component S_z of the spin of a spin-$\frac{3}{2}$ quanton. List the possible results of this experiment.

Q8B.9 Imagine that we do an experiment to evaluate the component S_z of the spin of a spin-1 quanton. List the possible results of this experiment.

Q8B.10 Imagine that we put 24 spin-0 pions in a box. After these pions settle to their lowest possible energy levels, what is the value of n for the highest box energy level occupied by a pion?

Synthetic

Q8S.1 Imagine that an electron is trapped in a box whose length is L. What is the *smallest* value that L can have if the box produces visible light?

Q8S.2 Imagine that an electron is trapped in a box whose length is 1.2 nm. Draw a spectrum chart (like figure Q8.2) showing all its *visible* emission lines.

Q8S.3 Imagine that an electron is trapped in a box whose length is 0.80 nm. Draw a spectrum chart (like figure Q8.2) showing all its *visible* emission lines.

Q8S.4 Verify equation Q8.7.

Q8S.5 Imagine that the only visible photons that an electron in a box of unknown length emits have wavelengths of 689 nm and 413 nm. Identify the transitions and find the length of the box. (*Hint:* Make a guess as to what the first transition is, use that to calculate the ground-state energy of the box, and then see if that guess correctly predicts the next transition. You will not have to make many guesses.)

Q8S.6 Imagine that the only visible photons that an electron in a box of unknown length emits have wavelengths of 620 nm and 443 nm. Identify the transitions and find the length of the box. (*Hint:* Make a guess as to what the first transition is, use that to calculate the ground-state energy of the box, and then see if that guess correctly predicts the next transition. You will not have to make many guesses.)

Q8S.7 An He^+ ion has one electron, just as a hydrogen atom does, but has *two* protons in its nucleus instead of just one.
(a) Find a formula for the wavelength λ of an emission line for this ion in terms of the magnitude of the hydrogen ground-state energy $|E_1| = 13.6$ eV, $n_{initial}$, and n_{final}. (*Hint:* Review the discussion of the Bohr model in chapter Q7 and determine how having two protons in the nucleus affects the results for a_0 and E_n.)
(b) Show that all transitions to $n = 5$ from $n \geq 12$ produce visible emission lines, and find the limiting wavelength of this series of lines as $n \to \infty$.
(c) There are nine other emission lines within the visible region (700 nm to 400 nm). Find the wavelengths of these nine lines.

Q8S.8 It is now almost possible (using integrated-circuit technology) to manufacture a "box" that traps electrons in a region only a few nanometers wide (see Reed, "Quantum Dots," *Scientific American*, **268**, 1, pp. 118–123). Imagine that we make an essentially one-dimensional box with a length of 3.0 nm. If we put 10 electrons in such a box and allow them to settle into the lowest energy states they can, consistent with the Pauli exclusion principle, what will be the value of n for the highest energy level occupied? What will be the total energy of the electrons (ignoring their electrostatic repulsion)? Would your answer be different if the electrons were bosons instead of fermions?

Q8S.9 The potential energy function for a proton or neutron in an atomic nucleus can be crudely modeled as being a "box" that is about 4 fm $= 4 \times 10^{-15}$ m wide. Imagine that we put 12 neutrons into such a box. Find the minimum *total* energy of the 12 neutrons in this case. (*Hint:* Only two neutrons can occupy each energy level. Explain why.) How much lower would this minimum total energy become if we changed one-half of the neutrons to protons?

Q8S.10 Here is another danger about literally considering a spinning electron to be a rotating ball of charge. Assume that the electron is a uniform sphere of radius R and total spin angular momentum $L = \hbar/2$. Experiments show that if an electron is not a point particle, its radius is certainly smaller than 10^{-18} m. Show that this means that if the electron really is a spinning ball, points on the electron's equator must be moving *much* faster than light. (Quantum field theory does *not* require an electron to be an actual spinning object in the classical sense. The electron behaves *as if it were* a spinning object because its full quantum wavefunction must behave in a certain way when it is rotated by 360° if it is to be both self-consistent and consistent with relativity. This imposes restrictions on the wavefunction that make it similar to the wavefunction of a quanton that is actually moving around some origin.)

Q8S.11 In the description in section Q8.6, we assumed that the magnetic field in a Stern-Gerlach experiment everywhere points in the $-z$ direction and has a magnitude that increases with increasing z.
(a) Argue that such a field would violate Gauss's law for the magnetic field.
(b) Figure Q8.11 shows an end view of an electromagnet that might be used in a modern version

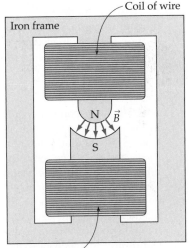

Figure Q8.11

An end view of the magnet used in a modern version of the Stern-Gerlach experiment to apply the external magnetic field. The gap in a typical experiment might be several millimeters across, and the magnet about 10 cm long in the direction perpendicular to the drawing. (Adapted from A. P. French and E. F. Taylor, *An Introduction to Quantum Physics*, Norton, 1978.)

of the Stern-Gerlach experiment. The poles of this magnet are shaped like sections of two cylinders that are about 10 cm long in the direction perpendicular to the plane of the drawing. The gap between the poles is a few millimeters across. The magnetic field near the center of this gap (the lowest part of the gap) is consistent with the field described in section Q8.6. Assume that the effective radius of the north pole is essentially 1.0 cm, the effective radius of the curved inner surface of the south pole is 1.5 cm, and the magnetic field points essentially perpendicular to the surface of the pole faces, as shown in the drawing. Assume also that mag($\vec{B}$) is about 5.0 T on the surface of the north pole. Use Gauss's law to estimate the average value of dB_z/dz at the lowest part of the gap. (Assume that the magnetic field has no component perpendicular to the plane of the drawing.)

Q8S.12 Why must we do a Stern-Gerlach experiment with neutral atoms instead of bare electrons? In this problem, we will consider just one of the difficulties. Imagine that we have an electron beam with a kinetic energy of 100 eV moving in the $+x$ direction through a region of space where the magnetic field points in the $+z$ direction and its strength goes from $B \approx 0.5$ T to $B \approx 1.5$ T in a vertical distance of 1.0 cm.

(a) Use a dipole model for the electron's equivalent magnet to argue that the magnitude of the deflection force on an electron will be given by

$$F_{\text{net}} = \mu \left| \cos\theta \, \frac{dB_z}{dz} \right| \qquad (Q8.11)$$

where B_z is the component of the magnetic field in the z direction, θ is the angle that the electron's spin angular momentum makes with the z axis, and μ is some constant of proportionality related to how strong a magnet the electron is.

(b) The magnitude of μ for an electron turns out to be 9.3×10^{-24} J/T. Show that the maximum magnetic force on an electron in this experimental situation would be about 9.3×10^{-22} N.

(c) Compute the speed of a 100-eV electron, and show that the simple Lorentz force $F_{em} = evB$ on an electron moving through the middle of such a magnetic field has a magnitude about 1 billion times larger than the result for part (a).

(d) This would be fine if we could correct for the effect. To roughly how many decimal places would we have to know the magnetic field strength at every point along the electron's curving trajectory to be able to isolate the effect we are looking for? Does this seem practical?

Q8S.13 Consider the case of an oxacarbocyanine molecule with N bonds between the two nitrogen atoms. The

bond length between atoms in the central chain is still $b \approx 0.14$ nm, but the different end groups in this molecule (compared to the molecule in example Q8.1) allow electrons to penetrate roughly $0.5b$ beyond the nitrogen atom at each end. This actually varies weakly with N, but for the sake of simplicity, let us assume that it does not.

(a) Show that the lowest-energy photon that the electrons along the molecule's central chain can absorb will have a wavelength of roughly

$$\lambda \approx \frac{8mc^2b^2}{hc}(N+1) \qquad (Q8.12)$$

(b) Predict these wavelengths for dye molecules with $N = 6$, $N = 8$, and $N = 10$. Compare with the observed wavelengths: roughly 485 nm, 582 nm, and 687 nm, respectively. (The absorption lines are broad, but have peaks at these wavelengths.)

Q8S.14 Imagine that we can model a certain cyanine dye molecule as having 10 electrons confined to a one-dimensional box of length L. Imagine that we use an external source of light to illuminate a sample of this molecule with photons having just the right energy to excite electrons from the $n = 5$ level to the $n = 7$ level of the box.

(a) When such excited electrons decay back toward their lowest available energy state, in addition to reradiating photons of the same energy as those from the external source, they can radiate photons of two and *only* two other longer wavelengths (this is the phenomenon known as *fluorescence*). Using an energy-level diagram showing possible transitions, identify the two transitions that produce these photons and explain why other transitions (say, from the $n = 7$ to the $n = 1$ level) are impossible.

(b) Assume that of the photons discussed in part (a), the ones with the longest wavelength have a wavelength of 600 nm (in the red-orange region of the spectrum). What is the wavelength of the original photons from the external source? Are the photons from the external source visible?

Rich-Context

Q8R.1 When excited, electrons in a certain substance give off photons having a variety of wavelengths. The longest two wavelengths emitted are 830 nm (in the near infrared) and 496 nm (blue-green). Are the electrons in this substance bound by a springlike interaction, or are they trapped in a box? Carefully defend your response.

Q8R.2 The magnetic field of an iron bar is almost entirely created by aligned electron spins (one electron spin per iron atom). Imagine that we suspend an

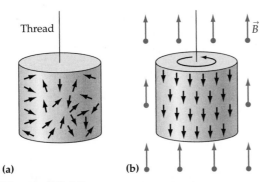

Thread

$\vec{B}$

(a)

(b)

Figure Q8.12

(a) Before an external magnetic field is applied, spinning electrons in an iron cylinder are randomly oriented. (b) After a strong vertical magnetic field is applied, the electrons that provide the magnetism in iron become oriented opposite to the direction of the applied field. To conserve angular momentum, the iron cylinder must rotate.

unmagnetized iron cylinder by a thin thread as shown in figure Q8.12a, and assume that the iron bar is initially completely at rest and not rotating. Imagine then that we suddenly immerse the bar in a strong magnetic field. Initially, the electron spins are randomly aligned, but after the magnetic field is applied, the spins will all antialign themselves with the field, as shown in figure Q8.12b. This means that the component of the total electron spin along the z axis changes when the magnetic field is applied. If electron spin is genuinely angular momentum (as the newtonian model implies), then since the z component of the total angular momentum of the iron bar is conserved, the bar will begin rotating in the opposite direction to compensate for the change in the electron spin angular momenta. Assuming that the spin angular momentum of a single electron has a magnitude of $\frac{1}{2}\hbar$, and assuming that one electron per iron atom becomes aligned with the

field, estimate how fast a bar 1.0 cm in diameter would rotate after the field is applied, and specify how much time would be required for the bar to complete 1 rev at this rate. Express your answer in days. (*Hint:* To solve this problem, you need only a periodic table, the formula for the moment of inertia for a cylinder, and some basic information about angular momentum that you can find in chapter C13.)

Comment This twisting of a ferromagnetic object in suddenly applied magnetic field is called the *Einstein-de Haas effect*. Although the effect is small, experiments show quite clearly that it exists, demonstrating that electron spin is a genuine form of angular momentum.

Advanced

Q8A.1 The spectrum of real diatomic molecules is a product of transitions that involve both rotational *and* vibrational energy levels. The formula for the vibrational-rotational energy levels of such a molecule is given by

$$E_{n\ell} = \hbar\omega\left(n + \frac{1}{2}\right) + \frac{\ell(\ell+1)\hbar^2}{2I} \qquad (Q8.13)$$

where I is the molecule's rotational moment of inertia and where $n = 0, 1, 2, \ldots$ and $\ell = 0, 1, 2, \ldots$. It turns out that conservation of angular momentum and other restrictions require that $\Delta n = 1$ for an allowed vibrational transition *and also* $\Delta\ell = \pm 1$ during such a transition. Draw an energy-level diagram for this situation, and find the possible energies for emitted photons if $\hbar\omega = 0.358$ eV and $\hbar^2/2I = 0.00135$ eV (the values for the HCl molecule). Draw a spectrum chart that has energy as the horizontal axis and vertical lines representing the spectral lines. Only worry about spectral lines for $\ell_{init} \leq 6$ (in practice, the lines get significantly dimmer beyond this point).

ANSWERS TO EXERCISES

Q8X.1 The energy of the photon produced by a $3 \rightarrow 2$ transition is

$$E_{ph} = E_3 - E_2 = (1.5\,\text{eV})(3^2 - 2^2)$$

$$= 5(1.5\,\text{eV}) = 7.5\,\text{eV} \qquad (Q8.14)$$

The wavelength of this photon is thus

$$\lambda = \frac{hc}{E_{ph}} = \frac{1240\,\text{eV}\cdot\text{nm}}{7.5\,\text{eV}} = 165\,\text{nm} \qquad (Q8.15)$$

Similarly, the energy of the photon produced by a $3 \rightarrow 1$ transition is

$$E_{ph} = (1.5\,\text{eV})(3^2 - 1^2) = 8(1.5\,\text{eV})$$

$$= 12\,\text{eV} \qquad (Q8.16)$$

$$\Rightarrow \quad \lambda = \frac{hc}{E_{ph}} = \frac{1240\,\text{eV}\cdot\text{nm}}{12\,\text{eV}}$$

$$= 103\,\text{nm} \qquad (Q8.17)$$

Q8X.2　In the case of the harmonic oscillator, the photon energy associated with a transition from the energy level numbered n_{init} to that numbered n_{final} is

$$E_{\text{ph}} = E_{\text{init}} - E_{\text{final}} = \hbar\omega\left(n_{\text{init}} + \tfrac{1}{2}\right) - \hbar\omega\left(n_{\text{final}} + \tfrac{1}{2}\right)$$

$$= \hbar\omega(n_{\text{init}} - n_{\text{final}}) \qquad \text{(Q8.18)}$$

Therefore, the photon wavelength is

$$\lambda = \frac{hc}{E_{\text{ph}}} = \frac{hc}{\hbar\omega(n_{\text{init}} - n_{\text{final}})} \qquad \text{(Q8.19)}$$

Since $\hbar = h/2\pi$, $hc/\hbar\omega = 2\pi c/\omega = c/f$, where f is the oscillator's classical frequency in cycles per second. However, it is more common to state the value of $\hbar\omega$ for a quantum oscillator (which is an energy) than to state its classical frequency, so the form of the equation given above is generally the most useful.

Q8X.3　Plugging in the numbers, we get

$$\frac{2(hc)a_0}{ke^2} = \frac{2(1240\ \text{eV} \cdot \text{nm})(0.053\ \text{nm})}{1.44\ \text{eV} \cdot \text{nm}} = 91.3\ \text{nm}$$
$$\text{(Q8.20)}$$

Q8X.4　Looking at the energy-level chart for the hydrogen atom in figure Q8.1c, we see that jumps from any energy level to the $n = 1$ ground level will involve a greater energy difference than any jump to the $n = 2$ or $n = 3$ level. The transition to the $n = 1$ level that produces the photon with the lowest energy (and thus the longest wavelength) will be the $2 \to 1$ transition. According to equation Q8.7, the wavelength of this photon is

$$\lambda(2 \to 1) = (91.3\ \text{nm})\left(\frac{2^2 1^2}{2^2 - 1^2}\right) = 122\ \text{nm} \quad \text{(Q8.21)}$$

which is in the far ultraviolet. All *other* transitions to the $n = 1$ level will produce photons with still greater energies (and thus *still* shorter wavelengths). Therefore, *no* transition to the $n = 1$ level produces a visible photon.

Q8X.5　According to equation Q8.3, the energy of the photon emitted in a transition between energy levels numbered n_{init} and n_{final} will be

$$E_{\text{ph}} = E_{\text{init}} - E_{\text{final}} = -13.6\ \text{eV}\left(\frac{1}{n_{\text{init}}^2} - \frac{1}{n_{\text{final}}^2}\right)$$

$$= +13.6\ \text{eV}\left(\frac{1}{n_{\text{final}}^2} - \frac{1}{n_{\text{init}}^2}\right) \qquad \text{(Q8.22)}$$

The longest-wavelength transition to the $n = 2$ final state will be the $3 \to 2$ transition. Using the equation above and equation Q8.7, we can see that the

energy and wavelengths of the photon produced by this transition are

$$E_{\text{ph}}(3 \to 2) = 13.6\ \text{eV}\left(\frac{1}{2^2} - \frac{1}{3^2}\right) = 1.89\ \text{eV}$$
$$\text{(Q8.23a)}$$

$$\lambda(3 \to 2) = (91.3\ \text{nm})\left(\frac{3^2 2^2}{3^2 - 2^2}\right) = 657\ \text{nm}$$
$$\text{(Q8.23b)}$$

Similarly $\quad E_{\text{ph}}(4 \to 2) = 2.55\ \text{eV}, \quad \lambda(4 \to 2) = 486\ \text{nm}$, $E_{\text{ph}}(5 \to 2) = 2.86\ \text{eV}, \lambda(5 \to 2) = 435\ \text{nm}$, and $E_{\text{ph}}(6 \to 2) = 3.02\ \text{eV}, \lambda(6 \to 2) = 411\ \text{nm}$. The $7 \to 2$ transition produces a photon with a wavelength of 398 nm, which is just beyond the visible range, so no transitions to $n = 2$ with $n_{\text{init}} > 6$ produce visible photons. It also turns out that all transitions to the $n = 3$ level produce photons with wavelengths too long to be visible (see problem Q8B.5). Therefore, the four transitions from $n = 3, 4, 5,$ and 6 to $n = 2$ are the only hydrogen transitions that produce visible photons. Figure Q8.5 gives a spectrum chart for hydrogen.

Q8X.6　The magnetic field exerts a torque $\vec{\tau}$ on the electron's equivalent magnet that seeks to align that magnet with the local field direction, which we will assume in this case to be the $+z$ direction. An electron's equivalent magnet is antiparallel to its spin angular momentum vector $\vec{S}$ (see figure Q8.7), so the torque seeks to twist the electron's spin $\vec{S}$ so that it is *antialigned* with the $+z$ direction. Since the torque wants to rotate $\vec{S}$ so that it and the $-z$ direction close like the blades of a pair of scissors, the axis around which the torque wants to rotate $\vec{S}$ (and thus the direction of the torque vector $\vec{\tau}$) is necessarily perpendicular to both $\vec{S}$ and the z axis, just as the axis around which scissors blades rotate is perpendicular to both blades. Now, torque is defined to be the rate of change of angular momentum, so in this case, $\vec{\tau} \equiv d\vec{S}/dt$. During any infinitesimal time interval, therefore, the torque will cause the electron's spin angular momentum $\vec{S}$ to change by $d\vec{S} = \vec{\tau}dt$. Since we know that the torque is at every instant perpendicular to both $\vec{S}$ and the z axis, this torque will neither change the magnitude of $\vec{S}$ (since $d\vec{S}$ has no component in the direction of $\vec{S}$) nor its projection S_z on the z axis (since $d\vec{S}$ has no component in the z direction). This in turn implies that the angle $\theta = \sin^{-1}(S_z/S)$ between $\vec{S}$ and the z axis cannot change. Therefore, the electron *cannot* flip over in response to this torque. (Instead of closing the angle between $\vec{S}$ and the $-z$ direction, the torque instead causes the tip of $\vec{S}$ to precess around the field direction, just as the gravitational torque on a spinning top causes its spin to precess. See section C14.1 for more discussion of this.)

Q8X.7 According to newtonian mechanics, the spin angular momentum $\vec{S}$ of a symmetric spinning object is related to its angular velocity of rotation $\vec{\omega}$ by the equation $\vec{S} = I\vec{\omega}$, where I is the object's moment of inertia. The moment of inertia of a spinning sphere of radius R and mass m is $I = \frac{2}{5}mR^2$ (see chapter C9). As a first approximation, let us model both the proton and electron as uniform charged spheres. Since the proton and electron have the same magnitude of angular momentum (both are spin-$\frac{1}{2}$ quantons), we have

$$1 = \frac{I_e \omega_e}{I_p \omega_p} = \frac{\frac{2}{5}m_e R_e^2 \omega_e}{\frac{2}{5}m_p R_p^2 \omega_p} \quad \Rightarrow \quad \frac{\omega_e}{\omega_p} = \frac{m_p}{m_e}\left(\frac{R_p}{R_e}\right)^2$$

(Q8.24)

where the e subscript refers to the electron and the p subscript to the proton. If the proton and electron have roughly the same radius, then

$$\frac{\omega_e}{\omega_p} \approx \frac{m_p}{m_e} \approx 2000 \qquad \text{(Q8.25)}$$

This means that the electron spins roughly 2000 times faster than the proton. Since each has the same charge, and since the magnetic field that each creates in this model depends on the magnitude of the circulating current that each represents, the spinning electron represents a circulating current that is 2000 times larger than that of the proton, implying that (other things being equal) the electron's magnetic field should be roughly 2000 times larger than that of the proton. (The experimentally observed ratio is about 660, so this semiclassical model is too crude to yield exactly the right result, but does give us the right order of magnitude.)

Q9

Understanding Atoms

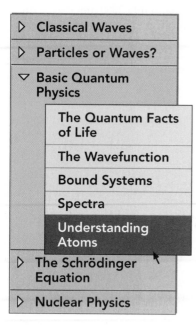
Chapter Overview

Introduction

In this closing chapter of the unit subdivision on basic quantum mechanics, we look at how what we have learned so far helps us understand the quantum physics of atoms.

Section Q9.1: Radial and Angular Waves

Our goal in this section is to learn to interpret (not derive) the features of the three-dimensional energy eigenstates of the hydrogen atom. Imagine for the moment that an electron in a hydrogen atom can move *only* radially. Our work in chapter Q7 suggests that the electron's energy eigenfunction must have an integer number n_r of half-wavelengths between $r = 0$ and the outer turning point where $V(r) = E$. (The absolute square of such a wavefunction will have n_r "bumps.") These radial eigenfunctions (by luck) have the same energies as found in the Bohr model.

The rule that the wavefunction must match itself as it goes once around the atom still applies to the three-dimensional energy eigenfunctions, so an integer number ℓ of full wavelengths still must fit around any circle around the nucleus. This implies (as we saw in chapter Q7) that the electron's orbital angular momentum magnitude must be an integer multiple of $\hbar$: $L = \ell\hbar$.

A general hydrogen energy eigenfunction has both radial and angular waves. The energy associated with a general energy eigenfunction is $E = -E_0/n^2$, where $E_0 = 13.6$ eV and $n = n_r + \ell$. A general wavefunction is therefore completely specified by four quantum numbers:

1. n, which specifies the energy level ($n = 1, 2, 3, \ldots$)
2. ℓ, which specifies the magnitude of the electron's orbital angular momentum
3. m, which specifies the z component of that angular momentum ($L_z = m\hbar$)
4. m_s, which specifies the z component of the electron's spin ($S_z = m_s\hbar$)

For a given n, the possible values of ℓ are $0, 1, \ldots, n - 1$. The value of m ranges from $-\ell$ to $+\ell$ in integer steps ($2\ell + 1$ possible values) and m_s is either $+\frac{1}{2}$ or $-\frac{1}{2}$.

Section Q9.2: The Periodic Table

Interactions between electrons make eigenfunctions for multielectron atoms very difficult to calculate exactly. A simplistic qualitative model is to imagine that the electrons in lower energy states than a given electron form a spherical cloud that shields part of the nuclear charge from that electron. This model suggests that electron energy eigenfunctions should be hydrogenlike, though with somewhat different energies. Also, electrons in eigenfunctions with low values of ℓ have a larger probability of being found deep inside this cloud compared to electrons with eigenfunctions with the same value of n but higher ℓ. This means that lower-ℓ electrons feel more of the unshielded attraction of the nucleus, and this has the effect of lowering the energy of such states compared with those with the same n but higher ℓ. This energy-level arrangement (see figure Q9.3) determines the structure of the periodic table.

One can specify how many electrons occupy a given energy level by using **spectroscopic notation,** for example, $2p^5$. In this notation, the big number specifies the value of n, the letter the value of ℓ (s, p, d, f mean $\ell = 0, 1, 2, 3$, respectively), and the superscript the number of electrons in that level [the maximum possible number is $2(2\ell + 1)$].

Section Q9.3: Selection Rules

Photons turn out to have spin $s = 1$. [Relativistic effects imply that the projection of this spin on a given axis has two values ($\pm\hbar$) instead of the expected three.] An emitted photon therefore necessarily carries away some angular momentum, and conservation of angular momentum implies that this can only come from either the emitting electron's orbital angular momentum (most likely) or a spin flip (very unlikely). If the angular momentum comes from orbital angular momentum, then that orbital angular momentum must change by exactly 1 unit, so such transitions are only possible if $\Delta\ell = \pm 1$. This is an example of a **selection rule** that specifies "allowed" transitions. Transitions that violate this rule are not *strictly* forbidden, but are much less likely. Therefore, if an electron is in an excited state whose decay is forbidden by a selection rule (a **metastable** state), it will last much longer than it typically would in an excited state.

Section Q9.4: Stimulated Emission and Lasers

A passing photon with the right frequency can "shake loose" a photon from an atom in an excited state (this is essentially a resonance effect). We call this phenomenon **stimulated emission.** The new emitted photon has a wavefunction whose wavelength and phase are exactly the same as those of the initiating photon. We call such photons **coherent.**

A **laser** generates light by stimulated emission. To build a laser:

1. Find an atom with an appropriate metastable state.
2. Find a way to put more atoms in that state than in lower available states (this is called a **population inversion**).
3. Trap any photons that are emitted between two mirrors for a while so that they have plenty of opportunity to stimulate emission.

(The common helium-neon laser cleverly uses a metastable state of helium to keep neon atoms in an appropriate excited state for stimulated emission.)

Because photons created by stimulated emission have exactly the same wavelength, direction, and phase, lasers have many applications that require light that is monochromatic and coherent and/or light that can be very tightly focused.

Q9.1   Radial and Angular Waves

When we developed the Bohr model of the hydrogen atom in chapter Q7, we made a number of approximations and assumptions that are really wrong (even though they happen to yield the right energy levels). The correct way to find the hydrogen energy eigenfunctions in quantum mechanics is to solve the fully three-dimensional Schrödinger equation, a task that is quite a bit beyond the level of this course (we will explore the much simpler *one-dimensional* Schrödinger equation in chapters Q10 and Q11, and that will turn out to be quite enough!). My goal in this section is not to derive but rather to present the results and show you how we can *interpret* the qualitative features of these energy eigenfunctions.

Radial waves

The main problem with the Bohr model is that it ignores the possible *radial* motion of the electron (we imagined the electron to orbit the nucleus in a circular orbit). For the moment, let's make the *opposite* error and ignore the electron's angular motion *around* the nucleus. A graph of the potential energy of an electron bound by a proton as a function of r is displayed in figure Q9.1a. A classical bound electron (whose total energy E is fixed and less than zero) moving along the r axis would be trapped between $r = 0$ (clearly the electron cannot be found at negative values of r) and the classical turning point at the value of r where $E = V(r) = -ke^2/r$. As in all the one-dimensional cases that we discussed in chapter Q7, the electron's wavefunction must fit an integer number of half-wavelengths between these two boundaries.

The possible energy eigenfunctions with the three lowest possible energies for purely *radial* motion of the electron thus look something like those shown in figure Q9.1b, which are actual solutions of the Schrödinger equation assuming purely radial motion. (The absolute squares of these energy eigenfunctions are shown in figure Q9.1c.) Note that both the amplitude and

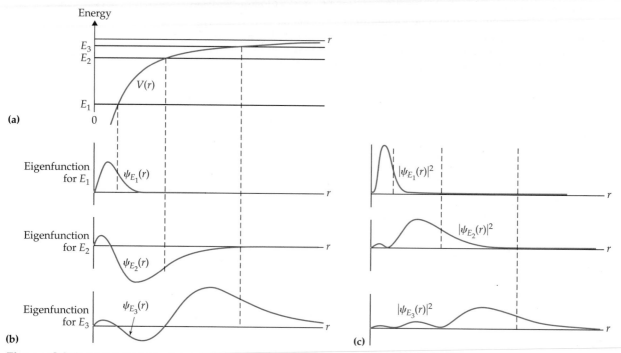

Figure Q9.1
(a) A graph of the potential energy of an electron in a hydrogen atom. (b) The energy eigenfunctions for the three lowest energy states of radial motion. (c) The absolute squares of those wavefunctions.

the effective wavelength of these energy eigenfunctions get smaller as we get closer to the nucleus; also instead of going strictly to zero at the outer turning point, the wavefunction turns into a decaying exponential there. I will explain the qualitative reasons for these features in chapter Q11.

These peculiarities should not obscure in your mind the basic fact that the absolute squares of these quantized radial energy eigenfunctions have an integer number of "bumps." It turns out that the energy associated with the nth radial energy eigenfunction is *also* $E_n = -ke^2/2a_0n^2$, the same result as predicted by the Bohr model (except that n here is *not* the number of full wavelengths of the quanton's wavefunction that fit *around* the atom, but rather is the number of bumps in the absolute square of the wavefunction as we go *away* from the atom!).

Now let us review the picture of angular motion presented in the Bohr model. In that model, we ignored the freedom that the electron had to move in the radial direction (imagining it instead to orbit at a fixed radius) but required that the electron wavefunction match itself as we go exactly once around the orbit. This led to the requirement that the wavefunction go through ℓ complete sinusoidal cycles as we go around the atom. We *also* saw that this in turn meant that the magnitude of the electron's orbital angular momentum in this model is given by

Angular waves

$$L = \ell\hbar \qquad (Q9.1)$$

(See exercise Q7X.6.)[†] Note that if a wavefunction goes through ℓ sinusoidal cycles around the atom, the *square* of the wavefunction will have 2ℓ bumps arranged around the atom.

Now let us consider both radial *and* angular motion. The requirement that the wavefunction match itself as we go once around the atom still holds, so the general energy eigenfunction must *still* have an integer number ℓ of complete sinusoidal cycles as we go around the atom. The square of the general energy eigenfunction for the hydrogen atom will therefore have both an integer number n_r of radial bumps *outward* from the atom and an even integer number 2ℓ of bumps *around* the atom at any given radius. (This qualitative result applies to essentially any quanton that is bound by a potential energy that is a function of r alone.)

General energy eigenfunctions have both radial and angular waves

It turns out that for a given *hydrogen* energy level $E_n = -ke^2/2a_0n^2$, we can trade one or more radial bumps for one or more complete wavelengths (*pair* of bumps) *around* the atom and remain at the same n. The only restriction is that the number of radial bumps n_r must be at *least* 1 (though ℓ can be zero):

$$n = n_r + \ell \qquad (n_r = 1, 2, \ldots \text{ and } \ell = 0, 1, \ldots) \qquad (Q9.2)$$

So, for example, for $n = 3$, we could have three radial bumps and no angular bumps ($n_r = 3$ and $\ell = 0$), two radial bumps and two angular bumps ($n_r = 2$ and $\ell = 1$), or one radial bump and four angular bumps ($n_r = 1$ and $\ell = 2$). All three of these completely different energy eigenfunctions have the *same* energy $E_3 = ke^2/18a_0$! (This turns out to be a *special* feature of any system where the potential energy $V \propto 1/r$.)

Figure Q9.2 shows the squares of the three energy eigenfunctions that I have just described, plotted as a function of position in the xy plane. The nucleus is at the center of each graph, and the graph goes out to $\pm 20a_0$ in both the x and y directions. (Actually, the graphs show the square of only the *real*

[†]Note that for technical reasons beyond us at present, the Bohr model is not quite right about this. When one measures the actual *magnitude* of a quanton's angular momentum (instead of its projection on an axis), one finds that the possible values are actually $L = \sqrt{\ell(\ell+1)}\,\hbar$. However, this does not affect any of the other results discussed in this chapter.

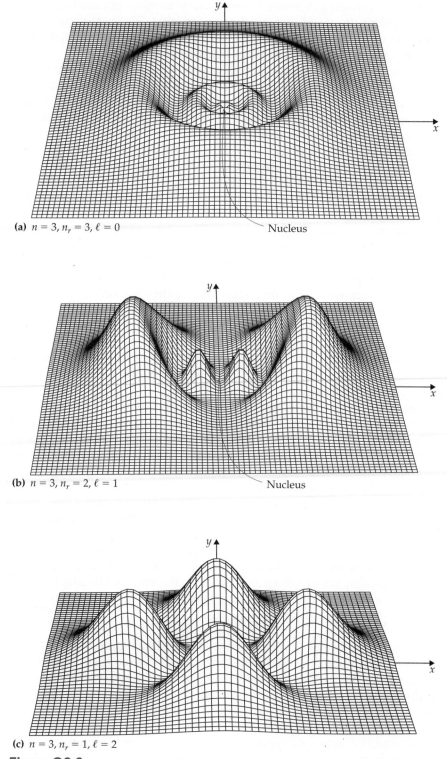

(a) $n = 3, n_r = 3, \ell = 0$ — Nucleus

(b) $n = 3, n_r = 2, \ell = 1$ — Nucleus

(c) $n = 3, n_r = 1, \ell = 2$

Figure Q9.2

Plots of the square of the (real part of the) electron energy eigenfunctions in hydrogen (evaluated in the xy plane) for $n = 3$ and various values of ℓ. Note how as ℓ increases, the number of radial bumps n_r decreases.

part of these energy eigenfunctions: adding the square of the imaginary part washes out the angular bumps, obscuring the fact that the complex wave-function really is "waving" an integer number of times as we go around the nucleus. The square of just the real part thus better illustrates what is really going on.)

This figure, in combination with equation Q9.2, helps us understand why the simplistic Bohr model happens to give the right answers. The Bohr model essentially calculates the energies associated with the energy eigen-states having one radial bump and maximal angular momentum ($\ell = n - 1$). As figure Q9.2 illustrates, modeling such eigenstates as being an electron in a circular orbit of some radius is really not so far from the mark. The fact that other possible mixtures of radial and angular motions satisfying equation Q9.2 with n fixed have the same energy as the states with maximal angular momentum means that the Bohr model essentially gets *all* the hydrogen energy levels right.

Why the Bohr model works so well

Exercise Q9X.1

What are the possible values of n_r and ℓ if $n = 2$? $n = 4$?

We call any energy level that has two or more completely different energy eigenfunctions associated with it a **degenerate** energy level. (I'm not making this up!) We see that the $n = 3$ energy level has at least *three* associated energy eigenfunctions. In fact it has even more than that. Figure Q9.2 assumes that the electron's orbital angular momentum is perpendicular to the xy plane. As we saw in chapter Q8, though, a quanton's orbital angular momentum whose magnitude is specified by ℓ can have a quantized z component L_z ranging in steps of $\hbar$ from $+\ell\hbar$ to $-\ell\hbar$. There is a different energy eigenfunction associated with each of the possible values of L_z. (It is hard to illustrate these eigenfunctions clearly because one really needs to plot them as a function of x, y, and z, and this would require a four-dimensional graph.) You can show that in general there are $2\ell + 1$ such eigenfunctions for a given value of ℓ. So there are really *three* eigenfunctions for $n_r = 2, \ell = 1$ and *five* eigenfunctions for $n_r = 1, \ell = 2$ (only one of which is shown in figure Q9.2 in each case) for a total of *nine* eigenfunctions for $n = 3$ (including the single $n_r = 3, \ell = 0$ eigenfunction).

Sets of eigenfunctions with the same energy

The complete energy eigenstate of an electron in a hydrogen atom is therefore completely specified by four integer **quantum numbers:** n (specifying the energy level), ℓ (specifying the number of angular cycles of the wavefunction around the atom and also the number $n_r = n - \ell$ of radial bumps in the squared wavefunction), m (specifying one of the $2\ell + 1$ possible orientations of the wavefunction in space: $L_z = m\hbar$), and m_s (specifying one of the two possible electron spin orientations: $S_z = m_s\hbar$ where $m_s = \pm\frac{1}{2}$).

Quantum numbers for hydrogen eigenstates

Exercise Q9X.2

Show that the number of possible values of L_z (or S_z) corresponding to a given value of ℓ (or s) is indeed $2\ell + 1$ (or $2s + 1$).

Exercise Q9X.3

How many distinct electron quantum states are associated with the $n = 4$ level of hydrogen?

Q9.2   The Periodic Table

Hydrogenlike eigenstates

Now consider the bare nucleus of a more complicated atom with Z protons. The possible energy eigenfunctions for the *first* electron that we add to such an atom have the same *shape* as the hydrogen atom eigenfunctions but a different scale of length a_0 (the stronger attraction of the nucleus will pull the electron wavefunctions inward, making the radial characteristics of the eigenfunction smaller by a factor of Z). We will call these rescaled energy eigenfunctions **hydrogenlike** eigenfunctions.

When we add a second electron, however, the problem becomes *much* more complicated, because in addition to interacting with the nucleus, the electrons interact with each other. This added factor makes a complete analytic determination of the electron eigenfunctions for even the helium atom impossible.

Model for dealing with multielectron atoms

We can, however, get around this problem to a limited extent by making the following approximation. Imagine that we have already added N electrons to our atom. Let us *assume* that these N electrons form a roughly spherical cloud around the nucleus. To the extent that the next electron lies *outside* this cloud, its eigenfunction will essentially be the same as the hydrogenlike eigenfunction with the same n for a *bare* atom with $Z - N$ protons (since the inner electron cloud effectively cancels the charge of N of the atom's protons). If, on the other hand, the hydrogenlike eigenfunction implies that the electron has a significant probability of being *inside* this cloud, then the electron "feels" more of the uncanceled attraction of the nucleus, and this lowers the energy of that eigenfunction somewhat (but we assume that it doesn't much change its shape). This approximation turns out to work pretty well mostly because the wavefunctions with different values of n don't really overlap very much (particularly in multielectron atoms). The sets of electrons with different values of n thus essentially form a series of nested *shells* (see figure Q9.1 to see these shells for $\ell = 0$ wavefunctions having $n = 1, 2,$ and 3).

Energies of states with the same n but different ℓ differ in multielectron atoms

As we can see from figure Q9.2, as ℓ gets smaller, the energy eigenfunction has more bumps closer to the nucleus, which implies that an electron having that eigenfunction has a greater probability of being inside the cloud formed by the other electrons, which in turn implies that the energy associated with that eigenfunction is smaller than we would expect from the hydrogen model. This means that no longer will all energy eigenfunctions having the same n have also the same energy: eigenfunctions with smaller values of ℓ will have a somewhat lower energy than those with larger values of ℓ. The values of m and m_s (which specify orientation in space) are still irrelevant for the energy, though.

Spectroscopic notation

Before quantum mechanics was completely developed, physicists doing atomic spectroscopy had already developed a notation for describing the different hydrogenlike energy levels in an atom. Basically, the notation consisted of a *number* (representing the value of n) followed by a lowercase letter (representing the value of ℓ). The correspondence between the letters and values of ℓ is as follows:

letters	s	p	d	f
ℓ	0	1	2	3

So, for example, $3p$ refers to an energy eigenstate with $n = 3$ and $\ell = 1$. Physicists and chemists call this **spectroscopic notation** (the letters originally described characteristics of spectral lines, such as "sharp" and "principal").

Exercise Q9X.4

What are n and ℓ for the state designated $5f$?

Figure Q9.3 shows the relative energies of the different hydrogenlike eigenstates in a multielectron atom. I have labeled each level, using the spectroscopic notation, and shown the number of electrons that this level can hold (2 times $2\ell + 1$) in parentheses. Note that for each value of n, the s level ($\ell = 0$) has the lowest energy, the p ($\ell = 1$) level is next lowest, the d ($\ell = 2$) level is next lowest, and so on. Note also, though, that the highest $n = 3$ levels are actually higher than the lowest $n = 4$ levels and so on for higher n (!). The levels linked by the tinted boxes are actually so close in energy that which is higher and which is lower depend on the specific atom involved.

To determine the lowest-energy arrangement of electrons for an atom with Z protons (and thus Z electrons), we simply fill in the levels on the chart, starting with the $1s$ level, until we have used up all Z electrons. For example, hydrogen has one electron, so it goes in the $1s$ level. Helium has two electrons, so both go to the $1s$ level (with opposite spins). The big energy gap between the $1s$ and $2s$ states contributes to helium's chemical inactivity: there is not a lot of flexibility for shifting electrons around in the formation of a chemical bond. Lithium has three electrons; the first two fill the $1s$ level and the third occupies one of the $2s$ slots. This makes lithium chemically similar to hydrogen (which also had a single electron in an s state). Beryllium has four electrons, so they fill the $1s$ and $2s$ states. The comparatively small gap between the filled $2s$ level and the $2p$ level means that beryllium is much more reactive than helium (which also had a filled s level).

How the periodic table reflects energy-level structure

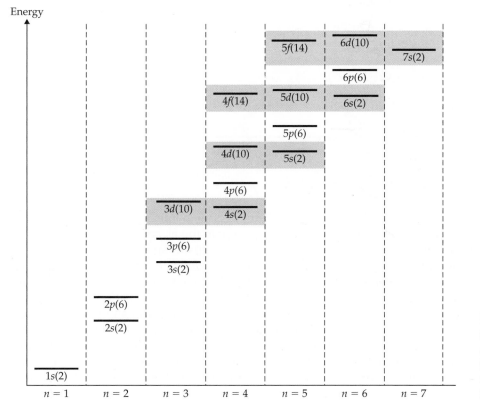

Figure Q9.3
General energy-level diagram for a multielectron atom. (Adapted from Cohen-Tannoudji et al., *Quantum Mechanics*, Wiley, New York, 1978, p. 1415.)

As we continue to fill the states with electrons, we find that the two *left-most* columns of the periodic table correspond to chemically similar atoms with one and two electrons (respectively) in an *s* level, the six rightmost columns of the periodic table correspond to atoms with a sequentially increasing number of electrons in *p* levels, the long rows in the middle correspond to the atoms we get as we fill a *d* level, and the two even longer rows below the main table correspond to atoms we get as we fill the *f* level. The last column on the right corresponds to chemically inactive gases because these atoms have filled outer *p* levels (or *s* level in the case of helium) and a big energy gap to the next level.

Because the levels in the tinted boxes are so close together, the electrons might not fill the states as regularly as this would suggest. For example, nickel ($Z = 28$) has the $1s$, $2s$, $2p$, $3s$, $3p$, and $4s$ levels filled and 8 electrons in the $3d$ level. (We can write this electron configuration compactly as $1s^2 2s^2 2p^6 3s^2 3p^6 4s^2 3d^8$.) You would expect that copper ($Z = 29$) would have 9 electrons in the $3d$ level, but instead it has 10 electrons in that level and only 1 in the $4s$ level, even though on the diagram the $4s$ level is lower than the $3d$ level. There are many other examples of this kind of disorder among states linked by the tinted boxes.

Even given these small irregularities, the basic pattern is clear: the structure of the periodic table reflects the character of the quantum-mechanical energy levels in a multielectron atom. Note how this is all ultimately a consequence of the Pauli exclusion principle: if the principle were *not* true for electrons, all electrons in all atoms would be in the $1s$ level, and all atoms would thus behave pretty much alike. Thus it is ultimately the Pauli exclusion principle that makes chemistry interesting (and biology possible)!

Exercise Q9X.5

Use figure Q9.3 to predict the electron configuration for oxygen ($Z = 8$) and aluminum ($Z = 13$).

Exercise Q9X.6

How many distinct electron energy eigenstates are there corresponding to $n = 5$?

Q9.3   Selection Rules

Photons are spin-1 bosons

Consider the discussion of the 21-cm hydrogen line in chapter Q8. We stated that this line arises from a transition where the electron in a hydrogen atom in the ground-state energy eigenfunction flips its spin so that it becomes antialigned with the spin of the proton. Let us define our *z* axis to coincide with the proton's spin axis. This means that the *z* component of the electron's angular momentum decreases by $\hbar$. The electron's wavefunction remains in the lowest energy eigenfunction $n = 1$, $\ell = 0$ during this transition, so this means that the atom's total angular momentum changes by $\hbar$ in this transition. This means that since angular momentum is conserved, the photon must carry away a spin angular momentum whose *z* component is $+\hbar$. This means that the photon *must* be a boson and that its spin magnitude quantum number must be *at least* $s = 1$.

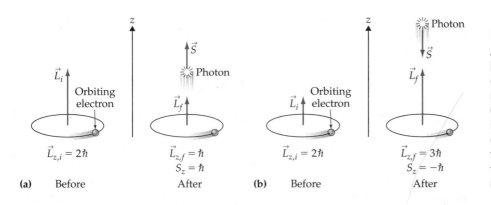

Figure Q9.4
A schematic diagram illustrating how emitting a photon that carries away one unit of angular momentum can lead to either (a) a decrease or (b) an increase in the magnitude of an electron's orbital angular momentum. (Note that the arrow labeled $\vec{S}$ describes the photon's spin angular momentum, not its velocity.)

It turns out that a variety of other puzzles about atomic spectra can be explained if we assume that the photon is a spin-1 quanton. This means that since an atomic transition process that emits a photon must conserve angular momentum, the angular momentum of the electron involved in a transition *must* change to compensate for the angular momentum removed by the photon (and *cannot* change by more than the photon can take away).

It also turns out that spin flips are *much* less probable than changes in orbital angular momentum when both are possible. This is so because a spin flip transition only creates a change in the electron's relatively weak *magnetic* field instead of changes in its electric field; it is *much* less likely in a given time to produce a photon than a transition involving a change in orbital angular momentum would. The spin flip transition that gives rise to the 21-cm hydrogen line is only noticeable at all because there is a lot of hydrogen in the universe, this interstellar hydrogen is so cold it cannot get into any higher energy levels, and it is so rarified that collisions between atoms do not deexcite the atoms before they have a chance to emit the 21-cm photon.

Therefore, transitions between atomic states giving rise to normal spectral lines almost always involve changes in the electron's *orbital* angular momentum. Let the direction of the electron's *initial* angular momentum define the $+z$ direction in space. According to equation Q8.8, the spin angular momentum of the emitted photon, since it has $s = 1$, has three possible projections with respect to that axis: $S_z = +\hbar$, 0, and $-\hbar$. However, there is a relativistic effect that suppresses the middle result in quantons moving at or very near the speed of light, so there are really only two projections that count: $S_z = +\hbar$ and $-\hbar$. This means that the emitted photon can carry away an angular momentum of $\hbar$ either in the *same* direction as the electron's initial orbital angular momentum or *opposite* to that direction. As figure Q9.4 illustrates, this means that the electron's orbital angular momentum must either decrease or increase by $\hbar$ to conserve angular momentum. Therefore virtually all atomic emission lines involve transitions where an orbiting electron's angular momentum magnitude quantum number ℓ changes by 1 unit: $\Delta\ell = \pm 1$.[†]

Photons really do carry genuine angular momentum, as R. A. Beth showed in 1936. A beam of photons was prepared so that their spin angular momentum vectors all pointed parallel to their direction of motion. These photons were allowed to fall on a metal plate that was free to rotate around an axis parallel to the beam. The photons hitting the plate were absorbed and transferred their angular momentum to the plate, which subsequently

An experiment showing that photons carry angular momentum

[†]As usual, this model is somewhat oversimplified, but is useful if not taken too literally.

Conservation of angular
momentum in photon
emission processes

How a selection rule can create
a metastable state

began to rotate. The effect is pretty subtle (see problem Q9R.1), but the plate really does rotate at a rate that is consistent with the model.

The restriction $\Delta\ell = \pm 1$ discussed above is an example of what is called a **selection rule.** Such selection rules are rarely absolute: often a transition that is "forbidden" by a selection rule is simply far less *probable* than one that is "allowed." For example, a spin flip *can* occur (making a $\Delta\ell = 0$ transition like the one that produces the 21-cm hydrogen line possible), or multiple photons *might* be emitted (allowing transitions where $|\Delta\ell| > 1$), but such processes are *much* less probable than the straightforward transitions allowed by the selection rule. For example, the typical lifetime of an electron in an excited state with an "allowed" transition to a lower state is tens of nanoseconds, but the average lifetime of the excited state of hydrogen that leads to the 21-cm line is on the order of 10^7 *years*. We can only see this transition at all because galaxies contain a *lot* of hydrogen that is not emitting anything else near this wavelength.

In many atoms, selection rules can help us identify what are called **metastable** excited states. A metastable excited state is an energy level whose decay to *any* lower level is forbidden by one or more selection rules. For example, imagine that we somehow excite an electron in a hydrogen or helium atom to the 2*s* level. The only lower level that the electron could decay to is the 1*s* level, but since both the 1*s* and the 2*s* levels have the same ℓ, the transition would violate the $\Delta\ell = \pm 1$ selection rule. The 2*s* level in these atoms is thus a metastable state. The electron in the 2*s* level *can* decay via a spin flip transition, but (as discussed above) this is much less likely to occur during a given time interval than an ordinary transition would be. So the electron will remain in the metastable 2*s* state much longer than it would remain in a 2*p* state, for example.

Toys and materials that glow in the dark take advantage of very long-lived metastable states in certain atoms and molecules: once electrons in these materials are energized by exposure to bright light (which kicks them into excited energy levels, some of which decay to the metastable state), the electrons remain in the metastable state for many minutes (instead of tens of nanoseconds as is common in allowed transitions). The glow that you see is the visible photons that the electrons produce when they finally decay from the metastable state.

Another important selection rule is $\Delta n = \pm 1$ for transitions between vibrational energy states in a diatomic molecule (this rule is also related to the angular momentum of the photon). Since the energy states of a simple harmonic oscillator are equally spaced, this might lead us to expect that the spectrum of a diatomic molecule would consist of a *single* line with $E_{ph} = \hbar\omega$. In fact, the spectrum of a diatomic molecule consists of a number of equally spaced lines due to transitions between *rotational* energy levels (see problem Q8A.1).

Exercise Q9X.7

Imagine that an electron is in the 4*s* energy level of hydrogen. What levels might it decay to? What lower levels can it *not* decay to?

Exercise Q9X.8

Imagine that we excite one of the two outer 5*s* electrons of strontium to a higher energy level. Are there any higher levels that could be metastable? (Assume that the vertical placement of levels in figure Q9.3 is exactly correct.)

Q9.4   Stimulated Emission and Lasers

An interesting aspect of the resonance phenomenon in quantum mechanics is that the presence of a passing photon whose frequency is resonant with a certain transition between two energy levels for an electron in a system can trigger the transition from the high to the low level as well as cause a transition from the low level. The periodic disturbance represented by the passing photon essentially shakes a second photon loose from the excited state much as shaking an apple tree at its resonant frequency can dislodge an apple.

Stimulated emission

We call this process of a passing photon of the right frequency facilitating the emission of another photon **stimulated emission.** One of the interesting features of this process is that the second photon's wavefunction not only is identical in wavelength and energy to the wavelength and energy of the passing photon but also oscillates exactly in step with it. We therefore say that light produced by this process is **coherent** as well as being monochromatic. Figure Q9.5 is a cartoon illustration of the process.

The phenomena of stimulated emission and metastable states make **lasers** possible. To construct a laser, you need to do three things:

How to make a laser

1. Find an atom or molecule with a suitable pair of energy states, the upper one of which is a metastable state. (I will call the atom or molecule involved simply an *atom* in what follows.)
2. Figure out a way to create a **population inversion** in a collection of such atoms so that there are more atoms in the metastable state than there are in the lower state of our chosen pair. (The metastable nature of the upper state makes this practical: if we can just get an appreciable number of atoms into that state, they will stay put for a while.)
3. Finally, make sure that any photons moving through this population have plenty of opportunity to stimulate emission from any atoms in the metastable state.

We can usually do the latter by placing mirrors (the front one half-silvered) at opposite ends of the tube containing the atoms, so that any emitted photons bounce back and forth a number of times (and thus stimulate emission from many more atoms) before they finally escape through the half-silvered front mirror. See figure Q9.6 for a schematic diagram for such a laser. (The word *laser* is an acronym for *light amplification by stimulated emission of radiation,* a phrase that nicely summarizes the process.)

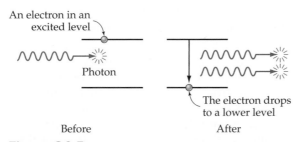

An electron in an
excited level

Photon

The electron drops
to a lower level

Before   After

Figure Q9.5

A simplified illustration of the process of stimulated emission. An incoming photon whose energy is equal to the energy difference between an excited level and a lower level can stimulate an electron to make that transition. This process emits a photon that has exactly the same energy, wavelength, and phase as the incoming photon.

Figure Q9.6

A schematic diagram of a typical laser. The mirrors trap the light for a while so that it can sweep back and forth through the atoms many times, stimulating emission of more photons.

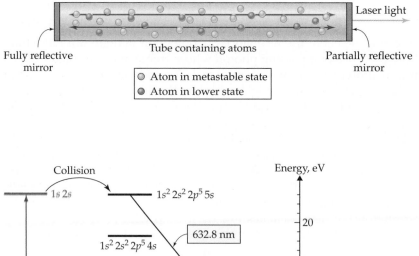

Laser light

Fully reflective mirror

Tube containing atoms

Partially reflective mirror

○ Atom in metastable state
● Atom in lower state

Figure Q9.7

Simplified energy-level diagrams for helium and neon showing the important energy levels for a normal 632.8-nm helium-neon laser.

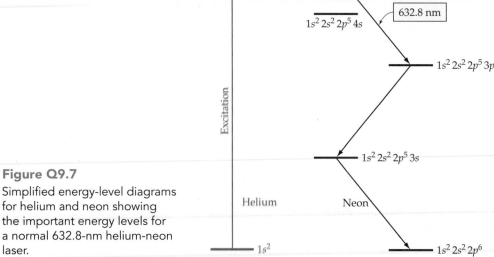

The helium-neon laser

Exercise Q9X.9

Explain why a population inversion is required to create a laser. (*Hint:* A photon with the right energy to stimulate a transition from the upper level to the lower level also has the right energy to be *absorbed* by an electron in the lower level, kicking it to the upper level.)

The very common helium-neon laser illustrates a somewhat less direct way of creating the neccessary population inversion. The gas in a helium-neon laser tube contains about 10 parts helium to 1 part neon at about one-thirtieth of normal atmospheric pressure. An electric current flowing through the tube causes free electrons to collide with the helium atoms. These collisions excite electrons in the helium atoms to a variety of excited energy levels, but most decay very rapidly back to the ground state. In the process of decaying after a collision, though, some helium electrons find their way into the metastable 2s state. As time passes, more and more helium electrons find their way into this state until the system reaches a steady state where a significant fraction of the helium atoms are caught in this metastable state.

This metastable state happens to have an energy of 20.61 eV above the ground level. (An energy-level diagram for this situation is shown in figure Q9.7.) The 5s excited state for an outer electron of neon happens to be 20.66 eV above its normal 2p state. If a neon atom collides with a helium atom in the metastable state, the helium electron can drop back to the ground level and the energy released (plus a little bit from the collision itself) can lift the neon electron to the 5s level. Such an electron then decays to the 3p level (emitting a photon with a wavelength of 632.8 nm) and then to the 3s level and finally back to the ground state.

Since collisions with helium atoms constantly replenish the number of neon atoms with an electron in the 5s state, while the electrons in the 3p level of neon quickly decay to lower levels, there are in the steady state more neon atoms in the 5s level than in the 3p level, creating the population inversion for the transition that produces a photon with a wavelength of 632.8 nm. This means passing photons of this wavelength are more likely to stimulate photon emission from neon atoms in the 5s state than to be absorbed by neon atoms in the 3p state. In a laser tube like that shown in figure Q9.5, this leads to continuous emission of laser light at that wavelength.

The characteristics and utility of laser light

The light produced by any such laser process is monochromatic, coherent (all the photon wavefunctions are exactly in step), and highly directional. All three of these properties make it possible to focus laser light to an unusually small point, making extreme intensities possible. The fact that laser light is monochromatic and coherent also makes it especially easy to display interference effects with laser light. These special properties of laser light make it useful for a wide variety of applications, ranging from reading compact disks and laser surgery (which are possible because laser light can be tightly focused) to creating holograms and making precision position measurements (which are possible because laser light is so monochromatic and coherent).

TWO-MINUTE PROBLEMS

Q9T.1 If all the $n = 3$ energy eigenstates in a multielectron atom are filled with electrons, how many electrons are there in those states?
- A.   2
- B.   4
- C.   8
- D.   16
- E.   18
- F.   Other (specify)

Q9T.2 An electron in a $4d$ energy eigenstate has what orbital angular momentum quantum number?
- A. $\ell = 0$
- B. $\ell = 1$
- C. $\ell = 2$
- D. $\ell = 3$
- E. $\ell = 4$
- F.   Other (specify)

Q9T.3 One of the spectroscopic symbols listed below specifies an energy eigenstate of the hydrogen atom that does not exist. Which is the impossible state?
- A. $6f$
- B. $8p$
- C. $3f$
- D. $7s$
- E. $5d$
- F. $12f$

Q9T.4 Each of the sets of symbols below is *supposed* to describe the ground-state (lowest-energy) electron configuration of some atom. Which one is blatantly wrong? The symbol [Ar] is shorthand for the argon configuration $1s^2 2s^2 2p^6 3s^2 3p^6$.
- A. $[Ar]4s^1$
- B. $[Ar]3d^3 4s^2$
- C. $[Ar]3d^5 4s^1$
- D. $[Ar]3d^{10} 4s^1$
- E. $[Ar]3d^{12} 4s^2 4p^5$
- F.   All are possible!

Q9T.5 Selection rules "forbid" certain transitions between energy levels because those transitions
- A.   Do not effectively create photons for some reason
- B.   Would violate energy conservation
- C.   Do not have the right resonant frequency
- D.   Other (specify)

Q9T.6 An electron in a hydrogen atom is in a $5d$ energy level. Which of the states below could it *not* decay to?
- A. $2p$
- B. $4f$
- C. $3p$
- D. $2s$
- E. $4p$
- F.   All are possible

HOMEWORK PROBLEMS

Basic Skills

Q9B.1 The square of a hydrogen energy eigenfunction with $n = 4$ and $\ell = 2$ has how many radial bumps? How many complete sinusoidal waves *around* the atom does this eigenfunction make?

Q9B.2 The square of a hydrogen energy eigenfunction with $n = 5$ and $\ell = 1$ has how many radial bumps? How many complete sinusoidal waves *around* the atom does this eigenfunction make?

Q9B.3 Use figure Q9.3 to predict the ground-state electron configuration of a manganese atom (Z = number of protons = 25).

Q9B.4 Use figure Q9.3 to predict the ground-state electron configuration of a chlorine atom (Z = number of protons = 17).

Q9B.5 Argue that the selection rule $\Delta n = 1$ implies that to the extent that a diatomic molecule can be treated as a harmonic oscillator, it would only emit photons having a *single* wavelength, no matter what n_{init} might be (if transitions between rotational states were not also a factor).

Synthetic

Q9S.1 Prove that the nth energy level of the hydrogen atom comprises $2n^2$ distinct electron energy eigenstates. (*Hint:* Argue that

$$\text{Number of states} = 2 \sum_{\ell=0}^{n-1} (2\ell + 1) \qquad \text{(Q9.3)}$$

and then evaluate this sum.)

Q9S.2 Use figure Q9.3 to predict the ground-state electron configuration of cadmium ($Z = 48$).

Q9S.3 Imagine that we excite the lone outer $4s$ electron of potassium to a level above that $4s$ state. Using figure Q9.3, identify an excited state that is metastable.

Q9S.4 Imagine that we excite one of the outermost $5p$ electrons of xenon to an excited state above the $5p$ level. Using figure Q9.3, identify an excited state that is metastable.

Q9S.5 Calculate the energy for the third electron in Li, assuming that the inner two electrons completely shield two out of the three proton charges. Compare this to the measured value of -5.4 eV. Can you suggest a reason why the actual value might differ from your calculated result?

Q9S.6 Consider the energy-level pair in the neon atom that creates the laser photons in a helium-neon laser. The upper level of this pair is continually re-populated by collisions with helium atoms. The bottom level in this pair is *not* the ground state. Why is it advantageous for creating a population inversion if the bottom level is not the ground state? Explain carefully.

Q9S.7 Photons carry momentum. Conservation of momentum implies that an atom initially at rest must recoil a bit *backward* when it emits a photon. Also an atom initially at rest that absorbs a photon will afterward move slowly in the direction of the absorbed photon's motion. Considering this, does the photon emitted by an atom during a certain transition from a higher energy level to a lower energy level have *exactly* the same wavelength as the photon that the atom needs to absorb to get it to go from the lower level to the upper level? If so, explain exactly how the momentum issue is irrelevant. If not, explain how you know this and which photon (the emitted or the absorbed photon) has to have the higher energy.

Rich-Context

Q9R.1 Even though we cannot interpret a quanton's spin as arising from actual rotation around an axis, its spin does represent real and measurable angular momentum! Consider a photon which has spin 1. Imagine that we prepare beam of *circularly polarized* visible photons that have their spins aligned with the direction of motion, and we allow these photons to fall on a 100-g circular metal plate 10 cm in diameter whose axis is parallel to the beam and which is free to rotate around that axis. As the photons are absorbed by the metal plate, the plate will have to rotate to conserve angular momentum. If the light has the intensity of sunlight (about 1000 W/m^2), estimate how long it would take to get the plate rotating at a rate of 1 turn per second. If you could create a beam of the same intensity consisting of microwave photons ($\lambda \approx 1$ cm) instead of visible photons, would it accelerate the plate faster or slower? (In spite of how small the effect is, R. A. Beth quantitatively verified its existence for visible light in 1936.)

Advanced

Q9A.1 Consider the following Bohr model of the helium atom.
(a) First, adjust the Bohr model of the hydrogen atom to take account of the increased charge of the nucleus, but completely ignore the electrostatic repulsion between the two electrons. Show that the energy required to remove either one of the two electrons (the *ionization energy* for helium) is 54.4 eV according to this model. This is not very consistent with the observed ionization energy of 24.5 eV.

(b) How might we include the Coulomb repulsion energy of the electrons? Let's assume that the main effect of this repulsion is to keep the electrons as far away from each other as possible, so that the electrons orbit in the same paths that they would follow alone but stay on opposite sides of the nucleus. If this is true, the separation between the electrons will be constant, and we can treat their interaction by adding a constant electrostatic potential energy term to the basic orbital energy predicted by the Bohr model.

Show that the energy required to liberate an electron according to this model is now 27.2 eV, which is much better. *Hint:* The initial state of this atom occurs when the two electrons are in the ground state. The final state occurs when one electron has been removed to infinity and the other atom is *alone* in the lowest energy level. Find the difference in energy between these two states. (This problem was adapted from Taylor and French, *An Introduction to Quantum Physics*, Norton, New York, 1978, pp. 577–578.)

ANSWERS TO EXERCISES

Q9X.1 When $n = 2$, the possible values of n_r and ℓ are [2, 0] and [1, 1], respectively. When $n = 4$, n_r and ℓ may have values [4, 0], [3, 1], [2, 2], and [1, 3], respectively.

Q9X.2 In general, the possible values for L_z range from $-\ell\hbar$ to $+\ell\hbar$ in steps of $\hbar$ and the possible values for S_z range from $-s\hbar$ to $+s\hbar$. For low values of ℓ or s, one can verify this by direct counting: for $\ell = 0$, there is $2 \cdot 0 + 1 = 1$ possible value ($L_z = 0$); for $s = \frac{1}{2}$, there are $2(\frac{1}{2}) + 1 = 2$ values ($S_z = \pm\frac{1}{2}\hbar$); and so on. In general, if ℓ is an integer (as in the case of orbital angular momentum), L_z will have ℓ possible values between $\hbar$ and $\ell\hbar$, the same number between $-\hbar$ and $-\ell\hbar$, and one more value ($L_z = 0$) for a total of $2\ell + 1$ different values. The same is true for integer values of s. If s is $s_0 - \frac{1}{2}$, where s_0 is an integer (as is possible for fermion spins), there will be one fewer value because both the top and bottom values will be shifted in by $\frac{1}{2}\hbar$, which reduces the distance between these extremes by $\hbar$. This means that the number of values is also $(2s_0 + 1) - 1 = 2s_0 = 2(s + \frac{1}{2}) = 2s + 1$ in this case.

Q9X.3 When $n = 4$, ℓ can be 0, 1, 2, or 3 (see exercise Q9X.1), and there are $2\ell + 1$ possible L_z values for each value of ℓ. This means that there is one $\ell = 0$ state, and there are three $\ell = 1$ states, five $\ell = 2$ states, and seven $\ell = 3$ states, for a total of 16. If we count the two possible electron spin orientations as distinct states, then there are 32 possible energy eigenstates in the $n = 4$ set of energy eigenstates.

Q9X.4 The corresponding quantum numbers are $n = 5$ and $\ell = 4$.

Q9X.5 The ground-state electron configuration of oxygen is $1s^2 2s^2 2p^4$, and the ground-state configuration for aluminum is $1s^2 2s^2 2p^6 3s^2 3p^1$.

Q9X.6 Electrons in the $n = 5$ level can have the following values of ℓ: $\ell = 0, 1, 2, 3, 4$. There is one possible L_z value for $\ell = 0$; there are three possible values for $\ell = 1$, five for $\ell = 2$, seven for $\ell = 3$, and nine for $\ell = 4$ ($2\ell + 1$ in each case). There are also two possible spin orientations for each of these cases, so there are a total of $2(1 + 3 + 5 + 7 + 9) = 2(25) = 50$ total distinct electron states in the general category of $n = 5$.

Q9X.7 A hydrogen electron in a 4s level must decay to a lower energy level by conservation of energy. The $\Delta\ell = \pm 1$ selection rule implies that the electron can decay to either the $3p$ or $2p$ level ($\Delta\ell = +1$) but cannot decay to the $3s$, $2s$, or $1s$ levels because $\Delta\ell$ would be zero for such a transition. While the $3d$ level in hydrogen has the same energy as the $3p$ and $3s$ levels, the electron cannot decay to that level, because $\Delta\ell = 2$ for such a transition. This exhausts the possibilities.

Q9X.8 A strontium electron in the $4d$ level *might* be metastable if it is really higher than the $5s$ level in strontium, as indicated in figure Q9.3 (remember that the relative placement of levels connected by gray may vary from atom to atom). An electron in the $4d$ level could not decay back to the $5s$ level without violating the $\Delta\ell = \pm 1$ selection rule, and all other lower energy levels are filled with other strontium electrons. No other excited state shown in figure Q9.3 has any hope of being metastable.

Q9X.9 The same photon that can *stimulate* a transition from the upper to the lower level of our chosen pair of energy levels can also be *absorbed* by an atom in the lower level (causing it to move to the upper level). If there are more atoms in the lower state than in the upper state, then photons moving through the atoms will more likely be absorbed than stimulate the production of new photons, so any initial beam of photons will be *extinguished* by absorption instead of amplified by stimulated emission. The breakeven point occurs when there is the same number of atoms in each level: then the probability that a photon will be absorbed by an encounter with an atom is the same as the probability that it will stimulate emission of a new photon, so the beam strength will remain the same.

Q10

The Schrödinger Equation

Chapter Overview

Introduction

This chapter introduces a two-chapter unit subdivision on the *Schrödinger equation*, which provides a general method for calculating energy eigenfunctions and associated energy values. In this chapter, we will develop the equation from the de Broglie relation and discuss various ways to solve the equation. In chapter Q11, we will develop some intuitive methods for sketching and interpreting energy eigenfunctions.

Section Q10.1:　Generalizing the de Broglie Relation

Our goal in this chapter is to generalize the de Broglie relation $\lambda = h/p$ to cases where the quanton's momentum p is not necessarily constant. Consider the following qualitative example. Imagine that our quanton moves in one dimension in a region where its potential energy $V(x)$ increases as x increases. This means that $K = E - V(x)$ decreases as x increases, so a newtonian particle moving to the right will slow down as it moves to the right, meaning that its momentum decreases. Now the de Broglie relation implies that $\lambda = h/p$. So might a quanton's wavelength λ increase as x increases?

Section Q10.2:　Local Wavelength

To clarify what we really mean by a "changing wavelength," we need to define a function's wavelength at every point along the axis. We can define the **local wavelength** of a function $f(x)$ by noting the following:

1.　The wavelength of a normal wavelike function $f(x)$ is linked to its *curvature*.
2.　The function's second derivative d^2f/dx^2 quantifies that curvature.
3.　However, d^2f/dx^2 of a wavelike function increases with amplitude, while λ does not.
4.　The quantity $(d^2f/dx^2)/f$ is amplitude-independent and so could be linked with λ.
5.　The units work out only if $\lambda^2 \propto f/(d^2f/dx^2)$.
6.　The constant of proportionality must be $-4\pi^2$ for this to work with a sinusoidal wave.

Therefore, a plausible definition of the local wavelength of an arbitrary function $f(x)$ is

$$[\lambda(x)]^2 = \frac{-4\pi^2 f(x)}{d^2f/dx^2} \tag{Q10.8}$$

Section Q10.3:　Finding the Schrödinger Equation

We can use this to find the Schrödinger equation as follows. Imagine that a quanton's energy E is well defined (this means that its wavefunction is an energy eigenfunction). The de Broglie relation and the newtonian definition of kinetic energy imply that $E - V(x) = K = p^2/2m = h^2/2m\lambda^2$. If we plug the definition of local wavelength into this, we get the **Schrödinger equation:**

$$-\frac{\hbar^2}{2m}\frac{d^2\psi_E(x)}{dx^2} + V(x)\psi_E(x) = E\psi_E(x) \qquad \text{(Q10.12)}$$

Purpose: This is the equation that an energy eigenfunction $\psi_E(x)$ must satisfy to be consistent with the generalized de Broglie relation. Given the quanton's mass m and potential energy function $V(x)$, one can use this equation to find both the quanton's possible energy eigenfunctions and their corresponding energies E.

Symbols: $\hbar \equiv h/2\pi$, where h is Planck's constant.

Limitations: This equation applies only to nonrelativistic quantons moving in one dimension. The potential energy function $V(x)$ must be time-independent.

Section Q10.4: Solving the Schrödinger Equation

The Schrödinger equation is a simple **differential equation.** Solving a differential equation is different from solving for a variable, because you are solving for a *function*, not just a single value. A solution to a differential equation solves that equation for *all* values of x. One cannot solve such an equation simply by isolating the function on one side. Indeed, there are *no* general methods for solving differential equations other than (intelligent) trial and error.

For the purposes of this course, it is enough if you can simply recognize when a trial solution works and when it does not. This is generally easiest to see if you put the Schrödinger equation in the form

$$-\frac{\hbar^2}{2m}\frac{d^2\psi_E(x)}{dx^2} - [E - V(x)]\psi_E(x) = 0 \qquad \text{(Q10.14)}$$

A valid solution will make the left side zero for all x.

Section Q10.5: A Numerical Algorithm

We can solve the Schrödinger equation *numerically* if we divide the x axis into a discrete set of points x_i separated by a fixed Δx. We can then represent $\psi_E(x)$ as a list of values $\psi_E(x_i)$. Finding a numerical solution means generating this list of values. The section discusses how in this situation, the Schrödinger equation becomes approximately

$$\psi_E(x_{i+1}) = 2\psi_E(x_i) - \psi_E(x_{i-1}) - \frac{8\pi^2 mc^2 \Delta x^2}{(hc)^2}[E - V(x_i)]\psi_E(x_i) \quad \text{(Q10.29)}$$

If we know $V(x)$, E, m, Δx, and two starting values $\psi_E(x_0)$ and $\psi_E(x_1)$, we can use this equation repeatedly to calculate the value of $\psi_E(x_i)$ at all other points. But how do we choose $\psi_E(x_0)$ and $\psi_E(x_1)$?

Section Q10.6: Using SchroSolver

One of the basic features of any quantum wavefunction is that it must be normalizable, which means that the area under the curve of $|\psi|^2$ must be finite, which in turn means that $\psi_E(x)$ must go to zero as $x \to \pm\infty$.

SchroSolver is a computer program that implements the algorithm discussed in section Q10.5. It initializes $\psi_E(x_0)$ and $\psi_E(x_1)$ by choosing $\psi_E(x_0) = 0$ and $\psi_E(x_1) =$ something small at points x_0 and x_1 well outside the quanton's classically allowed region, where its position probability should be very small. Choosing different values of $\psi_E(x_1)$ simply amounts to multiplying the solution by different overall constants, so it really does not matter much what we choose for $\psi_E(x_1)$. The program then evaluates $\psi_E(x)$ for any energy E you specify, but only certain values of E for a bound quanton produce solutions that go to zero as $x \to +\infty$. This shows why the energy values of a bound system must be quantized.

Q10.1   Generalizing the de Broglie Relation

The goal is to generalize
$p = h/\lambda$ to nonfree quantons

In chapters Q7 through Q9, we have seen that one can learn a great deal about energy eigenfunctions and energy levels by using very simple models. However, in a number of places, I have also had to appeal to results from the full theory of quantum mechanics, which provides methods for calculating energy eigenfunctions in general circumstances.

Our ultimate goal in this chapter is to find a general method of determining the energy eigenfunctions (and corresponding energy levels) of a quanton moving in one dimension in a region where its potential energy $V(x)$ varies with position. As it was in chapter Q7, our special interest here will be *bound* systems, where the quanton is trapped between two barriers in the potential energy function. While a method limited to one-dimensional motion will not be able to handle intrinsically three-dimensional situations, it will deepen our understanding of what three-dimensional eigenfunctions might look like and how they might be found in principle.

How might the shape of such a quanton's energy eigenfunction be linked to the newtonian behavior of an analogous particle moving in response to the same potential energy function? Consider a classical particle with energy E moving subject to the hypothetical potential energy function shown in figure Q10.1a. As the particle moves toward the right, the difference between its potential energy and its total energy decreases, implying that its kinetic energy $K = E - V(x)$ decreases: the particle will thus slow down as it moves to the right. If the particle is nonrelativistic, its kinetic energy is related to its momentum as follows:

$$K = \frac{1}{2}mv^2 = \frac{(mv)^2}{2m} = \frac{p^2}{2m} \quad \Rightarrow \quad p = \sqrt{2mK} \qquad (Q10.1)$$

Therefore, as the object moves toward the right in the potential energy function shown in figure Q10.1, the magnitude of its *momentum* will also decrease.

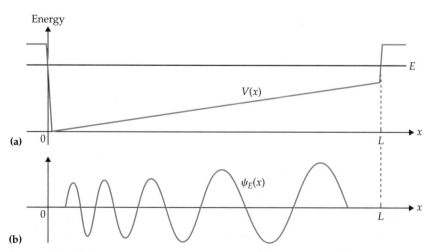

Figure Q10.1

(a) A sketch of a hypothetical ramplike potential energy function. If a classical particle were moving to the right in this allowed region, its kinetic energy K and thus the magnitude of momentum p will decrease. (b) If the de Broglie relation is true even when the quanton in question is subject to a non-constant $V(x)$, its energy eigenfunction might look something like that shown here. Note that the wavelength of this function increases to the right where the classical magnitude of the quanton's momentum would decrease.

Now, we have seen that the de Broglie relationship

$$\lambda = \frac{h}{p} \qquad\qquad (Q10.2)$$

Perhaps wavelength varies as momentum varies

links the effective wavelength of the wavefunction of a beam of quantons with the classical momentum of those quantons. In chapter Q6 we saw that we could use the postulates of quantum mechanics and this equation to explain the results of a low-intensity two-slit interference experiment. Now, although the experimental evidence that we have so far only supports this equation in the case where the quantons are *free* (and thus subject to *no* potential energy function), let's explore the implications of the idea that this relationship might be true *even when the quanton is subject to a nonconstant potential energy function V(x).*

If this hypothesis is true, then the wavefunction of a quanton moving in response to the potential energy function shown in figure Q10.1a might look qualitatively like the function sketched in figure Q10.1b. Since the quanton's classical momentum *decreases* as x gets larger, the wavelength of its quantum wavefunction should *increase* as x gets larger. (Note that because we are assuming that the quanton's energy E has a single, well-defined value, the wavefunction we are sketching here must be an energy eigenfunction.)

Exercise Q10X.1

The graph below shows the potential energy $V(x)$ for a harmonic oscillator. On the empty axis below the plot of $V(x)$, sketch a possible energy eigenfunction of a quanton trapped in the classically allowed region. (Choose a fairly high value of n so that the wavefunction has many bumps.)

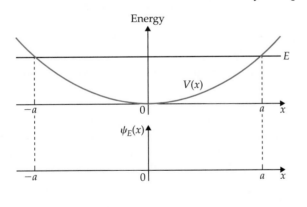

Q10.2   Local Wavelength

It is all well and good to talk qualitatively about the wavelength of a function "increasing" or "decreasing," but how can we *quantify* the wavelength of a wave whose wavelength isn't constant? Specifically, what we'd like to do is to be able to calculate the wavelength of a quanton's wavefunction *at a specific point*, so that we can use the de Broglie relation to link it to the momentum that the quanton would classically have at that same position. But how can we define the "local wavelength" of a wavefunction *at a point*?

How do we define wavelength at a point?

The usual definition of *wavelength* does not help us here. Previously, we have defined the wavelength to be the distance between a feature of the waveform (such as a crest) and the next repetition of that feature. If we apply this to a wave whose wavelength is changing, this prescription does not give

Figure Q10.2
The wavelength of a function seems to be inversely related to the curvature of its wavefunction: the shorter the wavelength, the greater the curvature of the wavefunction.

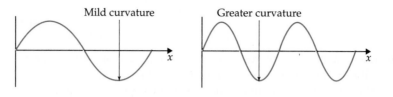

Mild curvature   Greater curvature

Relationship between wavelength and curvature

the wavelength *at a point*, but a kind of the *average* wavelength for the interval between the two crests used. We need something a little bit more precise.

Consider figure Q10.2: perhaps you can see that the wavelength of a wave seems to be inversely related to the curvature of the function $f(x)$ that describes it: other things being equal, the shorter the wavelength of a wave, the greater its curvature of the function. Moreover, as the wavelength goes to infinity, the curvature of the wave function goes to zero.

The curvature of a function $f(x)$ at a point is described by its second derivative d^2f/dx^2 evaluated at that point [the first derivative df/dx describes the *slope* of $f(x)$ at that point, while the second derivative describes the rate at which that slope changes, which describes the function's curvature]. So perhaps we can express the wavelength of a function at a point in terms of some inverse power of its second derivative at that point.

This isn't the whole story, though, because as figure Q10.3 shows, the amplitude of a wavefunction also affects its curvature even if the wavelength remains the same: as the amplitude grows, the curvature also grows. It is at least plausible that the curvature grows in direct proportion to the amplitude: if this is so, then the ratio

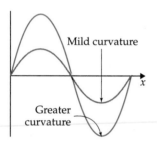

Mild curvature

Greater curvature

Figure Q10.3
Curvature increases with amplitude even as wavelength remains the same.

$$\frac{d^2f/dx^2}{f(x)} = \text{measure of curvature} \qquad (Q10.3)$$

gives a measure of the function's curvature that (like the wavelength) is *independent* of the amplitude of the wave (e.g., if the amplitude of wave doubles, the curvature as expressed by the second derivative also doubles, so the ratio stays the same).

So how is this measure of curvature related to the function's wavelength? We can determine this by examining the equation's *units*. First, note that the second derivative of a function $f(x)$ with respect to x has whatever the units of the function are, divided by two powers of *length* units. How do we know this? The first derivative of the function is defined to be

$$\frac{df}{dx} \equiv \lim_{\Delta x \to 0} \frac{f(x + \Delta x) - f(x)}{\Delta x} \qquad \text{so} \qquad \text{units of} \left[\frac{df}{dx}\right] = \frac{\text{units of } [f(x)]}{[\text{length}]}$$
$$(Q10.4)$$

This clearly has units of whatever the units of $f(x)$ are, divided by one power of length, since Δx has units of length. The second derivative is the derivative of the first derivative, so

$$\text{Units of} \left[\frac{d^2f}{dx^2}\right] = \frac{\text{units of } [df/dx]}{[\text{length}]} = \frac{\text{units of } [f(x)]}{[\text{length}]^2} \qquad (Q10.5)$$

by equation Q10.4. Therefore, the units of the ratio given by equation Q10.3 are

$$\text{Units of} \left[\frac{d^2f/dx^2}{f(x)}\right] = \frac{\text{units of } [f(x)]/[\text{length}]^2}{\text{units of } [f(x)]} = \frac{1}{[\text{length}]^2} \qquad (Q10.6)$$

[It is nice that the unknown units of $f(x)$ so nicely cancel for us!] So if this ratio is to be linked with the wavelength, the *only* way for the units to come out right is for

$$\lambda^2 = C \frac{f(x)}{d^2 f/dx^2} \qquad \text{(Q10.7)}$$

where C is a unitless constant whose magnitude cannot be determined by an argument based on units. In fact, as we will see in exercises Q10X.2 and Q10X.3, in order for us to get the right answer for a *sinusoidal* wave (whose wavelength is defined independently of this formula!), we must have $C = -4\pi^2$:

$$[\lambda(x)]^2 = \frac{-4\pi^2 f(x)}{d^2 f/dx^2} \qquad \text{or} \qquad \frac{1}{[\lambda(x)]^2} = \frac{-d^2 f/dx^2}{4\pi^2 f(x)} \qquad \text{(Q10.8)}$$

Final definition of *local wavelength*

Since we can evaluate both $f(x)$ and its second derivative at any position x, this formula provides a means of calculating the wavelength of a function at a given point x, even if that wavelength *itself* is a function of x. We will take this formula as defining the **local wavelength** $\lambda(x)$ of a wavefunction at the position x.

Exercise Q10X.2

Note that for an ordinary wavelike wavefunction, like those shown in figure Q10.3, when $f(x)$ is *negative*, its second derivative is *positive* and vice versa [e.g., when $f(x)$ is negative, the curve is *concave up*, so the function's slope is becoming more positive]. Explain why this means the constant C in equation Q10.7 must be *negative*.

Exercise Q10X.3

We have seen that the function $f(x)$ describing a sinusoidal wave with wavelength λ can be written

$$f(x) = A \sin\left(\frac{2\pi x}{\lambda}\right) \qquad \text{(Q10.9)}$$

Show that if we plug this into equation Q10.7, the right and left sides of the equation are equal *if and only if* $C = -4\pi^2$.

Exercise Q10X.4

Summarize the crucial steps of the argument in this section in your own words, guided by the questions below.

1. Why do we want to find the wavelength of a wavefunction $f(x)$ at a point?
2. What made us think that the second derivative $d^2 f/dx^2$ might help?
3. Why did we decide that the ratio $(d^2 f/dx^2)/f(x)$ might be even more useful?
4. How did we decide that the ratio should be proportional to $1/\lambda^2$ instead of $1/\lambda$ or λ?
5. How did we determine that constant of proportionality?

Q10.3   Finding the Schrödinger Equation

In section Q10.2, we found a formula which we think might describe the *local wavelength* $\lambda(x)$ of a function. The de Broglie relation suggests that the local wavelength of a quanton's wavefunction at a point might be related to the momentum it would have at that point if it were a newtonian particle. Let's see what happens if we bring these two ideas together mathematically.

Consider a quanton moving in one dimension in a region where it has a potential energy $V(x)$ due to interactions with its surroundings. Imagine that the quanton is nonrelativistic and has well-defined value of energy E (so that its wavefunction is an energy eigenfunction). Then, according to our proposed generalization of the de Broglie relation (see equations Q10.1 and Q10.2),

The Schrödinger equation as a generalized de Broglie relation

$$\frac{h}{p} = \lambda \quad \Rightarrow \quad \frac{h}{\lambda} = p = \sqrt{2mK} \quad \Rightarrow \quad \frac{h^2}{2m}\frac{1}{\lambda^2} = K = E - V(x)$$

$$\text{(Q10.10)}$$

Our hypothesis is that position-dependent wavelength λ given by this equation is the wavelength of the quanton's wavefunction [which is an energy eigenfunction $\psi_E(x)$ in this case]. According to equation Q10.8, the inverse square of the local wavelength of this wavefunction is given by

$$\frac{1}{\lambda^2} = \frac{-d^2\psi_E/dx^2}{4\pi^2\psi_E(x)} \qquad \text{(Q10.11)}$$

By plugging this into equation Q10.10 and rearranging things a bit, you can show that

The Schrödinger equation

$$-\frac{\hbar^2}{2m}\frac{d^2\psi_E(x)}{dx^2} + V(x)\psi_E(x) = E\psi_E(x) \qquad \text{(Q10.12)}$$

Purpose: This is the equation that an energy eigenfunction $\psi_E(x)$ must satisfy to be consistent with the generalized de Broglie relation. Given the quanton's mass m and potential energy function $V(x)$, one can use this equation to find both the quanton's possible energy eigenfunctions and their corresponding energies E.

Symbols: $\hbar \equiv h/2\pi$, where h is Planck's constant.

Limitations: This equation applies only to nonrelativistic quantons moving in one dimension. The potential energy function $V(x)$ must be time-independent.

We call this very important equation the (time-independent) **Schrödinger equation** (the qualification in parentheses distinguishes this equation from another equation, the *time*-dependent *Schrödinger equation*, that we will not study in this text).

Exercise Q10X.5

Fill in the missing algebraic steps that lead from equations Q10.10 and Q10.11 to equation Q10.12.

Q10.4   Solving the Schrödinger Equation

What does this equation tell us? The Schrödinger equation, I have attempted to argue, represents a natural extension of the de Broglie relation that links the wavelength of a quanton's wavefunction to the momentum that the quanton would have if it were a newtonian particle. The purpose of the

Schrödinger equation is this: given the potential energy function $V(x)$ and the quanton's definite energy E, we can solve for the quanton's energy eigenfunction $\psi_E(x)$ corresponding to that energy level E.

The Schrödinger equation is an example of a simple **differential equation**. Solving a differential equation is a bit different from solving an ordinary algebraic equation for an unknown, for three reasons. (1) The solution of a differential equation is a *function* [in this case, the energy eigenfunction $\psi_E(x)$], *not* a variable. (2) A valid solution makes the right side of the equation equal to the left side at *all* values of x, not just a particular value. (3) We *cannot* solve such an equation by simply isolating the function on one side, the way we would a variable. If we do this with the Schrödinger equation, for example, we get

Solving differential equations is different from solving algebraic equations

$$\psi_E(x) = \frac{1}{E - V(x)}\left(\frac{-\hbar^2}{2m}\frac{d^2\psi_E}{dx^2}\right) \qquad \text{(Q10.13)}$$

We cannot use this to find $\psi_E(x)$, though, because to determine the right side of the equation, we have to know what $\psi_E(x)$ is. Unfortunately, this function is what we are trying to *find*.

Indeed, unlike the case of an algebraic equation, there is no universal technique for solving a general differential equation other than trial and error! A mathematics course in differential equations fundamentally teaches you to be an intelligent guesser as you look for solutions.

In this text, I will not ask you to solve such an equation, by guessing or any other means. It will be enough if you can simply distinguish between a wavefunction that satisfies the equation and one that does not. In checking for a valid solution, you may find it helpful to put the Schrödinger equation in the following form:

It is enough here to recognize a valid solution

$$\frac{-\hbar^2}{2m}\frac{d^2\psi_E}{dx^2} - [E - V(x)]\psi_E(x) = 0 \qquad \text{(Q10.14)}$$

If a given guess for $\psi_E(x)$ makes the left side of this equation equal to zero *for all values of x*, then $\psi_E(x)$ is a solution to the Schrödinger equation.

Example Q10.1   A Bad Guess

Problem   Consider a quanton with energy E that is moving in a region where its potential energy function $V(x)$ is the harmonic oscillator potential energy function $V(x) = \frac{1}{2}m\omega^2x^2$ (where m and ω are constants). In this region, assume that $\psi_E(x) = A\sin kx$ (where A and k are constant values to be determined). Is this hypothetical wavefunction a solution to the Schrödinger equation?

Solution   The first and second derivatives of the trial function are

$$\frac{d\psi_E}{dx} = \frac{d}{dx}(A\sin kx) = A\cos kx\frac{d}{dx}(kx) = kA\cos kx \qquad \text{(Q10.15)}$$

$$\frac{d^2\psi_E}{dx^2} = \frac{d}{dx}\left(\frac{d\psi_E}{dx}\right) = kA\frac{d}{dx}(\cos kx) = kA(-k\sin kx) = -k^2A\sin kx$$

$$\text{(Q10.16)}$$

Plugging equation Q10.16 into the Schrödinger equation (Q10.14), we get

$$+\frac{\hbar^2k^2}{2m}A\sin kx - \left(E - \tfrac{1}{2}m\omega^2x^2\right)A\sin kx = 0 \qquad \text{(?)} \qquad \text{(Q10.17)}$$

Dividing through by $A \sin kx$, we find that

$$+\frac{\hbar^2 k^2}{2m} - E + \tfrac{1}{2}m\omega^2 x^2 = 0 \qquad \text{(?)} \qquad \text{(Q10.18)}$$

Evaluation   Can this be true for all x? Absolutely not! The value of the left side of this equation clearly varies as x varies; and no matter what the values of k, m, and ω might be, the left side of this equation simply cannot be zero for all x. Therefore our trial solution $\psi_E(x) = A \sin kx$ is *not* a valid solution to the Schrödinger equation for this particular potential energy function.

Example Q10.2   A Better Guess

Problem   Let's make another guess. Perhaps $\psi_E(x) = Ae^{-bx^2}$, where A and b are constants to be determined. See if this function solves the Schrödinger equation for $V(x) = \tfrac{1}{2}m\omega^2 x^2$ (where m and ω are known constants).

Solution   The first derivative of the trial function is

$$\frac{d\psi_E}{dx} = A\frac{d}{dx}\left(e^{-bx^2}\right) = Ae^{-bx^2}\frac{d}{dx}(-bx^2) = -2bxAe^{-bx^2} \quad \text{(Q10.19)}$$

according to the chain rule. Using the chain and product rules, we find the *second* derivative of the trial function to be

$$\frac{d^2\psi_E}{dx^2} = \frac{d}{dx}\left(\frac{d\psi_E}{dx}\right) = -2bA\frac{d}{dx}\left(xe^{-bx^2}\right)$$

$$= -2bA\left[\frac{dx}{dx}e^{-bx^2} + x\frac{de^{-bx^2}}{dx}\right] = -2bA\left[1 \cdot e^{-bx^2} + xe^{-bx^2}\frac{d(-bx^2)}{dx}\right]$$

$$= -2bAe^{-bx^2}[1 - 2bx^2] = 2b[2bx^2 - 1]Ae^{-bx^2} \qquad \text{(Q10.20)}$$

(Whew!) Plugging this into Schrödinger's equation (equation Q10.14), we get

$$\frac{\hbar^2 b}{m}[1 - 2bx^2]Ae^{-bx^2} - \left[E - \tfrac{1}{2}m\omega^2 x^2\right]Ae^{-bx^2} = 0 \quad \text{(?)} \quad \text{(Q10.21)}$$

Dividing through by Ae^{-bx^2} and regrouping terms, we find that

$$\left[\frac{\hbar^2 b}{m} - E\right] + \left[\frac{m\omega^2}{2} - \frac{2\hbar^2 b^2}{m}\right]x^2 = 0 \quad \text{(?)} \qquad \text{(Q10.22)}$$

At first glance, it looks as if we have the same problem here as before: the left side clearly varies as x varies, so it cannot be zero for all x. But wait! What if the two quantities in square brackets were zero? Then the left side would be zero no matter *what* the value of x was! Making the second bracket zero means that we have to choose the value of our "to be determined" constant b so that

$$0 = \frac{m\omega^2}{2} - \frac{2\hbar^2 b^2}{m} \quad \Rightarrow \quad 0 = \frac{m^2\omega^2}{4\hbar^2} - b^2 \quad \Rightarrow \quad b = \frac{m\omega}{2\hbar} \quad \text{(Q10.23)}$$

Making the first bracket zero means that we have to have

$$0 = \frac{\hbar^2 b}{m} - E \quad \Rightarrow \quad E = \frac{\hbar^2 b}{m} = \frac{\hbar^2 m\omega}{m \cdot 2\hbar} = \frac{\hbar\omega}{2} \qquad \text{(Q10.24)}$$

So this trial function *does* satisfy the Schrödinger equation *if* b and E have these values.

Evaluation   We have just found, then, that $\psi_E(x) = Ae^{-bx^2}$ (with $b = m\omega/2\hbar$) *is* an energy eigenfunction corresponding to the energy level $E = \frac{1}{2}\hbar\omega$! Note that equation Q7.16 implies that $E = \frac{1}{2}\hbar\omega$ is in fact the ground-state energy of the harmonic oscillator, which is useful corroborating evidence that we have done things right. The function we have found is therefore the energy eigenfunction describing the oscillator's ground state.

Exercise Q10X.6

(Important!) Note that the calculation in example Q10.2 did *not* determine the overall amplitude A of our trial function. Show that if $\psi_E(x)$ is a solution to the Schrödinger equation, then $A\psi_E(x)$ is *also* a solution, where A is a constant. This means that the Schrödinger equation only determines $\psi_E(x)$ up to an overall constant multiple. (We can determine A, if necessary, by choosing the value that makes the wavefunction normalized.)

Exercise Q10X.7

Is the wavefunction $\psi_E(x) = Ae^{-bx}$ another solution to the Schrödinger equation when $V(x) = \frac{1}{2}m\omega^2x^2$? If so, what are the corresponding values of E and b that make the solution work?

Exercise Q10X.8

Is the wavefunction $\psi_E(x) = Axe^{-bx^2}$ another solution to the Schrödinger equation when $V(x) = \frac{1}{2}m\omega^2x^2$? If so, what are the corresponding values of E and b that make the solution work? [To save you time, here is the second derivative of this function: $d^2\psi_E/dx^2 = (4b^2x^2 - 6b)Axe^{-bx^2}$.]

Q10.5   A Numerical Algorithm

As you can see, even just checking whether a function satisfies or doesn't satisfy the Schrödinger equation can be a lot of mathematical work. Wouldn't it be nice to have a machine such that we can just supply a potential energy function $V(x)$ and an energy E and it would grind out the desired solution $\psi_E(x)$?

We can in fact construct just a "Schrödinger solver" by using a computer (or even a programmable calculator). In what follows, I will describe the method in general terms (as if you were going in fact to do the calculations), but it is easy to automate the calculations by using a programmable calculator, a spreadsheet program, or a computer programming language.

This method will generate a *numerical* solution to the Schrödinger equation. That is, rather than yielding a *mathematical* function such as $\psi_E(x) = Ae^{-bx}$, it will yield a *list* of the function's numerical values $\psi_E(x_i)$ at selected points x_i along the x axis. If we graph these values as a function of x, we can get a sense of what our solution $\psi_E(x)$ looks like. While it would be nicer to have a mathematical function, this numerical list and graph are often the best we can do. (For example, in three-dimensional situations involving complicated

What it means to solve an equation numerically

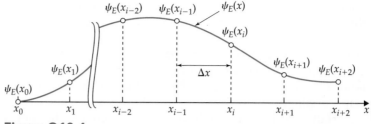

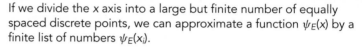

Figure Q10.4
If we divide the x axis into a large but finite number of equally spaced discrete points, we can approximate a function $\psi_E(x)$ by a finite list of numbers $\psi_E(x_i)$.

atoms or molecules, *no one* can solve the Schrödinger equation mathematically: numerical solutions using methods like the one I will describe are the *only* solutions anyone can produce.)

The first step in calculating a numerical solution to an equation like the Schrödinger equation is to divide the x axis into a finite number of points, each separated from its neighbors by some fixed distance Δx. These points will be the positions x_i where we will evaluate the wavefunction (see figure Q10.4).

The next step is to figure out how we can approximate the second derivative $d^2\psi_E(x)/dx^2$ of the wavefunction using only the values of $\psi_E(x)$ at the discrete points x_i. Consider a *specific* point x that is one of these discrete points x_i. As we have discussed before, the ratio

A numerical version of the Schrödinger equation

$$\frac{\psi_E(x + \Delta x) - \psi_E(x)}{\Delta x} \approx \frac{d\psi_E}{dx} \text{ halfway between } x + \Delta x \text{ and } x \quad \text{(Q10.25)}$$

We might call this halfway point $x + \frac{1}{2}\Delta x$. Similarly, the ratio

$$\frac{\psi_E(x) - \psi_E(x - \Delta x)}{\Delta x} \approx \frac{d\psi_E}{dx} \text{ halfway between } x \text{ and } x - \Delta x \quad \text{(Q10.26)}$$

We might call this halfway point $x - \frac{1}{2}\Delta x$. Now, the *second* derivative is defined to be the derivative of the *first* derivative. Approximating this derivative by a ratio as above and using equations Q10.25 and Q10.26, you can show that

$$\frac{\psi_E(x + \Delta x) - 2\psi_E(x) + \psi_E(x - \Delta x)}{\Delta x^2} \approx \frac{d^2\psi_E}{dx^2} \text{ at point } x \quad \text{(Q10.27)}$$

Exercise Q10X.9

Verify equation Q10.27.

If you plug this into the Schrödinger equation and solve for $\psi_E(x + \Delta x)$, you will get

$$\psi_E(x + \Delta x) = 2\psi_E(x) - \psi_E(x - \Delta x) - \frac{2m\Delta x^2}{\hbar^2}[E - V(x)]\psi_E(x) \quad \text{(Q10.28)}$$

Exercise Q10X.10

Verify equation Q10.28.

Note that if x is one of our set of discrete points x_i, then $x + \Delta x$ is the *next* point in the set and $x - \Delta x$ is the *previous* point in the set. Therefore, we might write the equation above as follows:

$$\psi_E(x_{i+1}) = 2\psi_E(x_i) - \psi_E(x_{i-1}) - \frac{8\pi^2 mc^2 \Delta x^2}{(hc)^2}[E - V(x_i)]\psi_E(x_i) \quad (Q10.29)$$

where x_i here is the ith position in the set of points, $x_{i+1} = x_i + \Delta x$ is the next position in the set, and so on. In this equation, I have also replaced $\hbar$ by $h/2\pi$ and multiplied the top and bottom of the constant factor by c^2 to express that factor in terms of hc and mc^2, which have more convenient values than $\hbar$ and m in most applications. (You can check that this constant factor has units of inverse energy.)

Now look at what this equation is saying. If we know the potential energy function $V(x)$ and values of $\hbar$ and m and we specify the values of E and Δx, then given the two previous values on our list of values for $\psi_E(x_i)$, we can find the next value on the list. Given values for $\psi_E(x_0)$ and $\psi_E(x_1)$, then, we can generate the whole list of values $\psi_E(x_i)$. This amounts to finding the eigenfunction $\psi_E(x)$ that solves the Schrödinger equation for the given value of E!

I want you to notice three important things about this method. First, to get good accuracy, Δx should be small. But the smaller Δx is, the more points you need to cover a given stretch of the x axis. To get even reasonable accuracy in most situations, you need at least 100 points. Doing this many calculations by hand, you can see, would become *very* tedious. This is why this is a *computer* method: it is really practical only if you can automate the calculation by using a programmable calculator, a computer program, or a spreadsheet.

Second, it is not obvious with this procedure how we are to choose the starting values $\psi_E(x_0)$ and $\psi_E(x_1)$. It should be clear that this equation will generate different lists of values (i.e., different solutions) for different choices of these starting values. What are the *right* choices for these values? Finally, given $\psi_E(x_0)$ and $\psi_E(x_1)$, equation Q10.29 will always generate a solution for *any* value of E. But didn't we find in chapter Q7 that energy must be quantized? This means that there should not *be* any eigenfunctions for E except for certain discrete values!

The answers to these questions are related, as we will see now.

Three important points about the method

Q10.6   Using SchroSolver

The computer program SchroSolver (which is available in both Macintosh and Windows formats on the *Six Ideas* website) uses the algorithm described in section Q10.5 to compute solutions to the one-dimensional Schrödinger equation for an electron responding to a variety of possible potential energy functions.

Figure Q10.5 shows a screen shot of the program's window after it has calculated a trial solution to the Schrödinger equation for a "well" potential energy function. This potential energy function is like that for a box (see chapter Q7) except that the barriers on either side of the allowed region are not infinitely high. SchroSolver divides the horizontal x axis shown in the graph into 1000 discrete points, makes an appropriate choice for $\psi_E(x_0)$ and $\psi_E(x_1)$ for the two leftmost points, and then uses equation Q10.29 and the value of E specified in the box on the lower right to grind out the values of $\psi_E(x)$ at the remaining points along the x axis.

Now, one of the basic features of any quantum wavefunction, as we discussed in chapter Q6, is that *the total area under the wavefunction's absolute square must be finite* so that it is possible to normalize the wavefunction. Any wavefunction that does *not* satisfy this constraint is simply *not* a physically

Introduction to SchroSolver

How to choose $\psi_E(x_0)$ and $\psi_E(x_1)$

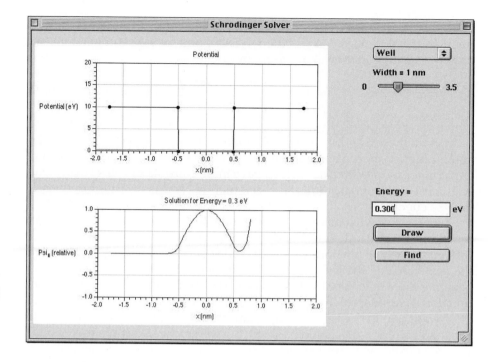

Figure Q10.5

This figure shows a screen shot of SchroSolver after it has generated a solution for $\psi_E(x)$ for the well potential energy function shown in the top graph.

reasonable quantum wavefunction. In the case of bound systems, this constraint implies that the wavefunction *must* go to zero as x goes to $\pm\infty$.

The program ensures that this is the case on the left by choosing $\psi_E(x_0)$ to be exactly zero. This will not be exactly correct, because most valid solutions to the Schrödinger equation will approach zero asymptotically as $x \to -\infty$ [see, e.g., the solution $\psi_E(x) = Ae^{-bx^2}$ that we found in example Q10.2 for the ground state of the harmonic oscillator]. However, if x_0 is sufficiently deep in the classically forbidden region, the probability of finding the quanton at x_0 will be so small as to be essentially zero, so this is a reasonable approximation.

How can we choose a value of $\psi_E(x_1)$? A logical choice, considering the discussion in the last paragraph, might be to choose $\psi_E(x_1) = 0$ as well. If we do this, though, then equation Q10.29 implies that *all* $\psi_E(x_i)$ will be zero, which is not a useful result! To avoid this problem, the program simply chooses $\psi_E(x_1)$ to be a very small positive value. It turns out that making different choices for this value simply generates solutions that have exactly the same shape but different amplitudes, in other words, solutions that differ by only a constant multiple. However, it is the *shape* of $\psi_E(x)$ that will be most interesting to us in what follows. Moreover, we saw in exercise Q10X.6 that the Schrödinger equation only allows us to determine the eigenfunction up to a constant multiple anyway, so our choice of $\psi_E(x_1)$ is arbitrary. (If necessary, we can determine the constant multiplying factor later by finding the value that makes the function normalized.)

These choices for $\psi_E(x_0)$ and $\psi_E(x_1)$ ensure that the function goes to zero on the left, but figure Q10.5 shows that the function appears to be going off to infinity on the right. (Note that the program cuts off the curve at the point where it is about to go off the graph.) If the function *does* go to infinity, this would violate the constraint that the area under $|\psi_E|^2$ be finite. Moreover, such a function would imply an infinite probability of finding the quanton in a classically forbidden region, something that is clearly absurd. Therefore, there is still something wrong with the solution shown in figure Q10.5.

Figure Q10.6 shows that this is so because the chosen value of the *energy* is not correct. If the energy value we type in is too high, the function goes off

Choosing the right energy value ensures good behavior on the right side

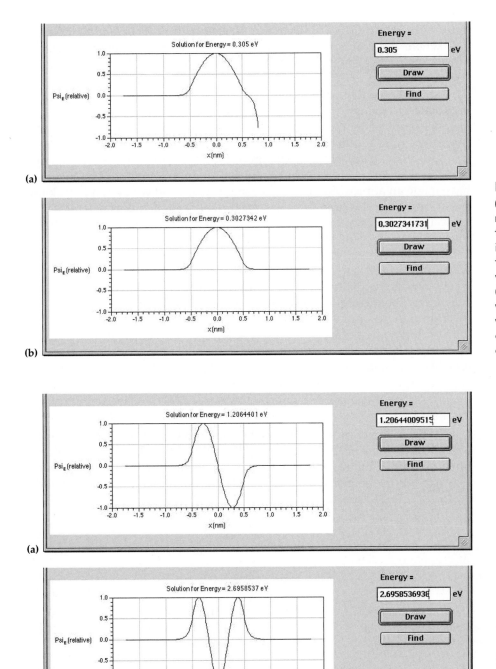

(a)

(b)

Figure Q10.6

(a) Choosing a value of *E* that is not correct leads to solutions that go off to positive or negative infinity as *x* becomes large. Such functions are *not* physically valid quantum wavefunctions. (b) However, with some effort, we can find a specific energy where the function goes to zero on both sides. This is a valid energy eigenfunction.

(a)

(b)

Figure Q10.7

Two more energy eigenfunctions (and their associated energy values) for the well potential energy function.

to infinity on one side of the *x* axis (see figure Q10.5). If we type in a value that is too low, the function goes off to infinity on the other side (see figure Q10.6a). By trial and error, though, we can find the value of the energy that is just right to cause the function to go to zero on the right as well as on the left (see figure Q10.6b).

By trial and error, one can also find other energies where the wavefunction is well behaved (see figure Q10.7). We see, therefore, that solutions to Schrödinger's equation are valid quantum wavefunctions only for certain *discrete* energy values: *this* is why energies must be quantized. Moreover, figures Q10.6 and Q10.7 indicate that such energies correspond to the energies where the corresponding wavefunctions have an integer number of bumps (half-wavelengths) between the classically forbidden regions, just as we found in chapter Q7.

I should note that even though we must supply the program with an energy stated to many decimal places to get the behavior we need, the precise energies required by the *program* may not necessarily be exactly equal to the precise energies required by the equivalent system in the real world. Equation Q10.29 is an *approximate* version of the Schrödinger equation that gets better and better as $\Delta x \to 0$, but is always an approximation. Moreover, we have made an approximation in choosing $\psi_E(x_0)$, and we do not even know the values of $\hbar$ and m to more than a handful of decimal places. Therefore, the answers that the program provides for the energy levels may look *precise*, but they are probably only *accurate* to about $\pm 1\%$.

The bottom line is that the condition that we must be able to normalize a wavefunction not only helps us choose appropriate values for $\psi_E(x_0)$ and $\psi_E(x_1)$ but also explains why the energies of a bound system are quantized. This solves the puzzles raised at the end of section Q10.5.

I strongly recommend that you download the SchroSolver program and play with it some before you read chapter Q11. Playing with the program will help you develop an intuitive feel for what the eigenfunctions should look like, and will help make the issues discussed in chapter Q11 clearer.

Exercise Q10X.11

Use the SchroSolver program to find the remaining bound-state energy levels of the 1-nm-wide well. State your results to three decimal places. (There are a finite number of bound energy levels for the well, since if the quanton's energy exceeds the height of the well's walls, the quanton is no longer bound.)

TWO-MINUTE PROBLEMS

Q10T.1 A quanton is in an energy eigenstate where its potential energy decreases to the right. As we go to the right, the wavelength of its energy eigenfunction
   A.   Increases
   B.   Remains the same
   C.   Decreases
   D.   Is undefined

Q10T.2 A quanton is moving in a classically allowed region. If the quanton had classically at point x_A twice the kinetic energy that it would have at point x_B, how would the wavelengths of an energy eigenfunction of the quanton compare at points A and B?
   A. $\lambda_A = 4\lambda_B$
   B. $\lambda_A = 2\lambda_B$
   C. $\lambda_A = \sqrt{2}\lambda_B$
   D. $\lambda_B = \sqrt{2}\lambda_A$
   E. $\lambda_B = 2\lambda_A$
   F. $\lambda_B = 4\lambda_A$
   T.   Other (specify)

Q10T.3 We can define the local wavelength of *any* continuous and doubly differentiable function, no matter how it is shaped, true (T) or false (F)?

Q10T.4 We know that the Schrödinger equation is specifically an equation for energy *eigenfunctions* because
   A.   Quantons quickly settle down into energy eigenstates
   B.   A bound quanton can only exist in an energy eigenstate
   C.   We derived it *assuming* that E had a definite fixed value
   D.   It is a generalization of the de Broglie relation
   E.   Other reason (specify)

Q10T.5 We can use the computer method to generate a (numerical) solution to Schrödinger's equation for a bound quanton with *any* energy E, T or F?

HOMEWORK PROBLEMS

Basic Skills

Q10B.1 Show that if $\psi_E(x) = A\cos bx$ or $A\sin bx$, then $d^2\psi_E/dx^2 = -b^2\psi_E(x)$.

Q10B.2 Show that if $\psi_E(x) = Ae^{\pm bx}$, then $d^2\psi_E/dx^2 = +b^2\psi_E(x)$.

Q10B.3 Show that if $\psi_E(x) = Axe^{-bx}$, then $d^2\psi_E/dx^2 = (b^2x - 2b)Ae^{-bx}$.

Q10B.4 Show that if $\psi_E(x) = Ax^2e^{-bx}$, then $d^2\psi_E/dx^2 = (b^2x^2 - 4bx + 2)Ae^{-bx}$.

Q10B.5 Use SchroSolver to find the possible bound-state energy levels of a well potential energy function whose classically allowed region is 0.75 nm wide. Specify the energy levels to three decimal places.

Q10B.6 Run SchroSolver and select "Ramp" from the pull-down menu in the upper right corner of the SchroSolver window. Then find the possible bound-state energy levels for this potential energy function. Specify the energy levels to three decimal places.

Q10B.7 Run SchroSolver and select "Oscillator" from the pull-down menu in the upper right corner of the SchroSolver window.
  (a) Find the first five possible bound-state energy levels for this potential energy function. Specify the energy levels to three decimal places.
  (b) Equation Q7.16 implies that $E_n = 2E_0(n + \frac{1}{2})$, where E_0 is the oscillator's energy when $n = 0$. Does this formula match your results pretty well?

Synthetic

Q10S.1 A quanton is moving in a region where $V(x) = V_0$ (a constant). Is the trial wavefunction $\psi_E(x) = A\cos bx$ a solution to the Schrödinger equation for $E > V_0$? If so, find b in terms of E and V_0. (*Hint:* See problem Q10B.1.)

Q10S.2 A quanton is moving in a region where $V(x) = V_0$ (a constant). Is the trial wavefunction $\psi_E(x) = A\cos bx$ a solution to the Schrödinger equation for $E < V_0$? If so, find b in terms of E and V_0. (*Hint:* See problem Q10B.1.)

Q10S.3 A quanton is moving in a region where $V(x) = V_0$ (a constant). Is the trial wavefunction $\psi_E(x) = Ae^{\pm bx}$ a solution to the Schrödinger equation for $E < V_0$? If so, find b in terms of E and V_0. (*Hint:* See problem Q10B.2.)

Q10S.4 A quanton is moving in a region where $V(x) = -ke^2/x$, where ke^2 is a constant. Is the trial wavefunction $-\psi_E(x) = Ae^{\pm bx}/x$ a solution to the Schrödinger equation in this case? If so, find b and E in terms of ke^2 and other constants. (*Hint:* See problem Q10B.2.)

Q10S.5 A quanton is moving in a region where $V(x) = -ke^2/x$, where ke^2 is a constant. Is the trial wavefunction $\psi_E(x) = Axe^{-bx}$ a solution to the Schrödinger equation in this case? If so, find b and E in terms of ke^2 and other constants. (*Hint:* See problem Q10B.3.)

Q10S.6 A quanton is moving in a region where $V(x) = -ke^2/x$, where ke^2 is a constant. Is the trial wavefunction $\psi_E(x) = Ax^2e^{-bx}$ a solution to the Schrödinger equation in this case? If so, find b and E in terms of ke^2 and other constants. (*Hint:* See problem Q10B.4.)

Q10S.7 Consider an electron in a region where its potential energy is $V(x) = \frac{1}{2}m\omega^2 x^2$, where $\hbar\omega = 1$ eV. Taking $\Delta x = 0.02$ nm, $E = 0.5$ eV, $\psi_E(x_0) = 1.00$, and $\psi_E(x_1) = 1.00$ (where $x_0 = 0$ and $x_1 = \Delta x = 0.02$ nm), use equation Q10.29 to find $\psi_E(x_i)$ for points x_2 through x_6.

Q10S.8 Consider the function $f(x) = \cos(b\sqrt{x})$, where $b = 6.28$ cm$^{-1/2}$.
  (a) Argue that this wave has crests at $x = 0$, 1 cm, 4 cm, 9 cm, and so on.
  (b) Sketch a graph of this function, and argue that it looks like a wave whose wavelength is increasing as x increases.
  (c) *Estimate* the local wavelength of this function at $x = 4.0$ cm by *averaging* the distances to adjacent crests on either side.
  (d) *Compute* the local wavelength of this function at $x = 4.0$ cm, using equation Q10.8, and compare with your answer for part (c).

Q10S.9 Consider the function $f(x) = \cos(bx^2)$, where $b = 0.0628$ cm^{-2}.
  (a) Argue that this wave has crests at $x = 0$, 10 cm, 14.1 cm, 17.3 cm, 20 cm, 22.4 cm, and so on.
  (b) Sketch a graph of this function, and argue that it looks like a wave whose wavelength is decreasing as x increases.
  (c) *Estimate* the local wavelength of this function at $x = 20$ cm by *averaging* the distances to adjacent crests on either side.
  (d) *Compute* the local wavelength of this function at $x = 20$ cm, using equation Q10.8, and compare with your answer for part (c).

Rich-Context

Q10R.1 Run SchroSolver and select "Ramp" from the pull-down menu in the upper right corner of the SchroSolver window.

(a) Find the possible bound-state energy levels for this potential energy function. (Specify the energy levels to three decimal places.)

(b) Carefully draw graphs of the first, third, and fifth energy eigenfunctions on your paper. (I would like you to *draw* these graphs by hand instead of simply printing them out, so that you pay full attention to all the features of these curves. For example, how does the amplitude change as x increases? How does the local wavelength change? What is the difference between the right side of the curve and the left side?)

(c) Use ideas in this chapter to explain, in your own words, *why* the local wavelength changes as x increases for the energy eigenfunctions of this potential energy.

(d) For the box potential energy function, $E_n = E_1 n^2$, where E_1 is the energy of the ground state. (1) Show that this formula does *not* accurately model the energy levels of the ramp. (2) What is the most important reason why it does not, do you think?

(e) Find a reasonably simple formula of a similar form (E_1 times some function of n) that better predicts the upper-level energies from the ground-state energy. Your formula does not have to be perfect (there are probably a number of reasonable answers).

(f) Then adjust the height of the ramp to 15 eV. Does your formula still work in this case? Comment on the results and/or improve your formula so that it works better in both cases.

Advanced

Q10A.1 Set up a spreadsheet to compute energy eigenfunctions for an electron in a region where its potential energy is $V(x) = \frac{1}{2}m\omega^2 x^2$, where $\hbar\omega = 1$ eV. Divide the x axis in the range of ± 2.0 nm into at least 100 points, and keep E below 10 eV. Set $\psi_E(x_0) = 0$ (where $x_0 = -2.0$ nm) and $\psi_E(x_1) = 0.01$, and generate solutions for $\psi_E(x_i)$ for several different values of E (including $E = 0.5$ eV and $E = 1.5$ eV). Graph these solutions if at all possible.

ANSWERS TO EXERCISES

Q10X.1 The wavefunction should have a *shorter* wavelength in the center (where the particle would classically be moving faster) and a longer wavelength on the sides.

Q10X.2 Since d^2f/dx^2 and $f(x)$ have opposite signs for a normal wave, and since $[\lambda(x)]^2$ must be positive, C has to be negative for the signs to work out.

Q10X.3 The derivatives of $f(x) = A\sin(2\pi x/\lambda)$ are

$$\frac{d}{dx} A\sin\frac{2\pi x}{\lambda} = A\left(\cos\frac{2\pi x}{\lambda}\right)\frac{2\pi}{\lambda} \qquad (Q10.30)$$

$$\frac{d^2f}{dx^2} = \frac{d}{dx} A\left(\cos\frac{2\pi x}{\lambda}\right)\frac{2\pi}{\lambda} = -A\left(\sin\frac{2\pi x}{\lambda}\right)\frac{4\pi^2}{\lambda^2} \qquad (Q10.31)$$

Therefore, in this case we have

$$\lambda^2 = C\frac{f(x)}{d^2f/dx^2} = C\frac{A\sin(2\pi x/\lambda)}{(4\pi^2/\lambda^2)A\sin(2\pi x/\lambda)}$$

$$= C\frac{-\lambda^2}{4\pi^2} \quad \Rightarrow \quad C = -4\pi^2 \qquad (Q10.32)$$

Q10X.4 You have to do this mostly for yourself. However, here are short answers to the questions posed:

1. Because our hypothesis is that the wavefunction's wavelength at a point might be linked through de Broglie's relation to the momentum that classical particle would have at that point.

2. Because the wavelength of a wave is linked to its curvature, which is linked to its second derivative.

3. Because this, like the wavelength, is an amplitude-independent measure of a wave's curvature.

4. Because this is the only way the units work.

5. By requiring that the definition yield the correct result for a sinusoidal wave.

Q10X.5 Plugging equation Q10.11 into Q10.10 yields

$$\frac{h^2}{2m}\frac{-d^2\psi_E/dx^2}{4\pi^2\psi_E(x)} = E - V(x) \qquad (Q10.33)$$

Since $\hbar = h/2\pi$, we can absorb $4\pi^2$ into the h^2 by making it $\hbar^2$. Doing this and adding $V(x)$ to both sides yield

$$-\frac{\hbar^2}{2m}\frac{d^2\psi_E/dx^2}{\psi_E(x)} + V(x) = E \qquad (Q10.34)$$

Multiplying both sides by $\psi_E(x)$ yields equation Q10.12.

Q10X.6 Define $\psi_E'(x) \equiv A\psi_E(x)$. If we plug this into equation Q10.14, we get

$$\frac{-\hbar^2}{2m}\frac{d^2\psi_E'}{dx^2} - [E - V(x)]\psi_E'(x)$$

$$= \frac{-\hbar^2}{2m}\frac{d^2(A\psi_E)}{dx^2} - [E - V(x)]A\psi_E(x) \qquad (Q10.35)$$

But by the constant rule for derivatives,

$$\frac{d^2(A\psi_E)}{dx^2} = \frac{d}{dx}\left(A\frac{d\psi_E}{dx}\right) = A\frac{d^2\psi_E}{dx^2} \qquad \text{(Q10.36)}$$

Therefore, equation Q10.35 becomes

$$\frac{-\hbar^2}{2m}\frac{d^2\psi_E'}{dx^2} - [E - V(x)]\psi_E'(x)$$

$$= A\frac{-\hbar^2}{2m}\frac{d^2\psi_E}{dx^2} - [E - V(x)]A\psi_E(x)$$

$$= A\left\{\frac{-\hbar^2}{2m}\frac{d^2\psi_E}{dx^2} - [E - V(x)]\psi_E(x)\right\} \qquad \text{(Q10.37)}$$

The quantity in curly braces is zero because $\psi_E(x)$ is a solution to the Schrödinger equation by hypothesis. Therefore we have

$$\frac{-\hbar^2}{2m}\frac{d^2\psi_E'}{dx^2} - [E - V(x)]\psi_E'(x) = 0 \qquad \text{(Q10.38)}$$

meaning that $\psi_E'(x) = A\psi_E(x)$ satisfies the Schrödinger equation also.

Q10X.7   The first and second derivatives of the trial function $\psi_E(x) = Ae^{-bx}$ are (using the chain rule)

$$\frac{d}{dx}Ae^{-bx} = Ae^{-bx}\frac{d(-bx)}{dx} = -bAe^{-bx} \qquad \text{(Q10.39)}$$

$$\frac{d^2\psi_E}{dx^2} = \frac{d}{dx}(-bAe^{-bx}) = +b^2Ae^{-bx} \qquad \text{(Q10.40)}$$

Plugging this into the Schrödinger equation, we get

$$-\frac{\hbar^2 b^2}{2m}Ae^{-bx} - \left(E - \tfrac{1}{2}m\omega^2 x^2\right)Ae^{-bx} = 0 \qquad \text{(Q10.41)}$$

$$\Rightarrow \quad -\frac{\hbar^2 b^2}{2m} - E + \tfrac{1}{2}m\omega^2 x^2 = 0 \qquad \text{(Q10.42)}$$

This cannot be made to equal zero for all x, so our trial solution does *not* satisfy the Schrödinger equation for this $V(x)$.

Q10X.8   Plugging the given second derivative into the Schrödinger equation, we get

$$-\frac{\hbar^2}{2m}[4b^2x^2 - 6b]Ae^{-bx^2} - \left[E - \tfrac{1}{2}m\omega^2 x^2\right]Ae^{-bx^2} = 0$$

$$\Rightarrow \quad \left[\frac{3\hbar^2 b}{m} - E\right] + \left[\frac{m\omega^2}{2} - \frac{2\hbar^2 b^2}{m}\right]x^2 = 0$$

$$\text{(Q10.43)}$$

This *can* be made to equal zero for all x if b and E have the right values so that the quantities in square

brackets are zero. Requiring the second bracket to be zero means that

$$0 = \left[\frac{m^2\omega^2}{4\hbar^2} - b^2\right] \quad \Rightarrow \quad b = \frac{m\omega}{2\hbar} \qquad \text{(Q10.44)}$$

Requiring the first to be zero implies that

$$0 = \left[\frac{3\hbar^2(m\omega/2\hbar)}{m} - E\right] \quad \Rightarrow \quad E = \tfrac{3}{2}\hbar\omega$$

$$\text{(Q10.45)}$$

This is the $n = 2$ level for the harmonic oscillator according to equation Q7.16.

Q10X.9   Since the second derivative of a function $f(x)$ is the derivative of the first derivative, for small Δx

$$\frac{d^2 f}{dx^2} \approx \frac{(df/dx)_{x+\frac{1}{2}\Delta x} - (df/dx)_{x-\frac{1}{2}\Delta x}}{\Delta x}$$

$$\approx \frac{\dfrac{f(x + \Delta x) - f(x)}{\Delta x} - \dfrac{f(x) - f(x - \Delta x)}{\Delta x}}{\Delta x}$$

$$= \frac{f(x + \Delta x) - 2f(x) + f(x - \Delta x)}{\Delta x^2} \qquad \text{(Q10.46)}$$

This most accurately reflects the derivative at the midpoint between $x + \frac{1}{2}\Delta x$ and $x - \frac{1}{2}\Delta x$, that is, simply at point x.

Q10X.10   Plugging equation Q10.27 into equation Q10.14, we get

$$-\frac{\hbar^2}{2m}\frac{\psi_E(x + \Delta x) - 2\psi_E(x) + \psi_E(x - \Delta x)}{\Delta x^2}$$

$$- [E - V(x)]\psi_E(x) = 0 \qquad \text{(Q10.47)}$$

Multiplying both sides by $2m\,\Delta x^2/\hbar^2$, we get

$$-\psi_E(x + \Delta x) + 2\psi_E(x) - \psi_E(x - \Delta x)$$

$$- \frac{2m\,\Delta x^2}{\hbar^2}[E - V(x)]\psi_E(x) = 0 \qquad \text{(Q10.48)}$$

Adding $\psi_E(x + \Delta x)$ to both sides yields equation Q10.28.

Q10X.11   I find the energies to be

$$E_4 = 4.737 \text{ eV}$$

$$E_5 = 7.250 \text{ eV}$$

$$E_6 = 9.861 \text{ eV (approximately)}$$

The next energy level would be above the wall height of 10 eV, so these are the only bound-state energies other than the ones shown in figures Q10.6 and Q10.7.

Q11

Energy Eigenfunctions

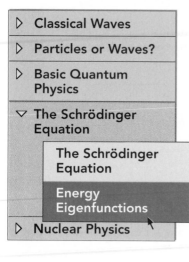
Chapter Overview

Introduction

In chapter Q10, we learned how to use Schrödinger's equation to determine a quanton's energy eigenfunctions and eigenvalues as it moves in response to a given potential energy function $V(x)$. These methods can be both difficult and abstract. In this section, we will learn how to sketch and interpret energy eigenfunctions *qualitatively*, which gives us greater insight into what these functions mean and how they behave.

Section Q11.1: How Energy Eigenfunctions Curve

We can learn quite a bit about how energy eigenfunctions must bend at a specific point if we rewrite the Schrödinger equation in the form

$$\frac{d^2\psi_E}{dx^2} = -\frac{2m}{\hbar^2}[E - V(x)]\psi_E(x) = 0 \tag{Q11.2}$$

This implies that in a classically *allowed* region, an eigenfunction $\psi_E(x)$ will curve *toward* the horizontal axis: the resulting curves are *wavelike*. In a classically *forbidden* region, $\psi_E(x)$ will curve *away* from the axis: the resulting curves are like exponentials. If the quanton is in a **square well** (where its potential energy is a small constant inside the classically allowed region and jumps discontinuously to a larger constant outside that region), the eigenfunctions are *exactly* sinusoidal inside and exponential outside. We also find that as long as a discontinuous jump in $V(x)$ is not infinite, the eigenfunction remains both continuous and smooth.

Section Q11.2: Why Bound-State Energies Are Quantized

We saw in chapter Q10 that only certain energy values produce physically reasonable energy eigenfunctions for bound quantons. We can now understand a bit better why this is so. A physically reasonable energy eigenfunction must go to zero as $x \to \pm\infty$. Since eigenfunctions curve away from the horizontal axis in a classically forbidden region, the only way that an eigenfunction can satisfy this constraint is by asymptotically approaching zero as $x \to \pm\infty$. This requires exactly the right energy value so that the function neither undershoots nor overshoots the axis. The eigenfunctions that work therefore have exponential-like tails that asymptotically approach the horizontal axis in both flanking classically forbidden regions. To make this work, the eigenfunction must also have an integer number of "bumps" inside the classically allowed region.

Section Q11.3: Tunneling

The exponential-like nature of the eigenfunction in classically forbidden regions means that a quanton can sometimes travel from one classically allowed region to another through a forbidden region: we call this phenomenon **tunneling**. Technological applications of this phenomenon include the scanning tunneling microscope (STM), tunneling diodes, Josephson junctions, SQUIDs, and so on.

Section Q11.4: Sketching Energy Eigenfunctions

Computer-generated solutions show that in a classically allowed region where the potential energy varies, the *amplitude* of the eigenfunction increases as the function's wavelength increases. We can understand this as follows. Remember that a short wavelength corresponds to a large classical speed. On the average, we are more likely to find a classical particle in regions where it is moving most slowly. This is reflected in quantum mechanics by the amplitude of the wavefunction becoming larger where the quanton would be moving most slowly.

The following is a summary of the basic rules for sketching an energy eigenfunction.

1. Solutions curve *toward* the axis (i.e., are wavelike) where $E > V(x)$ and curve *away* from the axis (i.e., are exponential-like) where $E < V(x)$.
2. The curvature of a solution increases with $|E - V(x)|$.
   a. Greater curvature means shorter wavelengths in classically allowed regions.
   b. Greater curvature means shorter "tails" in classically forbidden regions.
3. The amplitude of a wavelike part increases as the wavelength increases.
4. Solutions are continuous and smooth if jumps in $V(x)$ are finite.
5. Physically reasonable wavefunctions approach zero as $x \to \pm\infty$.
6. Acceptable solutions have an integer number of bumps in the classically allowed region, and the greater the number of bumps, the greater the energy.

Section Q11.5: The Covalent Bond

Consider the following simplified model of the covalent bond in the H_2^+ molecule. We will temporarily ignore the repulsive force between the protons and treat the electron as if it moved only along the line connecting the protons. Moreover, we will treat the potential energy of the electron as it interacts with each proton as if that proton were a square well. One can argue, using the qualitative reasoning we have learned in this chapter, that as the protons move apart, the electron's ground-state energy has to increase. If a system's energy increases as we separate its parts, the parts must be *attracting* each other. This illustrates how protons can attract each other by sharing an electron.

Q11.1   How Energy Eigenfunctions Curve

The Schrödinger equation

In chapter Q10, we discovered that a fairly natural generalization of the de Broglie relation for a quanton moving in one dimension is the *Schrödinger equation* (see equation Q10.14):

$$\frac{-\hbar^2}{2m}\frac{d^2\psi_E}{dx^2} - [E - V(x)]\psi_E(x) = 0 \qquad (Q11.1)$$

where $V(x)$ is the potential energy that the quanton would have if it were a classical particle at point x. Given a value of E, we can (at least in principle) solve this differential equation for the eigenfunction $\psi_E(x)$ that corresponds to that value of E. We also saw, however, that actually determining these solutions mathematically can be *very* difficult.

Our goals in this chapter

In this chapter, we will see that in spite of these difficulties, (1) we can fairly easily sketch a *qualitatively* accurate graph of a given energy eigenfunction and (2) we can learn some important things about a quantum system and its behavior from such sketches. Our goal in this chapter, then, is to explore some of the qualitative features and implications of the Schrödinger equation.

It is easier to see some of the implications of the Schrödinger equation if we express it in the following form:

$$\frac{d^2\psi_E}{dx^2} = -\frac{2m}{\hbar^2}[E - V(x)]\psi_E(x) \qquad (Q11.2)$$

Exercise Q11X.1

Argue that equation Q11.2 is equivalent to equation Q11.1.

This version of the equation shows how the second derivative $d^2\psi_E/dx^2$ of the quanton's energy eigenfunction $\psi_E(x)$ at a given point x is linked to the *value* of the eigenfunction at that x and the difference $E - V(x)$.

The second derivative $d^2\psi_E/dx^2$ at a given x, as we have discussed before, expresses the *curvature* of the eigenfunction $\psi_E(x)$ at that point: the larger the absolute value of $d^2\psi_E/dx^2$, the more strongly curved the wavefunction is. The *sign* of the second derivative expresses whether the function curves upward or downward: the function $\psi_E(x)$ curves (i.e., is concave) *upward* if $d^2\psi_E/dx^2 > 0$ and *downward* if $d^2\psi_E/dx^2 < 0$ (see figure Q11.1).

Qualitative implications of the Schrödinger equation

Equation Q11.2 has two important qualitative implications: (1) in a classically *allowed* region, an energy eigenfunction $\psi_E(x)$ always curves *toward* the horizontal axis, while (2) in a classically *forbidden* region, the eigenfunction curves *away* from the horizontal axis.

How do these statements follow from equation Q11.2? A region of the x axis is *allowed* for a classical particle if that particle's total energy E exceeds its potential energy $V(x)$ in that region (so that its kinetic energy $K = E - V$ is *positive*). We can see from equation Q11.2 that since $[E - V(x)]$ is positive in such a region by definition, the sign of $d^2\psi_E/dx^2$ is *opposite* to that of $\psi_E(x)$. Thus if the value of $\psi_E(x)$ is positive (i.e., *above* the axis) at a given x, then $d^2\psi_E/dx^2$ will be negative and the function will thus curve downward (*toward* the axis) there. Similarly, if the value of $\psi_E(x)$ is negative (*below* the axis), then $d^2\psi_E/dx^2$ will be positive and the function will thus curve upward (again *toward* the axis).

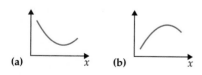

Figure Q11.1
(a) A function whose second derivative is *positive* curves upward. (b) A function whose second derivative is *negative* curves *downward*.

Exercise Q11X.2

Argue analogously that in a classically *forbidden* region, $\psi_E(x)$ will always curve *away* from the x axis.

A function $\psi_E(x)$ that always curves *toward* the x axis in a given region will be *wavelike* in that region: it is as if the function were attracted to the x axis, and even if the function overshoots the axis, it will always curve back toward it, implying that the function will oscillate about the x axis, as illustrated in figure Q11.2. On the other hand, a function $\psi_E(x)$ that always curves *away* from the x axis in a given region will be *exponential-like* in that region: the exponential function Ae^{ax} (where A and a are constants) is a function that always curves away from the x axis (whether a is positive or negative), as shown in figure Q11.3. This means that

1. $\psi_E(x)$ is *wavelike* in a classically allowed region [where $V(x) < E$].
2. $\psi_E(x)$ is *exponential-like* in a classically forbidden region [where $V(x) > E$].

Exercises Q11.3 and Q11.4 show that if $V(x)$ happens to be *constant* in a given classically allowed or forbidden region, the energy eigenfunction $\psi_E(x)$ is not just wav*elike* or exponential-like: it is *exactly* a sinusoidal wave if $V < E$ and *exactly* a sum of growing and decaying exponential functions if $V > E$.

Exercise Q11X.3

Show that the sinusoidal wave $\psi_E(x) = A\sin(bx + \phi)$ (where A, b, and ϕ are constants) is in fact an *exact* solution to the Schrödinger equation if $V(x) = V_0 = \text{constant} < E$. Determine how b is linked to E and V_0.

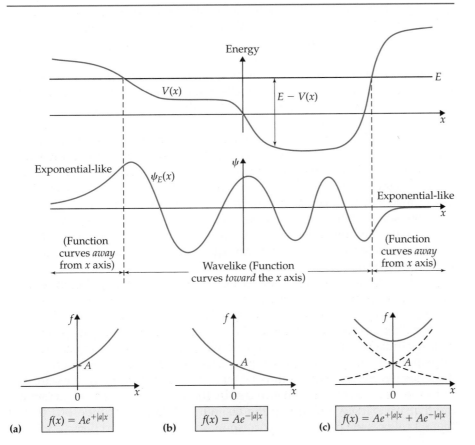

Figure Q11.2
Any energy eigenfunction $\psi_E(x)$ consistent with the Schrödinger equation will be wavelike in a classically allowed region and exponential-like in classically forbidden regions. (Note that the eigenfunction drawn here is like a growing exponential in the forbidden region on the left and like a decreasing exponential in the forbidden region on the right.)

Figure Q11.3
Graphs of (a) increasing and (b) decreasing exponential functions, and (c) the sum of such functions. Note that in each case, the function curves *away* from the x axis as x increases.

Exercise Q11X.4

Show that the sum of exponentials $\psi_E(x) = Be^{\beta x} + Ce^{-\beta x}$ (where B, C, and β are constants) is an *exact* solution to the Schrödinger equation if $V(x) = V_0 =$ constant $> E$. Determine how β is linked to E and V_0.

The degree of curvature depends on $|E - V(x)|$

Equation Q11.2 also implies that the absolute value of the second derivative [and thus the degree to which the function $\psi_E(x)$ is curved] at a given x depends fundamentally on the absolute value of difference $E - V(x)$. "Greater curvature" implies a *shorter wavelength* for the wavelike function in a classically allowed region (as we saw in chapter Q10) and *shorter, steeper exponential tails* in a classically forbidden region. This is also illustrated in figure Q11.2: note that the effective wavelength of the wavelike part and the exponential-like tail of $\psi_E(x)$ are both shorter on the right, since $|E - V(x)|$ is larger on that side.

Q11.2   Why Bound-State Energies Are Quantized

Physically reasonable solutions must remain finite as x goes to infinity

In chapter Q10, we saw (in the context of section Q10.6) that it is possible to find a mathematical solution $\psi_E(x)$ to the Schrödinger equation for *any* value of the quanton's energy E. We also saw that not all *mathematical* solutions $\psi_E(x)$ to the Schrödinger equation are *physically* reasonable. In fact, for bound quantons, we found by using SchroSolver that only certain specific and distinct (i.e., *quantized*) energy values lead to physically reasonable wavefunctions that go to zero as $x \to \pm\infty$.

We can understand this better conceptually if we consider the simple case of the *well* potential energy function $V(x)$ (sometimes called the **square well**), shown in figure Q11.4, which has a region of constant low potential energy flanked on either side by regions of constant high potential energy. This potential energy function is nice because if the quanton's definite energy E is between the top and bottom of the well, then the solutions to the Schrödinger equation are exactly exponentials outside the well and exactly sinusoidal functions inside the well.

The wavefunction $\psi_{E_1}(x)$ shown in figure Q11.4 corresponding to a quanton energy of E_1 is a physically reasonable solution to the Schrödinger equation: in the forbidden regions, the wavefunction has exponential tails that "lie down" on the x axis (i.e., asymptotically approach it), keeping the wavefunction finite even as it perpetually curves away from the axis. However, the solutions to the Schrödinger equation $\psi_a(x)$ and $\psi_b(x)$ that correspond to the slightly larger energy E_a and the slightly smaller energy E_b, respectively, cannot represent physical quantum states, because they don't curve away from the x axis at just exactly the right rate to "lie down" on the x axis in the right-hand forbidden region, and in fact end up going to positive or negative infinity as x grows large.

Solutions for only certain energies are reasonable

To see why the energy has to be exactly the right value for the solution to the Schrödinger equation to lie down on the x axis, consider the following analogy. Imagine that the curve of the wavefunction is the trajectory of an airplane that climbs away from the x axis in forbidden regions and dives back toward the x axis in the allowed regions, and that the plane curves away or toward the axis at a rate determined by the value of $|E - V(x)|\psi(x)$. Imagine that we start two planes at the far left at the same altitude $\psi = 0$, and plane 1 follows the path $\psi_{E_1}(x)$ corresponding to $E = E_1$ and plane A follows

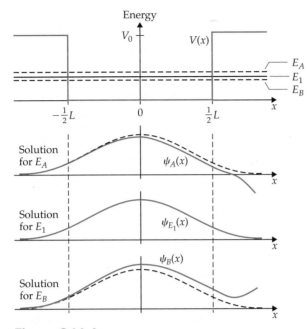

Figure Q11.4
The top graph shows the potential energy function $V(x)$ that physicists call the *square well*. The three graphs below show solutions to the Schrödinger equation for energies E_A, E_1, and E_B, respectively. The solution $\psi_{E_1}(x)$ corresponding to energy E_1 is a physically reasonable solution, but the solutions for E_A and E_B are *not* physically reasonable (even though these energies are close to E_1) because these solutions go to $-\infty$ or $+\infty$ as x becomes large. [The dotted curves in the graphs for $\psi_A(x)$ and $\psi_B(x)$ show $\psi_{E_1}(x)$ for the sake of comparison.]

the path $\psi_A(x)$ for $E = E_A$. Both planes curve away from the x axis as time increases, with plane 1 climbing a bit more rapidly since $|E_1 - V(x)| > |E_A - V(x)|$ in this region. When the planes reach the classically allowed region, however, the plane following path $\psi_A(x)$ dives back toward the x axis more rapidly than the one following path $\psi_{E_1}(x)$, since $|E_A - V(x)| > |E_1 - V(x)|$ in this region. This means that plane A ends up at the boundary of the right-hand forbidden region at a lower altitude and steeper dive angle than plane 1. Both planes begin to pull out of the dive as they move into the forbidden region, but while plane 1 is at the right altitude and dive angle to just barely pull out of the dive and land safely as x becomes large, plane A is too low and flying too steeply downward to level out before hitting the x axis. Once the function $\psi_A(x)$ passes below the x axis, the Schrödinger equation insists that the function veer away from the axis toward negative infinity as shown (this is where the plane analogy breaks down!).

Exercise Q11X.5

Using similar language, explain why the function $\psi_B(x)$ curves away to $+\infty$ as x becomes large.

The energies that work are isolated

We see that if the energy E is not *exactly* right (so that the exponential-like tails asymptotically approach the x axis in both forbidden regions), the solution to the Schrödinger equation for a bound quanton will not be a physically reasonable wavefunction. Even energy values very close to the precisely correct value E_1 in the case above will not yield solutions that correctly lie down on the x axis as $|x|$ becomes large. We see then that the energy E_1 that yields a reasonable energy eigenfunction is a well-defined and *isolated* value.

An energy eigenfunction (by definition) is that wavefunction that a quanton will have when we know that its energy is E. The solution to the Schrödinger equation for that value of E is supposed to be that energy eigenfunction. If the solution is not a physically reasonable wavefunction for a quanton to have, it follows that when we evaluate a quanton's energy, we will never find it to have that energy E. The only energy values that we can ever get for a bound quanton are those special and isolated values E_1, E_2, ... that correspond to physically reasonable wavefunctions whose exponential-like tails lie down on the x axis in *both* flanking classically forbidden regions. *This is why the energy of a bound quanton is quantized.*

Continuity of energy eigenfunctions

Note here that even though the potential energy in the well potential function jumps discontinuously at $x = \pm\frac{1}{2}L$, the solutions to the Schrödinger equation must remain both *continuous* and *smooth* through this boundary. As you can see from equation Q11.2, a discontinuous change in $V(x)$ will only discontinuously change the *second* derivative of $\psi_E(x)$. This means that the *slope* of a graph of the *first* derivative does change discontinuously at these points, but this does not mean that either the value of the function or its first derivative must change discontinuously. You might find the following analogy helpful as you think about this. The second derivative $d^2\psi_E/dx^2$ of a solution to the Schrödinger equation is analogous to the acceleration d^2x/dt^2 of a moving car. If you jam on the brakes, the acceleration of the car changes discontinuously but neither the car's x-velocity $v_x = dx/dt$ nor its position $x(t)$ changes discontinuously when you do this. By analogy, even when $d^2\psi_E/dx^2$ changes suddenly, neither the value of $d\psi_E/dx$ nor that of $\psi_E(x)$ changes discontinuously, meaning that the solution is *smooth* and *continuous*, respectively.

The energy E_1 shown in figure Q11.4 is only one of the energy values that yield physically reasonable wavefunctions for a quanton subject to the square well potential energy function. Other possible energy values and their associated wavefunctions are shown in figure Q11.5.

Exercise Q11X.6

Why are the exponential tails for $\psi_{E_3}(x)$ longer than those for $\psi_{E_1}(x)$?

Bound-state energy eigenfunctions have an integer number of bumps

The energy eigenfunctions corresponding to *any* one-dimensional bound quanton's set of possible discrete energies follow a predictable pattern with regard to their general features. The eigenfunction corresponding to the lowest possible energy (i.e., the one with the most gentle curvature in the classically allowed region) has exactly one wavelike bump in the classically allowed region. The eigenfunction corresponding to the next-higher energy has a sufficiently greater curvature to have exactly two bumps in the allowed region. The eigenfunction corresponding to the third definite energy state has three bumps, and so on. There are always an integer number of bumps essentially because the eigenfunction must approach zero in both the right and left classically forbidden regions, and there is no way to make this happen with a "bump and a half," for example, inside the classically allowed region. The energy eigenfunctions shown in figure Q11.5 for the square well

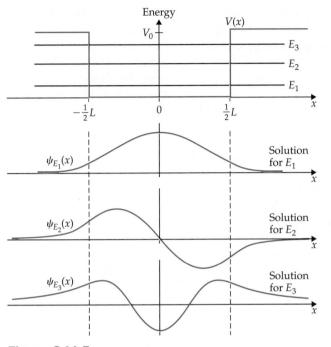

Figure Q11.5

Physically reasonable bound-state energy eigenfunctions have exponential-like tails that lie down on the x axis as |x| becomes large. This will only happen if the wavelike part of the eigenfunction in the classically allowed region has an integer number of bumps, as shown. The eigenfunction corresponding to the lowest energy E_1 has one bump, the eigenfunction corresponding to the next-highest energy E_2 has two bumps, and so on.

illustrate this principle (which applies to all bound quantons with a single classically allowed region flanked by two forbidden regions).

Exercise Q11X.7

Imagine that I claim that the wavefunction below is a physically reasonable solution to the Schrödinger equation for square well potential energy function for an energy $E_{1.5}$ between E_1 and E_2. The wavefunction has "a bump and a half" inside the classically allowed region (between the dashed lines). Why *can't* this actually be a solution to the Schrödinger equation?

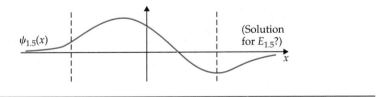

Q11.3   Tunneling

One of the most startling consequences of the Schrödinger equation is that even physically reasonable energy eigenfunctions for bound quantons all have exponential-like tails (rather than being strictly zero) in the classically

Quantons are not completely confined to classically allowed areas!

An experiment to display
this effect

forbidden regions that flank the quanton's classically allowed region. Since the absolute square of the quanton's wavefunction at a given x is proportional to the probability that an experiment to localize the quanton will yield the result x, this means that there is a nonzero probability that such an experiment will localize the quanton outside its classically allowed region!

We can actually test this assertion as follows. We know from our discussion of the photoelectric effect that it takes a certain energy W to liberate an electron from a metal (W is a metal-dependent constant that we called the *work function* in section Q3.5). This means that an electron's effective potential energy outside a given metal is W higher than inside the metal. This means that a graph of the electron's effective potential energy plotted across a bar of metal would be essentially the same as the square well potential energy shown in figure Q11.5 (with the height of the walls above the average electron energy being equal to W). If we put two pieces of metal a small but finite distance L apart and use a battery to raise the electric potential energy of electrons on the left relative to those on the right, then the total potential energy function that an electron sees might look something like that shown in figure Q11.6b.

Imagine that an electron is initially located in the left-hand metal bar. It will eventually settle down to an energy eigenfunction that might look something like the one shown in figure Q11.6c. Because of the exponential tail in the potential energy barrier (i.e., the vacuum) between the two pieces of metal, an electron with this energy eigenfunction will have a small probability of being located in the other piece of metal. Thus electrons originally in the left-hand metal bar have a small probability of ending up in the right-hand bar, even though they do not have sufficient energy to escape the left-hand bar classically. This would lead to a small but detectable current flowing from the left metal bar to the right. (Using the battery to raise the electric potential energy of the left bar relative to the right makes it harder for electrons to flow the other way across the gap: see problem Q11S.2.)

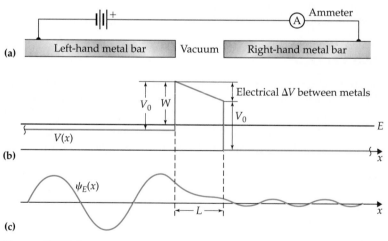

Figure Q11.6
(a) Two metal bars separated by a narrow length L of vacuum. A battery raises the electric potential energy of all electrons in the left bar by ΔV relative to those on the right. (b) An approximate potential energy graph for an electron with an energy W below the top of the potential energy barrier between the bars. (c) The solution $\psi_E(x)$ to the Schrödinger equation for an electron with the energy E shown. Note that an experiment to localize the electron has a nonzero probability of yielding a position in the right bar.

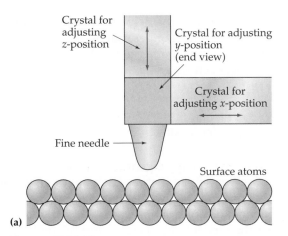

(a)

(b)

Figure Q11.7

(a) Schematic diagram of the needle of a scanning tunneling microscope in operation. (b) The results of a scan of the surface of a block of gold that clearly shows its individual atoms: brighter spots on the picture indicate where the needle had to be moved upward to keep the tunneling current constant.

Experiments like the one just described have been performed and yield results completely consistent with quantum mechanics. Analogous situations where quantons are observed to move through potential energy barriers that they could not penetrate classically occur in a variety of other contexts as well: physicists call the general phenomenon quantum-mechanical **tunneling**.

The specific situation shown in figure Q11.6 is the basis of the **scanning tunneling microscope** (STM), a device capable of imaging individual atoms on a metal surface. (G. Binning and H. Rohrer shared the 1986 Nobel Prize for inventing the STM in the early 1980s.) Figure Q11.7a shows a schematic diagram of an STM. A fine metal needle whose tip is only one or two atoms wide is affixed to a set of mutually perpendicular piezoelectric crystals. These crystals have the property that they expand or contract slightly when a voltage is applied to them; applying appropriate voltages to these crystals allows one to position the needle tip in three dimensions to an accuracy of about 0.01 Å ($\approx 10^{-3}$ nm). To image the atoms on the surface of a metal, one applies signals to the piezoelectric crystals to cause the needle tip to scan the surface of the metal about an atom's width above the metal. Because the electron wavefunction decays exponentially in the vacuum between the metal surface and needle, the rate at which electrons tunnel across the space between them depends very sharply on the distance between the needle tip and the atoms in the metal (see problem Q11R.1). By monitoring the tunneling current between the metal and the tip during the scanning process, one can register displacements of the surface toward or away from the tip that are much smaller than the size of an atom. Figure Q11.7b shows the results from such an STM scan (the bumps are individual atoms!).

The picture on this book's cover is an STM image of copper atoms manually arranged on a gold surface so that they create a stadiumlike arena. The waves are standing waves of free electrons whose waves bounce around in the arena, interfering with themselves.

The phenomenon of quantum-mechanical tunneling has a variety of other technological applications as well, including *tunneling diodes, Josephson junctions* (which someday might be used in very small, high-speed computers), and *superconducting quantum interference devices,* or SQUIDs (which can measure extremely weak magnetic fields). Tunneling also helps us resolve certain problems in nuclear physics, explaining how, for example, helium

The scanning tunneling microscope (STM)

Other applications and implications of tunneling

Labels in Figure Q11.7a:

Crystal for adjusting z-position

Crystal for adjusting y-position (end view)

Crystal for adjusting x-position

Fine needle

Surface atoms

nuclei can escape an atomic nucleus in alpha decay and how nuclear fusion occurs in the center of the sun, where the temperature is about a factor of 10 too small to drive protons close enough together to touch classically (tunneling allows the protons to interact even when separated by a significant electric potential energy barrier).

Q11.4   Sketching Energy Eigenfunctions

We *almost* know enough about what energy eigenfunctions look like qualitatively to be able to sketch a given eigenfunction fairly accurately, given a potential energy function $V(x)$. There is only one other feature of eigenfunctions that we ought to understand. Consider a region where the quanton's potential energy $V(x)$ is less than its definite energy E but is slowly decreasing, as shown in figure Q11.8a. Since $E > V(x)$, solutions to the Schrödinger equation will be *wavelike* in this region. We have also discussed that as $|E - V(x)|$ increases, solutions to the Schrödinger equation become more strongly curved, which for a wavelike solution means a shorter wavelength. Therefore, the effective wavelength of a solution to the Schrödinger equation in this case will have a wavelength that *decreases* as we go to the right, as shown in figure Q11.8b.

Why the *amplitude* of the wavefunction varies

It turns out that when we solve the Schrödinger equation in such situations, we also find that the *amplitude* of the wave decreases as its wavelength decreases. We can intuitively understand this as follows. Imagine for a moment that our quanton is a classical particle. Classically, $E - V(x) = K$, the quanton's kinetic energy, so as $E - V(x)$ increases, so do the particle's kinetic energy and thus its speed.

Now, imagine that we were to peek at this particle's position in this region at random times. What would we find? The *time* that a particle spends in the neighborhood of a given position is inversely proportional to its speed as it moves through that neighborhood, so the particle will spend proportionally more of its time near positions where it is moving slowly than near positions where it is moving quickly. Therefore the probability that a random peek will find the particle in a given neighborhood will be *smaller* where it is moving faster.

In quantum mechanics, faster classical motion corresponds to a greater classical momentum and (according to both the de Broglie relation $\lambda = h/p$

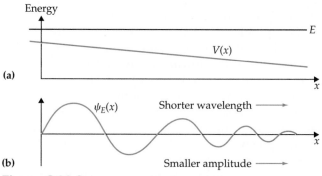

Figure Q11.8

(a) Graph of a region where the quanton's potential energy slowly decreases as x increases. (b) A solution to the Schrödinger equation in this region will have both a decreasing local wavelength and decreasing local amplitude as x increases.

and the Schrödinger equation that generalizes the de Broglie relation) thus a shorter *wavelength* for the quanton's wavefunction in quantum mechanics, as we have seen. Probability, on the other hand, corresponds to the square of the *amplitude* of the quanton's wavefunction: the greater the probability of locating a quanton at a point, the greater the amplitude that its wavefunction must have near that point. The classical argument in the previous paragraph implies that (in the classical limit) we should have a lower probability of locating the quanton in a neighborhood where it is moving quickly than one where it is moving slowly. Quantum mechanics will yield results consistent with this classical prediction only if the overall amplitude of the quanton's wavefunction gets smaller as we move to positions where the quanton's classical speed would increase. Therefore, in the situation shown in figure Q11.8a, it is plausible that the quanton's wavefunction should decrease in amplitude as we go to the right, as shown in figure Q11.8b. (This behavior is more and more clearly displayed as the number of bumps in the wavefunction increases.)

Now we know enough to create good qualitative sketches of a quanton's energy eigenfunctions if we know the potential energy function $V(x)$ for that quanton. Here is a summary of the rules that we have discovered in this chapter.

Summary of rules for sketching wavefunctions

1. Solutions to the Schrödinger equation curve *toward* the x axis in classically allowed regions and *away* from the x axis in classically forbidden regions. *Implication:* Solutions are *wavelike* in allowed regions and *exponential-like* in forbidden regions.
2. The curvature of solutions increases with $|E - V(x)|$. Note that $|E - V(x)|$ corresponds to the distance between the energy line and the potential energy curve on a potential energy graph. This implies that:
    a. In classically allowed regions, the *local wavelength* of the wavelike part of the solution gets shorter as $|E - V(x)|$ increases.
    b. In classically forbidden regions, the *length* of any exponential-like tail gets shorter as $|E - V(x)|$ increases.
3. The local amplitude of the wavelike part of a solution *decreases* as the value of $|E - V(x)|$ increases.
4. Solutions are always *continuous* and smooth (no kinks) [as long as any discontinuous jumps in $V(x)$ are finite].
5. Physically reasonable wavefunctions remain finite as $|x| \rightarrow \infty$.
    a. The exponential-like parts of physically reasonable solutions for bound quantons must actually decrease to *zero* as $|x| \rightarrow \infty$ in the classically forbidden regions that flank the central allowed region.
    b. Only certain quantized energy values E_n give rise to physically reasonable solutions (i.e., energy eigenfunctions) for bound quantons.
6. Energy eigenfunctions for bound quantons have an integer number of bumps in the central classically allowed region: the more bumps, the greater the corresponding energy. The energy eigenfunction corresponding to the quanton's lowest possible energy E_1 has one bump in its wavelike part, the eigenfunction corresponding to the quanton's second-lowest energy E_2 has two bumps, and so on.

Example Q11.1   Wavefunction Sketching

Problem   Sketch the energy eigenfunction corresponding to the seventh-lowest possible energy for a quanton whose potential energy as a function of x is shown in figure Q11.9a.

Figure Q11.9
(a) A graph showing the potential energy and total energy for a certain quanton. (b) A sketch of the energy eigenfunction corresponding to the seventh-lowest possible energy level for this quanton.

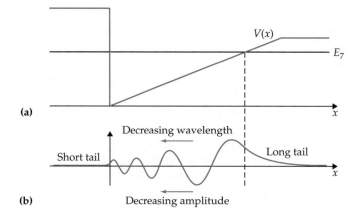

Solution Figure Q11.9b shows the solution to this problem. The energy eigenfunction should be wavelike within the classically allowed region and should have exponential-like tails that go to zero outside that region. Since it is supposed to correspond to the seventh-lowest energy level, it should have seven bumps in the classically allowed region. Toward the left-hand side of the allowed region, the difference between E_7 and $V(x)$ is large, so the curvature of the wave should be large, corresponding to a small effective wavelength. As we go toward the right, $E_7 - V(x)$ gets smaller, so the wave's curvature should decrease, increasing its effective wavelength. The amplitude of the wave should also increase as we go to the right. Notice also that the exponential tail in the left forbidden region should be shorter than that in the right forbidden region, since $|E - V(x)|$ is larger on the left than on the right.

Exercise Q11X.8

Sketch the energy eigenfunction corresponding to the fourth-lowest bound-state energy level of the potential energy function shown below. You may use the blank axes provided. Indicate the ways in which you have applied the listed rules.

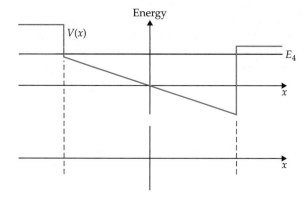

Q11.5   The Covalent Bond

In this section, we will use our eigenfunction sketching skills and a greatly simplified model to show qualitatively how "sharing" an electron can give rise to an attractive force that can bind two protons to form an H_2^+ molecule.

The basic principle illustrated here is of great importance in both chemistry and physics: not only does it show something about the fundamentally quantum nature of chemical bonds, but also it illustrates how quantons can attract each other by sharing or exchanging a third quanton, a concept that is very useful in quantum field theory and elementary particle physics.

To make the problem tractable, we will create a greatly simplified model of this ionized hydrogen molecule. For the moment, we will ignore the re- pulsive electrostatic force between the two protons and focus on the behavior of the shared electron. We will also ignore the three-dimensional character of the molecule and pretend that the electron can only move back and forth along the line connecting the two protons. Finally, rather than consider the complicated way in which the electron's electrostatic potential energy de- pends on position, we will treat the electron's potential energy as if it were a square well about as wide as a hydrogen atom (0.1 nm) centered at each pro- ton's position. This model may seem ridiculously approximate, but more realistic models simply increase complexity without really adding anything important to our understanding of the basic effect.

Figure Q11.10a shows the simplified potential energy function for the molecule and the electron's ground-state eigenfunction. Note that the wave- function should be symmetric because there is no physical reason that the electron should prefer one proton to the other and thus is equally probable to be found near either one. Note also how the electron's wavefunction in the middle forbidden region curves away from the axis: in this region, the wave- function is a *sum* of decreasing and increasing exponentials.

Figure Q11.10b and c shows how this ground-state eigenfunction *changes* as we pull the two protons apart. As the central forbidden region becomes wider, the probability of localizing the electron in this region must become smaller, so the *value* of the wavefunction in this region must become smaller. The colored lines in the lower diagrams show a full half-cycle of a sinusoidal wave that matches the curvature of the wavefunction inside the classically allowed regions. You can see that the width of this half-cycle decreases as the

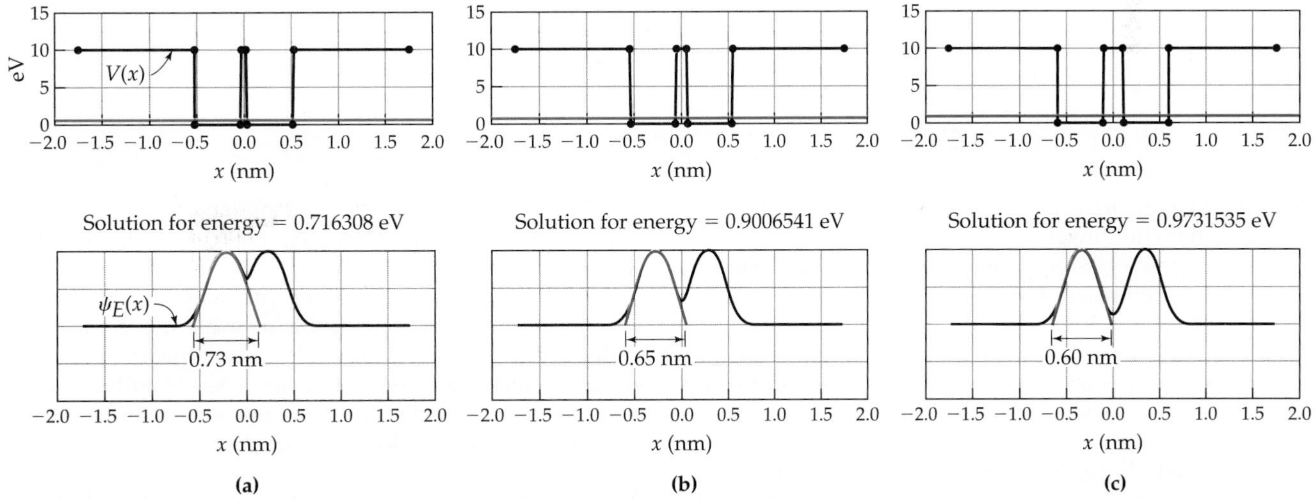

A simplified model of the ionized hydrogen molecule

Implications of the model

Figure Q11.10
These graphs show the potential energy function and ground-state electron eigenfunction (for increasing separations of the protons) for our simplified model of the H_2^+ ion. The top graph in each case shows the electron's potential energy function, and the bottom graph shows its ground-state eigenfunction. The colored curves indicate the best match of a half-cycle of a sinusoidal wave to the waveform in the left well. The indicated distances are therefore about one-half the wavelength of the electron's wavefunction in the classical region on the left. Note that these distances decrease as we pull the wells apart.

gap widens, so the quanton's wavefunction in this region must become shorter as the gap widens. Shorter wavelength implies greater energy, so *the electron's energy relative to the bottom of the well must increase as the protons are separated!*

Now, classically, we would say that if a system's energy increases when we separate its parts, these parts must *attract* each other. The way that the protons share the electron therefore essentially creates an interaction between the protons that is classically indistinguishable from an attractive force between them. It is this attractive force that holds the protons together in a hydrogen molecule.

Of course, if the protons get *too* close together, then their mutual electrostatic repulsion (which we have ignored up to now) overcomes the attractive effect of sharing the electron and pushes the protons apart. The stable distance between the protons is precisely the distance where the repulsive force between the protons balances the effective attraction of sharing the electron (this turns out to be about 0.105 nm for the H_2^+ molecule).

The power of qualitative reasoning

The point of this section is that the qualitative tools provided in this chapter make it possible to reason about how quantum systems must behave even when we lack mathematical or numerical solutions for the Schrödinger equation. Indeed, though we could have easily addressed this problem using SchroSolver, we can arrive at the same conclusion using purely qualitative reasoning.

Exercise Q11X.9

In the model considered in this section, consider a wavefunction that has only one bump in the left-hand classically allowed region and decreases exponentially to the right and left of this region (like the energy eigenfunction shown in figure Q11.10c except without the right-hand bump). Explain why this wavefunction could *not* be a solution to the Schrödinger equation in this case.

TWO-MINUTE PROBLEMS

Q11T.1 Imagine that the exponential tail for a solution to the Schrödinger equation where $E = E_A$ is *longer* in a given classically forbidden region than that for a solution where $E = E_B$. This means that

   A. $E_A > E_B$
   B. $E_A < E_B$
   C.   It is impossible to tell from the information provided.

Q11T.2 Must *all* physically reasonable wavefunctions remain finite as $|x| \to \infty$ (A), or does this only apply to solutions to the Schrödinger equation (B)?

Q11T.3 Must *all* physically reasonable wavefunctions be wavelike in classically allowed regions and exponential-like in forbidden regions (A), or does this only apply to solutions to the Schrödinger equation (B)?

Q11T.4 The possible energies of a quanton are only quantized if the quanton is bound, true (T) or false (F)?

Note: For each of the remaining problems in this section, select from the following possible responses. In some cases, multiple things may be wrong. If so, show the first thing that is wrong, and when your instructor asks, "Anything else?" show the second thing that is wrong, and so on. Show answer A when nothing else is wrong.

A.   The eigenfunction is correctly drawn (otherwise).

The eigenfunction drawn is incorrect because

B.   It curves toward the axis in a forbidden region or it curves away from the axis in an allowed region.
C.   Its wavy part doesn't have the right number of bumps.
D.   The amplitude of its wavy part is wrong.
E.   The wavelength of its wavy part is wrong.

F.   One of the exponential tails is the wrong length.

T.   Something else is wrong (specify).

Q11T.5 Consider a quanton whose potential energy is as shown in the graph below. The wavefunction shown in the second graph below is supposed to be the energy eigenfunction corresponding to the third-lowest possible energy for the quanton. What (if anything) is wrong with this eigenfunction as drawn? (See the note preceding this problem for information about how to answer.)

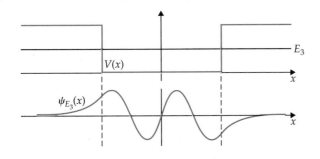

Q11T.6 Consider a quanton whose potential energy is as shown in the graph below. The wavefunction shown in the second graph below is supposed to be the energy eigenfunction corresponding to the fourth-lowest possible energy for the quanton. What (if anything) is wrong with this eigenfunction as drawn? (See the note preceding problem Q11T.5 for information about how to answer.)

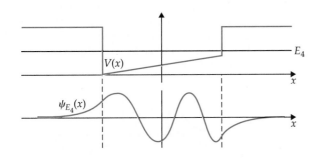

Q11T.7 Consider a quanton whose potential energy is as shown in the following graph. The wavefunction shown in the second graph below is supposed to be the energy eigenfunction corresponding to the fifth-lowest possible energy for the quanton. What (if anything) is wrong with this eigenfunction as drawn? (See the note preceding problem Q11T.5 for information about how to answer.)

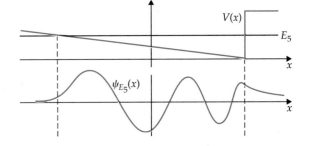

Q11T.8 Consider a quanton whose potential energy is as shown in the following graph. The wavefunction shown in the second graph below is supposed to be the energy eigenfunction corresponding to the sixth-lowest possible energy for the quanton. What (if anything) is wrong with this eigenfunction as drawn? (See the note preceding problem Q11T.5 for information about how to answer.)

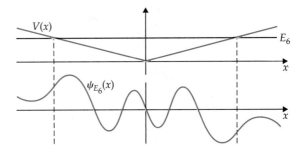

HOMEWORK PROBLEMS

Basic Skills

Q11B.1 Sketch the energy eigenfunction corresponding to the fourth-lowest bound-state energy for a quanton whose potential energy is shown in figure Q11.11.

Q11B.2 Sketch the ground-state energy eigenfunction for a quanton whose potential energy is as shown in figure Q11.11.

Q11B.3 Sketch the energy eigenfunction corresponding to the seventh-lowest bound-state energy for a quanton whose potential energy is shown in figure Q11.11.

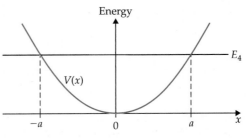

Figure Q11.11

The potential energy curve for problems Q11B.1 through Q11B.3.

Q11B.4 Sketch the energy eigenfunction corresponding to the fifth-lowest bound-state energy for a quanton whose potential energy is shown by the graph in figure Q11.12.

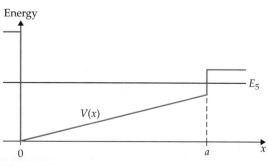

Figure Q11.12
The potential energy curve for problem Q11B.4.

Q11B.5 Sketch the energy eigenfunction corresponding to the fourth-lowest bound-state energy for a quanton whose potential energy is shown by the graph in figure Q11.13.

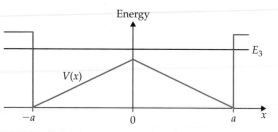

Figure Q11.13
The potential energy curve for problem Q11B.5.

Synthetic

Q11S.1 For each of the potential energy functions shown in figure Q11.14, sketch the ground-state energy eigenfunction for a quanton having that potential energy. Describe what you think will happen to the

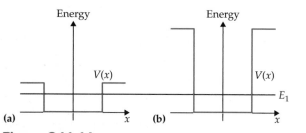

(a)   **(b)**

Figure Q11.14
What happens to the ground-state energy eigenfunctions as the walls become very high? (See problem Q11S.1.)

eigenfunction as the heights of the walls become extremely large, and compare your prediction to the ground-state eigenfunction for the box potential energy function shown in figure Q11.3.

Q11S.2 Consider the model of electron tunneling shown in figure Q11.6. In the case of an electron that was originally in the left-hand bar but has evolved to an energy eigenfunction with the energy value shown by the horizontal line, the amplitude of the electron's wavefunction in the right bar is small but essentially constant for very large distances away from the gap. This means that there is a finite probability of locating the electron a significant distance down the bar from the gap. Now imagine an electron originally in the *right*-hand bar with the same *kinetic* energy in that bar as the electron shown in figure Q11.6 has in the left bar. With the help of a sketch, argue that the probability of locating this electron a significant distance to the *left* of the gap is comparatively small, making the probability of tunneling from *right to left* much smaller than that of tunneling *left to right*. Explain how the battery sets up this asymmetry.

Q11S.3 Another possible energy eigenfunction for the double-well potential energy shown in figure Q11.10c is shown in figure Q11.15. Note that like the ground-state eigenfunction shown in figure Q11.10c, this eigenfunction also predicts that the quanton is no more likely to be in one well than the other, and this eigenfunction also has only one bump of the wavefunction in each well. Argue even so that this eigenfunction must correspond to a higher energy than the ground-state eigenfunction shown in figure Q11.10c.

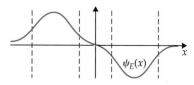

Figure Q11.15
An antisymmetric energy eigenfunction for the double-well potential energy function. (See problem Q11S.3.)

Q11S.4 We can use SchroSolver to verify the results discussed in section Q11.5. Run the program and select "Symmetric Well" from the pull-down menu. (Make sure that the pull-down menu above the energy text box says "Symmetric.")
(a) Plot a graph of the ground-state energy versus center-to-center separation of the wells.
(b) In a situation where the interaction between two objects is specified by a potential energy

function $V(x)$, the x-force that the interaction exerts on an object moving in the $+x$ direction is given by $F_x = -dV/dx$ (see section C11.1). In this case the total quantum energy changes E when we separate the wells, but this essentially behaves as a potential energy; so it makes sense that the magnitude of the force in this case would be $F_x = -dE/dx$, where dx is the increase in the center-to-center separation of the wells. Use this to estimate the force on the right well when the center-to-center distance is 1 nm.

(c) If you change the pull-down menu above the energy text box to read "Antisymmetric," SchroSolver will draw solutions to the Schrödinger equation where the wavefunction bumps in the two wells have opposite signs instead of the same sign. Show that the lowest-energy antisymmetric eigenfunction has a larger energy than the lowest-energy symmetric function, as claimed in problem Q11S.3. (This means that the symmetric eigenfunction is the true ground state of this system.)

(d) Plot the energy of the lowest antisymmetric state as a function of the center-to-center separation of the wells. Is there still an attraction between these wells, or do the wells repel each other?

Q11S.5 Consider an electron moving radially toward or away from a proton. Having a radial separation of less than zero is physically absurd; for this and other reasons, the electron's wavefunction must go strictly to zero at $r = 0$ as if there were an infinite barrier there. The potential energy function for the electron at other values of r due to its electrostatic attraction to the proton is as shown in figure Q11.16.

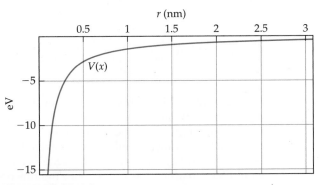

Figure Q11.16

The potential energy function for an electron interacting with a proton. (See problem Q11S.5.)

(a) Qualitatively sketch the first three energy eigenfunctions for the electron in this case, explaining the various features of the functions you draw (the number of bumps, amplitude/wavelength changes, tail lengths, and so on).

(b) Compare your answers with the results given in figure Q9.1 and/or the results displayed by SchroSolver (if the latter, print the graphs and attach them to your solution).

Q11S.6 Consider the "paddleball" potential energy function shown in figure Q11.17 (this is essentially an infinite barrier at $x = 0$ with a harmonic oscillator potential energy for $x > 0$).

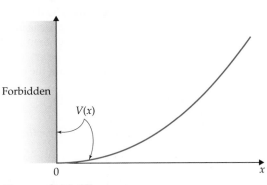

Figure Q11.17

The "paddleball" potential energy function. (See problem Q11S.6.)

(a) Qualitatively sketch the first three energy eigenfunctions for the quanton in this case, explaining the various features of the functions you draw (the number of bumps, amplitude/wavelength changes, tail lengths, and so on).

(b) How are these energy eigenfunctions related to the full harmonic oscillator eigenfunctions displayed in figure Q7.5?

Rich-Context/Advanced

Q11R.1 Consider the situation shown in figure Q11.6. Theoretical arguments imply that the tunneling current in this case is *roughly* proportional to $|\psi(\text{right})|^2/|\psi(\text{left})|^2$, where $\psi(\text{right})$ is amplitude of the electron's wavefunction just to the right of the gap and $\psi(\text{left})$ is its amplitude just to the left of the gap. Imagine that we increase the gap width from L to $L + \Delta L$. *Estimate* how big the change ΔL needs to be for the tunneling current to drop by an easily detectable amount (say, 5 percent). This will give an estimate of the smallest height variations that can be detected by a scanning tunneling microscope. (*Hints:* the typical value of W for many metals is about 4 eV. Assume for simplicity's sake that the potential energy is constant instead of slanted, so that $E - V = W =$ constant throughout the gap; and assume that the wavefunction in the gap is a purely *decreasing* exponential. Take advantage of the result of exercise Q11X.4.)

ANSWERS TO EXERCISES

Q11X.1 If we add $[E - V(x)]\psi_E(x)$ to both sides of equation Q11.1 and then multiply both sides by $-2m/\hbar^2$, we get equation Q11.2.

Q11X.2 In a classically forbidden region, $V(x) > E$ by definition, so the quantity $[E - V(x)]$ is *negative*. This means that $(-2m/\hbar^2)(E - V)\psi_E(x)$, and therefore $d^2\psi_E/dx^2$, has the same sign as $\psi_E(x)$ itself does. Thus when $\psi_E(x)$ is positive, its double-derivative is also positive, so $\psi_E(x)$ thus curves up, away from the x axis. If $\psi_E(x)$ is negative, then its double derivative is also negative, so $\psi_E(x)$ curves downward, which is again away from the axis.

Q11X.3 Using the chain rule, we find that

$$\frac{d\psi_E}{dx} = A\frac{d}{dx}\sin(bx + \phi) = Ab\cos(bx + \phi)$$

(Q11.3)

$$\frac{d^2\psi_E}{dx^2} = bA\frac{d}{dx}\cos(bx + \phi) = -b^2A\sin(bx + \phi)$$

(Q11.4)

The function $\psi_E(x)$ will therefore be a solution to the Schrödinger equation if (see equation Q11.1)

$$-\frac{\hbar^2}{2m}\frac{d^2\psi_E}{dx^2} - (E - V_0)\psi_E = 0$$

$$\Rightarrow \quad +\frac{\hbar^2}{2m}b^2A\sin(bx + \phi)$$

$$- (E - V_0)A\sin(bx + \phi) = 0$$

$$\Rightarrow \quad \left[\frac{\hbar^2b^2}{2m} - (E - V_0)\right]A\sin(bx + \phi) = 0$$

(Q11.5)

This will only be zero for all x if the quantity in square brackets is identically zero. This means that

$$\frac{\hbar^2b^2}{2m} = E - V_0 \quad \Rightarrow \quad b = \sqrt{\frac{2m(E - V_0)}{\hbar^2}}$$

(Q11.6)

So, as long as b has this value, then $\psi_E(x) = A\sin(bx + \phi)$ will be a solution to the Schrödinger equation. (Note that we must have $E > V_0$ for the square root to be meaningful.)

Q11X.4 Using the chain rule, we find that

$$\frac{d\psi_E}{dx} = \frac{d}{dx}(Be^{\beta x} + Ce^{-\beta x})$$

$$= \beta Be^{\beta x} - \beta Ce^{-\beta x}$$

(Q11.7)

$$\frac{d^2\psi_E}{dx^2} = \beta\frac{d}{dx}(Be^{\beta x} - Ce^{-\beta x})$$

$$= \beta^2 Be^{\beta x} + \beta^2 Ce^{-\beta x}$$

(Q11.8)

The function $\psi_E(x)$ will therefore be a solution to the Schrödinger equation if (see equation Q11.1)

$$-\frac{\hbar^2}{2m}\frac{d^2\psi_E}{dx^2} - (E - V_0)\psi_E = 0$$

$$\Rightarrow \quad -\frac{\hbar^2}{2m}\beta^2(Be^{\beta x} + Ce^{-\beta x})$$

$$- (E - V_0)(Be^{\beta x} + Ce^{-\beta x}) = 0$$

$$\Rightarrow \quad \left[-\frac{\hbar^2\beta^2}{2m} - (E - V_0)\right](Be^{\beta x} + Ce^{-\beta x}) = 0$$

(Q11.9)

This will only be zero for all x if the quantity in square brackets is identically zero. This means that

$$\frac{\hbar^2\beta^2}{2m} = V_0 - E \quad \Rightarrow \quad \beta = \sqrt{\frac{2m(V_0 - E)}{\hbar^2}}$$

(Q11.10)

So, as long as β has this value, then $\psi_E(x) = Be^{\beta x} + Ce^{-\beta x}$ will be a solution to the Schrödinger equation. (Note that we must have $V_0 > E$ for the square root to be meaningful.)

Q11X.5 Imagine that we start two planes at the far left at the same altitude $\psi = 0$, and plane 1 follows the path $\psi_{E_1}(x)$ corresponding to $E = E_1$ and plane B the path $\psi_B(x)$ for $E = E_B$. Both planes curve away from the x axis as time increases, with plane 1 climbing a bit less rapidly since $|E_B - V(x)| > |E_1 - V(x)|$ in this region. When the planes reach the classically allowed region, however, the plane following path $\psi_B(x)$ dives back toward the x axis less rapidly than the one following path $\psi_{E_1}(x)$, since $|E_B - V(x)| < |E_1 - V(x)|$ in this region. This means that plane B ends up at the boundary of the right-hand forbidden region at a higher altitude and shallower dive angle than plane 1. Both planes begin to pull out of the dive as they move into the forbidden region, but while plane 1 is at the right altitude and dive angle to just barely pull out of the dive and land safely as x becomes large, plane B is so high and is flying so shallowly downward that it levels out and begins to go upward long before it would have reached the x axis. Once the function $\psi_B(x)$ turns upward, the Schrödinger equation insists that the function continue to veer away from the axis toward positive infinity, as shown.

Q11X.6 Because $|E_3 - V_0| < |E_1 - V_0|$ in the classically forbidden regions, the tails for $\psi_{E_3}(x)$ should be less strongly curved and thus longer than the tails for $\psi_{E_1}(x)$.

Q11X.7 The function shown curves *toward* the x axis in part of the right-hand classically forbidden region. This is not consistent with the Schrödinger equation, which requires that it curve *away* from the axis in a forbidden region.

Q11X.8 The wavefunction should look as shown below (important features are marked).

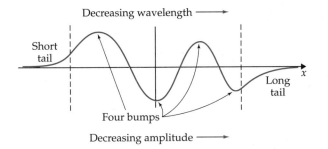

Q11X.9 If the wavefunction exponentially decreases to the right of the left-hand allowed region, then it will be exponentially decreasing in the right-hand allowed region. The Schrödinger equation, on the other hand, requires the function to be wavelike in the right-hand allowed region. Therefore, the wavefunction described cannot be a solution to the Schrödinger equation in this case.

Q12 Introduction to Nuclei

Chapter Overview

Introduction

This chapter opens a four-chapter unit subdivision on nuclear physics, which is not only a politically and socially important application of quantum mechanics but one that provides an opportunity to review material from previous units. This chapter provides essential background by discussing empirical facts about nuclear structure and radioactivity as well as by defining the crucial concept of *binding energy*.

Section Q12.1: Introduction to Nuclear Structure

Atomic nuclei are constructed of positively charged protons and uncharged neutrons, which we collectively call **nucleons.** The proton and neutron masses are nearly equal and are more than 1800 times larger than that of an electron.

We conventionally use Z to denote a nucleus's **atomic number** (the number of protons it contains); N, its **neutron number** (the number of neutrons it contains); and $A = Z + N$, its **mass number** (the total number of nucleons it contains). We use $^{A}_{Z}X$ (where X is the atom's chemical symbol) to specify a particular nucleus. **Isotopes** are nuclei that have the same Z but different N. We usually express nuclear masses in terms of the **unified mass unit,** where $1 \text{ u} \equiv (\frac{1}{12})$(mass of the $^{12}_{6}C$ atom).

Section Q12.2: The Size of the Nucleus

We can determine the radius of a given nucleus by firing charged quantons with increasing energy until changes in the scattering pattern indicate that the quantons are penetrating the nucleus. The experimental results imply that

$$r \approx r_0 A^{1/3} \qquad \text{where } r_0 = 1.2 \times 10^{-15} \text{ m} = 1.2 \text{ fm} \qquad \text{(Q12.5)}$$

Purpose: This links the radius r of a nucleus to its mass number A.
Symbols: r_0 is an empirically determined constant. (1 fm $\equiv 10^{-15}$ m.)
Limitations: This formula is approximate.
Note: The *volume* of a nucleus is roughly $\propto A$. The radius of a typical nucleus is about 100,000 times smaller than that of the corresponding atom.

Section Q12.3: The Strong Interaction

Empirically, the **strong interaction** that binds protons and neutrons into a tiny nucleus (in spite of the powerful electrostatic repulsion of the protons) appears to have the following characteristics:

1. It appears equally strong when acting between both types of nucleons.
2. It does not affect electrons (or any other leptons).
3. Its effective range is roughly 2 fm.

The last is so because nucleons are quark triplets that are *neutral* overall with respect to the strong interaction. The quarks in two nucleons almost have to intermingle to exert uncanceled forces on each other.

Section Q12.4: Binding Energy and Mass

We define a system's **binding energy** E_b to be the energy required to disperse that system to its constituent parts at rest at infinity. Special relativity implies that a system with positive binding energy thus has a mass that is slightly smaller than the total mass of its parts: we call this difference the system's **mass deficit** Δm.

$$E_b \equiv E_{\text{parts}} - E_{\text{sys}} \qquad\qquad (Q12.14a)$$

$$\Delta m \equiv m_{\text{parts}} - m_{\text{sys}} = \frac{E_b}{c^2} \qquad\qquad (Q12.14b)$$

Purpose: These equations define a system's binding energy E_b and mass deficit Δm.
Symbols: E_{sys} and m_{sys} are the system's rest energy and rest mass, respectively; E_{parts} and m_{parts} represent the sum of the rest energies and the sum of the rest masses of the system's constituent parts, respectively, when the parts are dispersed at rest at infinity; and c is the speed of light.
Limitations: These equations are definitions and so have no limitations. However, for E_b and Δm to be well defined, one needs to be clear about what the system's "constituent parts" are. For nuclei, the parts are nucleons.

This implies that the mass deficit of a nucleus is

$$\Delta m = Zm_{\text{H}} + Nm_n - m_{\text{atom}} \qquad\qquad (Q12.16)$$

Purpose: This specifies a nucleus' mass deficit Δm in terms of the mass m_{H} of the hydrogen atom, the mass m_n of a neutron, and the atom's mass m_{atom}.
Symbols: Z is the atom's atomic number and N is its neutron number.
Limitations: This expression ignores the negligibly small binding energies of the electrons in both the atom in question and the hydrogen atom.

Section Q12.5: Questions About Nuclear Stability

Some nuclei are stable, but others—**radioactive** nuclei—spontaneously shed quantons of various types. A chart of stable nuclei (see figure Q12.6) raises some questions:

1. Why do small nuclei have $Z \approx N$?
2. Why do more massive stable nuclei have $Z < N$?
3. Why are there no stable nuclei that have $Z > 83$?
4. What mechanisms do unstable nuclei use to move toward stability?
5. Can we find nuclear reactions that can tap nuclear energy?

Section Q12.6: An Historical Overview of Radioactivity

Empirical observations also suggest that there are three fundamental categories of radioactive decay processes: **alpha decay** processes that emit ^4_2He nuclei, **beta decay** processes that emit electrons (or positrons), and **gamma** processes that emit high-energy photons. Observations also suggest that the number of quanta radiated by a sample of radioactive nuclei decreases exponentially with time with a characteristic **half-life** (which is the time required for the decay rate to decrease by one-half). These observations raise these questions:

1. Why do unstable nuclei radiate only these kinds of particles?
2. Why does a sample's radioactivity decrease exponentially?

Q12.1   Introduction to Nuclear Structure

In about 1910, Hans Geiger and Ernst Marsden, who were graduate students working for the English physicist Ernst Rutherford, showed conclusively that atoms consisted mostly of empty space. Geiger and Marsden were conducting experiments in which they fired high-energy, positively charged particles at a piece of gold foil. They found that most of these particles simply went through the foil without being substantially deflected, as if the gold foil were empty space. However, a very few of these particles were highly deflected, indicating the presence of tiny, massive, and positively charged nuclei in the gold atoms. Quantitative measurements indicated that almost all the mass of an atom is contained in the nucleus, but also that the nucleus has a radius about 100,000 times smaller than that of the atom as a whole.

Subsequent painstaking work by a number of physicists gradually made atomic structure clear. This work showed that in their natural state, all atoms consist of a positively charged nucleus surrounded by one or more electrons occupying quantum-mechanical bound states in the electrostatic potential energy field created by the nucleus. The nucleus, in turn, is constructed of two types of quantons, protons and neutrons, each of which has a mass about 1800 times that of an electron. Collectively, we refer to protons and neutrons as **nucleons.**

The proton carries a positive charge equal in magnitude to the negative charge of an electron ($e \equiv q_{\text{proton}} = -q_{\text{electron}} = +1.602 \times 10^{-19}$ C) while the neutron has no net electric charge. In a *neutral* atom, then, the number of protons in the nucleus will be exactly equal to the number of electrons bound to the nucleus. The chemical properties of an atom are determined by the number and arrangement of its electrons (as we saw in chapter Q9). Therefore, neutral atoms having different numbers of protons will be chemically distinct and thus correspond to different chemical elements.

Nuclei are conventionally described using the following quantities:

1. The **atomic number** Z (the number of protons in the nucleus).
2. The **neutron number** N (the number of neutrons in the nucleus).
3. The **mass number** $A = Z + N$ (the total number of nucleons).

There is a convenient and conventional shorthand for describing the number of protons and neutrons in a nucleus that uses a symbol structured like this:

$$\,^{A}_{Z}\text{X} \qquad \text{where X is the standard chemical symbol for the element} \qquad \text{(Q12.1)}$$

So, for example, $\,^{12}_{6}\text{C}$ represents a carbon nucleus with 6 protons and 6 neutrons (12 nucleons total), $\,^{56}_{26}\text{Fe}$ represents an iron nucleus with 26 protons and 30 neutrons, and so on. This notation is actually a bit redundant: since the chemical properties of an element are completely determined by the number of protons in the nucleus, a carbon nucleus *always* has 6 protons and an iron nucleus always has 26 protons, so the Z subscript isn't really necessary. It is usually *convenient* to include it, though: it can save one from having to keep running back to a periodic table to get atomic numbers of strange elements.

Since the chemical properties of an atom are determined by the number of its electrons and thus the number of its protons, atoms having the same number of protons but different numbers of neutrons in their nuclei are therefore not significantly different from each other chemically. Thus all nuclei having the same Z are considered to be examples of the same chemical element: if these nuclei have different values of N, we call them different **isotopes** of that element. So, for example, $\,^{16}_{8}\text{O}$, $\,^{17}_{8}\text{O}$, and $\,^{18}_{8}\text{O}$ are all naturally occurring isotopes of oxygen (these isotopes have natural abundances of 99.758 percent, 0.038 percent, and 0.204 percent, respectively).

Nuclei are bound systems of protons and neutrons

Conventional numbers and symbols for describing nuclei

Since nuclear masses are tiny (on the order of 10^{-26} kg), it is not convenient to express these masses in kilograms. Instead, we conventionally express nuclear masses in terms of the **unified mass unit,** which is defined to be exactly one-twelfth of the mass of an atom of $^{12}_{6}C$. This implies that

$$1\,u = 1.660559 \times 10^{-27}\,\text{kg} \qquad (Q12.2a)$$

$$1\,uc^2 = 931.48\,\text{MeV} \quad \text{(equivalent rest energy)} \qquad (Q12.2b)$$

This unit is convenient because each nucleon has a mass of almost exactly 1 u:

$$m_p = \text{proton mass}\ \ = 1.007277\,u \qquad (Q12.3a)$$

$$m_n = \text{neutron mass} = 1.008665\,u \qquad (Q12.3b)$$

$$m_e = \text{electron mass} = 0.0005486\,u \qquad (Q12.3c)$$

Note that the mass of $^{12}_{6}C$ is actually less than the sum of the masses of its constituent protons, neutrons, and electrons: this is a manifestation of the relativistic principle that the mass of a system is not necessarily equal to the mass of its parts. (We'll discuss this more in section Q12.4.)

A conventional mass unit for atoms and nuclei

Q12.2   The Size of the Nucleus

How can one measure the size of something as tiny as an atomic nucleus? Rutherford, using Geiger and Marsden's data, was able to put an upper limit on the size of the nucleus in 1910. As discussed in section Q12.1, Geiger and Marsden fired **alpha particles** (bare helium nuclei) having a known kinetic energy K_0 at a gold foil and observed the number of particles scattered by the gold foil at various angles. The results observed were completely consistent with the results that one would expect if each particle were being scattered by a pointlike object having a charge of $+79e$ (the charge of the gold nucleus). *This means that the alpha particles never actually penetrated the gold nuclei.*

Imagine an alpha particle that approaches a gold nucleus head on, is brought instantaneously to rest by the nucleus' repulsive electric field, and then is scattered back the way it came (see figure Q12.1a through Q12.1c). Such an alpha particle will approach the nucleus more closely than an alpha particle that scatters off at some other angle (figure Q12.1d). (In both of the

How to measure the size of a nucleus

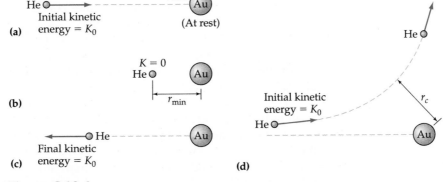

Figure Q12.1
(a), (b), (c) Stages of a head-on collision between a gold nucleus and an alpha particle. In (b), the particle has converted its initial kinetic energy entirely to electrostatic potential energy. (d) A glancing collision interaction between an alpha particle and the nucleus. Note that $r_{min} < r_c$.

collisions shown, we can consider the comparatively massive gold nucleus to remain essentially at rest.)

Exercise Q12X.1

Use conservation of energy principles to argue that r_{min} in figure Q12.1b must be smaller than r_c in figure Q12.1d.

Exercise Q12X.2

Show that in the case shown in figure Q12.1b, the distance of closest approach is given by

$$r_{min} = \frac{2Zke^2}{K_0} \qquad (Q12.4)$$

Exercise Q12X.3

The alpha particles used by Geiger and Marsden were known to have had an initial kinetic energy of about 7.0 MeV. Find r_{min}. (*Hint:* Remember that $ke^2 = 1.44\,\text{eV} \cdot \text{nm}$.)

The results of this kind of analysis convinced Rutherford that gold nuclei had radii smaller than roughly 3×10^{-14} m. Similar experiments with silver foil put an upper limit of about 2×10^{-14} m on the radius of the silver nucleus.

To *measure* the size of a nucleus, one has to use more energetic alpha particles (or electrons, which are more commonly used these days) that actually get close enough to get *inside* the nucleus. A quanton that gets inside a nucleus will be scattered in a different way than it would be by a point charge. Therefore, to measure the size of the nucleus, one simply adjusts the energy of the projectile particles until differences in the scattering pattern indicate that they are beginning to penetrate the nucleus. Since Rutherford's time, careful experiments of this type have established that nuclei are roughly spherical with radii given by the empirical formula

An empirical formula for the size of a nucleus

$$r \approx r_0 A^{1/3} \qquad \text{where } r_0 = 1.2 \times 10^{-15}\,\text{m} = 1.2\,\text{fm} \qquad (Q12.5)$$

Purpose: This formula describes the connection between the radius r of a nucleus and its mass number A.

Symbols: r_0 is an empirically determined constant. (1 fm = 1 femtometer $\equiv 10^{-15}$ m.)

Limitations: This formula is approximate.

Since 1 fm = 10^{-6} nm and atoms typically have radii on the order of tenths of nanometers, this means that the nucleus is roughly 100,000 times smaller than the atom in which it resides. (In an atom the size of a baseball stadium, the nucleus would be smaller than a typical marble.)

Note also that the *volume* of a nucleus, which is proportional to r^3, is proportional to A according to this formula, as if a nucleon takes up an essentially constant volume. We can therefore think of individual nucleons as being like incompressible balls as far as volume goes.

Exercise Q12X.4

Estimate the density of a typical atomic nucleus, and compare it with the density of water (1000 kg/m^3).

Q12.3   The Strong Interaction

The experimental evidence outlined in section Q12.2 implies that the positively charged protons in the nucleus are crammed together into an incredibly small space. Protons so close together will exert large electrostatic forces on each other that will tend to drive them apart. What is it that keeps the nucleus from instantly evaporating as a result of these strong repulsive forces?

While gravity is an attractive force that gets larger as objects get closer together, the gravitational force between two protons 1 fm apart is nearly 10^{36} times weaker than the electrostatic repulsion between those same protons. Nucleons are thus *not* bound in the nucleus by gravitational attraction.

Gravity cannot hold the nucleus together!

Exercise Q12X.5

Verify the statement about the relative strength of the gravitational and electrostatic forces acting between two protons. (Since both forces depend on the inverse square of the radius, the separation between the protons is actually irrelevant.)

The mere existence of nuclei indicates that some kind of very strong attractive interaction must exist between protons to hold them together despite their mutual electrostatic repulsion. Physicists call this interaction the **strong (nuclear) interaction** (the adjective *nuclear* is optional). As we saw in chapter C1, this interaction is one of the four fundamental interactions, analogous to but distinct from the gravitational and electrostatic interactions.

Experimental investigations of the strong interaction have shown that

Characteristics of the strong interaction

1. The strong interaction binds protons to protons, neutrons to neutrons, and protons to neutrons with roughly the same (large!) force.
2. The strong interaction does not seem to affect certain other kinds of quantons (specifically electrons).
3. The strong interaction has a range of 2 fm or so. Nucleons separated by more than this exert essentially *no* strong forces on each other.

The quark model of nucleons (see section C1.5) helps explain these observations. According to this model, the strong interaction only acts between quarks and does not affect leptons (such as the electron) because quarks carry a sort of strong-interaction "charge" that leptons do not. Nucleons (protons and neutrons) are structures of quarks that are strong-interaction-neutral, just as an atom is electrostatically neutral. Two nucleons can therefore interact according to the strong interaction only if they are close enough together that the distances between various pairs of quarks are significantly different, so that the interaction forces don't simply cancel. This is why the strong-interaction force between two nucleons drops quickly toward zero as the quantons are separated: the same is true of the electromagnetic interaction of two neutral atoms separated by more than a few atomic diameters.

Q12.4   Binding Energy and Mass

We can quantify how strongly nucleons are bound together in terms of the nucleus' *binding energy*. The **binding energy** of a system of bound quantons (or classical particles for that matter) is defined to be the energy required to completely disassemble the system into individual quantons or particles separated by an essentially infinite distance. If the total energy of the bound system (at rest) is E_{sys} and the energy of the disassembled parts (at rest at infinite separation) is E_{parts}, then the binding energy of the system E_b is such that

Definition of a system's binding energy E_b

$$E_b \equiv E_{parts} - E_{sys} \quad \text{so that} \quad E_{sys} + E_b = E_{parts} \quad \text{(Q12.6)}$$

The binding energy is almost always positive and gets larger as the system becomes more tightly bound.

Example Q12.1   Electrostatic Binding Energy

Problem   Consider two plastic balls, each with a radius of $R = 0.5$ cm. Imagine that one has a uniformly distributed charge of $+Q = 100$ nC $= 10^{-7}$ C and the other has a uniformly distributed charge of $-Q$. These balls stick together, as shown in figure Q12.2a. What is the binding energy of this system?

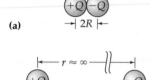

(a)

(b)

Figure Q12.2
(a) A system of two electrostatically bound plastic balls. (b) The same system with its parts dispersed to infinity.

Model   Spherical charge distributions behave as if their charge were concentrated at their centers. Therefore, we can compute the electrostatic potential energy of this bound system by using the equation $V_e(r) = kq_1q_2/r$ for point charges:

$$V_e(R) = \frac{k(+Q)(-Q)}{2R} = -\frac{9 \times 10^9 \text{ J} \cdot \text{m}/\text{C}^2 (10^{-7} \text{ C})^2}{10^{-2} \text{ m}} = -0.009 \text{ J} \quad \text{(Q12.7)}$$

(Note that the centers are separated by $2R$.) If the system is at rest (and not rotating), the ball's total kinetic energy is zero, so the total initial energy of the bound system is simply $E_{sys} = V_e(R) = -0.009$ J. The energy E_{parts} of the two balls at rest and separated by an infinite distance is

$$E_{parts} = V_e(\infty) = \frac{k(+Q)(-Q)}{\infty} = 0 \quad \text{(Q12.8)}$$

Solution   So the binding energy of this system is therefore

$$E_b = E_{parts} - E_{sys} = 0 - (-0.009 \text{ J}) = +0.009 \text{ J} \quad \text{(Q12.9)}$$

Evaluation   The units in equation Q12.7 are correct, and the result seems plausible for the relatively small charges involved.

Figure Q12.3
System of three electrostatically bound balls.

Exercise Q12X.6

Now consider *three* charged balls like those described in example Q12.1 (all with radius $R = 0.5$ cm), two with charge $+Q$ and one with charge $-Q$, where $Q = 10^{-7}$ C. Imagine that the three balls are stuck together in a line with the negative ball in the middle, as shown in figure Q12.3. Find the binding energy of this system. (*Hint:* The total potential energy of the bound system is the sum of the potential energies of each individual pair of charges: $V_{tot} = V_{AB} + V_{AC} + V_{BC}$.)

We *could* compute the binding energy of a nucleus in the same way if we knew the formula for the potential energy of the strong interaction, or if we could take a nucleus and actually disassemble it, nucleon by nucleon, keeping track of the energy involved. It turns out, however, that there is an *easier* way to determine the binding energy of a nucleus: all we need to know is its mass!

Remember from unit R that the mass of a system at rest is given by its total relativistic rest energy E_{rest}. In SI units, we write this relationship as follows:

$$E_{sys,rest} = m_{sys}c^2 \quad \text{or} \quad m_{sys} = E_{sys,rest}/c^2 \qquad \text{(Q12.10)}$$

How binding energy is related to mass

The total relativistic rest energy of a system includes not only the mass energies of all the objects in the system, but their kinetic and potential energies as well. So consider a bound system of a set of objects. The binding energy E_b is defined to be the energy that we have to add to the system to disassemble it so that its constituent parts are at rest but separated by an infinite distance. The total relativistic energy E_{parts} of the separated parts at rest is simply the sum of the individual objects' rest-mass energies, since they do not have any kinetic energy when they are at rest and since (in analogy with the electrostatic potential energy function) we will conventionally *define* the potential energy of the objects in the system to be zero at infinity. Since we have to *add* energy E_b to the bound system to disassemble it, the mass of the isolated objects will thus be *greater* than the mass of the bound system by E_b/c^2. The definition of binding energy is $E_{parts} = E_b + E_{sys}$, so dividing through by c^2, we get

$$m_{parts} = \sum m_{obj} = m_{sys} + \frac{E_b}{c^2} \quad \Rightarrow \quad m_{sys} = \sum m_{obj} - \frac{E_b}{c^2} \qquad \text{(Q12.11)}$$

This is illustrated in figure Q12.4.

The binding energy is therefore proportional to the bound system's **mass deficit** Δm, which we define to be the difference between the mass of the unbound system and that of the bound system:

$$\Delta m \equiv m_{parts} - m_{sys} = \left[\sum m_{obj} \right] - m_{sys} \qquad \text{(Q12.12)}$$

The mass deficit, like the binding energy, is usually a *positive* number, since the mass of the bound system is smaller than the mass of the unbound system. If you compare equations Q12.12 and Q12.11, you can see that

$$\Delta m \equiv \frac{E_b}{c^2} \quad \text{or} \quad (\Delta m)c^2 = E_b \qquad \text{(Q12.13)}$$

Thus the mass deficit of a bound system and its binding energy are essentially the same concept. The crucial thing to remember is that the more tightly bound a system is, the *larger* its binding energy is, but the *smaller* the system's mass is compared to the total mass of its parts.

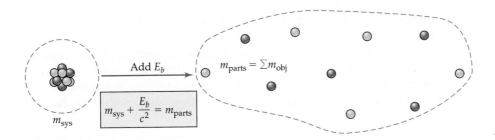

Figure Q12.4
An illustration of equation Q12.11.

In summary, the definitions of binding energy and mass deficit are as follows:

A summary of the definitions of binding energy and mass deficit

$$E_b \equiv E_{\text{parts}} - E_{\text{sys}} \tag{Q12.14a}$$

$$\Delta m \equiv m_{\text{parts}} - m_{\text{sys}} = \frac{E_b}{c^2} \tag{Q12.14b}$$

Purpose: These equations define a system's binding energy E_b and mass deficit Δm.

Symbols: E_{sys} and m_{sys} are the rest energy and the rest mass of the system, respectively; E_{parts} and m_{parts} represent the sum of the rest energies and the sum of the rest masses of the system's constituent parts, respectively, when the parts are dispersed at rest to infinite separation; and c is the speed of light.

Limitations: These equations are definitions and so have no limitations. However, for E_b and Δm to be well defined, one needs to be clear about what the system's "constituent parts" are. For nuclei, we consider the nucleons to be the constituent parts.

Using this approach to find the binding energy of nuclei

Using the mass deficit is convenient because it is very easy to measure the mass of an atom with a mass spectrometer to extraordinary accuracy. If we know the mass of an atom, we can easily compute its mass deficit (and thus its binding energy) by subtracting the masses of the individual nucleons and the atom's electrons. According to equation Q12.12, we have

$$\Delta m = Zm_p + Zm_e + Nm_n - m_{\text{atom}} \tag{Q12.15}$$

where m_{atom} is the mass of the atom, m_p is the mass of a proton, m_e is the mass of an electron, and m_n is the mass of a neutron. This formula actually gives the mass deficit for the entire atom (not just the nucleus), but the binding energy of the electrons in an atom turns out to be insignificant compared to that of the nucleus (as we will see shortly), so this mass deficit very accurately reflects the binding energy of the nucleus alone.

Instead of subtracting the mass of the protons and electrons separately, it is often easier to subtract Z times the mass of the hydrogen atom, since the mass of the hydrogen atom includes the mass of one proton and one electron (again, the binding energy of the electron in a hydrogen atom is 13.6 eV, which is 10^{-8} times smaller than the rest energy of a proton involved, so we can ignore it). The mass deficit of a nucleus can thus be written

A formula for the nuclear mass deficit

$$\Delta m = Zm_{\text{H}} + Nm_n - m_{\text{atom}} \tag{Q12.16}$$

Purpose: This equation specifies the mass deficit Δm for a nucleus in terms of the mass m_{H} of the hydrogen atom, the mass m_n of a neutron, and the atom's mass m_{atom}.

Symbols: Z and N are the atom's atomic number and neutron number, respectively.

Limitations: This expression is actually an approximation that ignores the binding energies of the electrons in both the atom in question and the hydrogen atom. Since electron binding energies are very small compared to nuclear binding energies, this is an excellent approximation.

Example Q12.2   The Mass Deficit for Carbon

Problem   The mass of the $^{12}_{6}C$ atom is defined to be 12.000000 u. What is the mass deficit for this nucleus? What is the binding energy?

Solution   According to equation Q12.16, we have

$$\Delta m = Zm_H + Nm_n - m_{atom}$$

$$= 6(1.007825\ u) + 6(1.008665\ u) - 12.000000\ u = 0.098940\ u \quad (Q12.17)$$

The conversion factor between the unified mass unit u and energy in mega-electronvolts is

$$1\ u = 931.5\ MeV/c^2 \qquad (Q12.18)$$

So the binding energy for this atom is

$$E_b = \Delta m\,c^2 = (0.098940\ \cancel{u}) \left(931.5 \frac{MeV}{\cancel{uc^2}} \right) \cancel{c^2} = 92.16\ MeV \quad (Q12.19)$$

Evaluation   As a comparison, the binding energy between two atoms in a molecule is on the order of a few electronvolts. Nuclear binding energies are typically 10^6 times larger than chemical binding energies: this is why you can get so much more energy from a kilogram of nuclear fuel than from a kilogram of chemical fuel.

Similarly, the energy required to completely strip a carbon atom of its electrons is on the order of about 1000 eV, which is about 10^4 times smaller than the binding energy we have just calculated. Therefore, the binding energy of an atom is equal to the binding energy of its nucleus to about four decimal places.

Exercise Q12X.7

The mass of the $^{40}_{20}Ca$ atom is 39.962591 u. Find its mass deficit and its binding energy.

The method of determining a system's binding energy by finding its mass deficit in principle would work for *any* kind of bound system; but in practice, the binding energy of most systems that are *not* nuclei is such a tiny fraction of its rest energy that this method is useless. It is useful for nuclei primarily because (1) we have techniques that make it possible to measure atomic masses to eight significant figures and (2) the nucleus is so tightly bound that its binding energy is a significant fraction of its rest energy.

Knowledge of the binding energy allows us to compare the strength of the strong and electromagnetic (EM) interactions, as example Q12.3 illustrates.

The relative strength of strong and EM interactions

Example Q12.3   The Strength of the Strong Interaction

Problem   The binding energy of the helium ($^{4}_{2}He$) atom is 28.3 MeV. Use this to estimate the strong-interaction potential energy between two nucleons. Compare this to the electrostatic potential energy of two protons separated by 1.0 fm (their approximate distance in the helium nucleus).

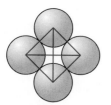

Figure Q12.5
There are six interacting pairs in a nucleus with four nucleons (count the lines connecting these pairs).

Model/Solution   The helium nucleus consists of four nucleons, and (as shown in figure Q12.5) this means that there are *six* distinct interacting pairs of nucleons. The binding energy will thus be the sum of the strong-interaction potential energies for each of these interacting pairs. If we assume that the potential energies of these pairs are about equal, we find that the strong-interaction potential energy between two nucleons in the helium atom is about

$$V_{\text{si}} \approx -\frac{28.3 \text{ MeV}}{6} = -4.7 \text{ MeV} \qquad (Q12.20)$$

(This potential energy is *negative* because the strong interaction is attractive, and so its potential energy must increase with distance. Since we have defined the potential energy to be zero at infinity, it will then have to be negative at any smaller r.)

The repulsive electrostatic potential energy of two protons with charge $+e$ that are 1.0 fm $= 10^{-15}$ m $= 10^{-6}$ nm apart is

$$V_e \approx \frac{ke^2}{r} = \frac{1.44 \text{ eV} \cdot \text{nm}}{10^{-6} \text{ nm}} \left(\frac{1 \text{ MeV}}{10^6 \text{ eV}}\right) = 1.4 \text{ MeV} \qquad (Q12.21)$$

Evaluation   We see that the binding energy of the strong interaction between two protons in this case is very roughly 3 times larger in magnitude than that for the electrostatic interaction, displaying (at least qualitatively) the greater strength of the strong interaction. The strong interaction between nucleons is actually much weaker than the strong interaction between individual quarks (just as the electrostatic forces that hold atoms together in a molecule are generally much weaker than those holding the electrons in the atom). The strong interaction between quarks is generally quoted as being about 100 times stronger than the electromagnetic interaction between the same.

Q12.5   Questions About Nuclear Stability

A chart of stable nuclei

The strong nuclear interaction, then, provides the "glue" that holds nucleons together in stable nuclei in spite of the electrostatic repulsion of their protons. Roughly 400 stable nuclei are known to exist. Hundreds of other nuclei have been studied that are unstable: these nuclei (which we also call **radioactive**) tend to eject quantons of various types, split into smaller pieces, or otherwise transform themselves so that they eventually settle down into a stable form.

Some interesting questions that this chart raises

Figure Q12.6 shows a plot of N (number of neutrons) versus Z (number of protons) for the known stable nuclei. While the strong nuclear interaction explains why nuclei stay bound together, it is not clear why some nuclei are stable and others are not. In particular, as we look at figure Q12.6, several interesting questions spring to mind:

1.   Why do stable nuclei tend to have $N \approx Z$ (especially light nuclei)? Why aren't there stable nuclei composed entirely of protons or neutrons?
2.   Why do heavier stable nuclei tend to have more neutrons than protons?
3.   Why are there apparently no stable nuclei with $Z > 83$?
4.   What makes unstable nuclei unstable in the first place, and what are the mechanisms by which they transform themselves into stable nuclei?

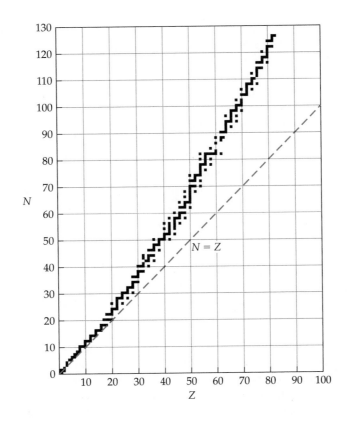

Figure Q12.6
Plot of N versus Z for stable nuclei. Each stable nucleus in this chart is represented by a square dot. Note that light nuclei tend to have $N \approx Z$, but for massive stable nuclei $N > Z$. Note also that there are no stable nuclei with $Z > 83$.

Another question is raised by the strength of the strong interaction and the compact size of the nucleus:

5. Is it possible to set up nuclear reactions that convert some of the very large strong-interaction potential energy (or even some of the electrostatic potential energy) between nucleons to useful forms of energy?

We will search for answers to all these questions in the next few chapters.

Q12.6   An Historical Overview of Radioactivity

The phenomenon of radioactivity itself raises some interesting questions. This section provides a brief overview of the history of radioactivity to provide the basis for those questions.

In 1895, Wilhelm Röntgen discovered that a beam of electrons directed at a metal surface produces a new kind of penetrating radiation that he called *X-rays* (we now understand that X-rays are simply high-energy photons). He detected these rays by observing that certain substances glow ("fluoresce") in their presence. He also observed that X-rays were able to fog a photographic plate even when it was wrapped in black paper, showing that X-rays were able to penetrate a substance opaque to ordinary light.

This result came to the attention of Henri Becquerel in 1896, who wondered if the inverse process might occur, that is, if substances induced to fluoresce by an intense source of light (such as sunlight) would produce X-rays. To test this hypothesis, he wrapped a photographic plate with black paper, placed a fluorescent material (which happened to be a uranium salt) on top of the wrapped plate, and exposed both to sunlight for several hours. When he developed the plate, he found that the plate was indeed fogged. After several such experiments, he described his results to a meeting of the Académie

Becquerel discovers
radioactivity (by accident)

des Sciences in Paris on February 24, 1896, reporting that the uranium salts seemed to be more effective than other fluorescent materials.

The following week, though, Becquerel discovered that the uranium compound was able to fog the plate even in the dark (suggesting that the fluorescence of the compound was irrelevant), and that pure uranium was able to do the same thing. Thus, the uranium was of its own accord producing some kind of radiation that was able to penetrate the wrapping paper and expose the plate. Though Becquerel still did not completely understand what was going on (he persisted in thinking that the fluorescence was somehow related), he had at this point discovered that uranium actively and spontaneously produces some kind of penetrating radiation (i.e., that uranium is **radioactive** in Marie Curie's later terminology).

The three general types of
radiation

Later investigations by Marie and Pierre Curie and others led to the discovery of other radioactive elements. Within a few years, Ernst Rutherford and others had discovered that the radiation produced by these elements was different from X-rays and consisted of particles of three fundamentally different types, which were arbitrarily given the names **alpha, beta,** and **gamma particles** (this was long before anyone understood what these particles really were). Figure Q12.7 shows some observed distinctions between these types of radiation.

In 1908, Rutherford and Royds were able to prove that the positively charged *alpha particles* were in fact $^4_2$He nuclei by collecting such particles in an evacuated glass tube and showing that the resulting gas had the same spectrum as helium. Somewhat later, *beta particles* were shown to be simply electrons, and *gamma particles* were identified as high-energy photons. In 1934, Irène and Frédéric Joliot-Curie showed that some radioactive atoms also decay by emitting *positrons* (antielectrons) instead of electrons.

Rutherford discovers that
radioactivity decreases
exponentially with time

In 1900 Rutherford also discovered that the rate at which a given radioactive sample emitted particles decreased exponentially in time. Subsequent investigations showed this to be true for *all* radioactive substances and that each different kind of radioactive nucleus had a characteristic **half-life** (the time required for the activity of a sample to decrease by a factor of 2).

These empirical results raise some interesting questions.

1. Why do radioactive nuclei emit photons, electrons, positrons, and $^4_2$He nuclei but not other quanta (such as protons, neutrons, or $^3_2$He nuclei)?
2. Why does a sample's radioactivity decrease exponentially with time?

We will find answers to these questions as well in the next few chapters.

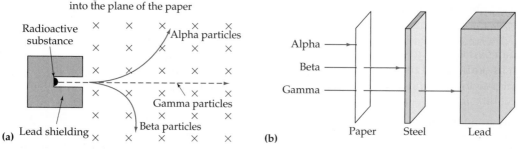

Figure Q12.7
The differences between the three types of radiation known in the first decade of the 1900s. (a) Their response to a magnetic field shows that alpha, beta, and gamma particles are positively charged, negatively charged, and neutral, respectively. (b) Alpha particles are generally unable to penetrate a sheet of paper, and beta particles are generally absorbed by a few millimeters of steel, but it takes many centimeters (even meters) of lead to stop gamma particles.

TWO-MINUTE PROBLEMS

Q12T.1 How many neutrons are in a $^{197}_{79}$Au nucleus?
A.   79
B.   179
C.   118
D.   Some other number (specify)

Q12T.2 About how many times larger in diameter is a $^{64}_{29}$Cu nucleus than a $^{27}_{13}$Al nucleus?
A.   2.37
B.   1.54
C.   1.33
D.   They are about the same size
E.   Some other factor (specify)

Q12T.3 The mass of the earth-moon system is somewhat less than the sum of the mass of the earth and the mass of the moon, true (T) or false (F)?

Q12T.4 The ground-state energy of an electron in a hydrogen atom is -13.6 eV. This means that the binding energy of that electron is
A. -13.6 eV
B. $+13.6$ eV
C.   Some other value (specify)

Q12T.5 If energy is *released* when the parts of a system are allowed to disperse to infinity, its binding energy is
A.   Positive
B.   Negative
C.   Zero
D.   Not defined in such a case

Q12T.6 According to the chart shown in figure Q12.6, how many stable isotopes of $_{30}$Zn exist?
A.   One
B.   Two
C.   Three
D.   Five
E.   There is no way to determine this from the chart.
F.   Some other number (specify)

Q12T.7 A certain radioactive substance emits quantons that, when placed in a magnetic field directed toward the observer, bend to the left side of their direction of travel. These radioactive nuclei are decaying via which kind of process?
A.   An alpha decay process
B.   A beta decay process
C.   A gamma decay process
D.   There is not enough information to specify

Q12T.8 The type of radiation that fogged Bequerel's photographic plates was probably
A.   Alpha radiation
B.   Beta radiation
C.   Gamma radiation
D.   Either alpha or beta radiation
E.   Either alpha or gamma radiation
F.   Either beta or gamma radiation
T.   There is no way of telling which kind of radiation fogged his plates.

HOMEWORK PROBLEMS

Basic Skills

Q12B.1 How many protons are in each of the following nuclei? How many neutrons?
(a) $^9_4$Be
(b) $^{13}_6$C
(c) $^{197}_{79}$Au
(d) $^{40}_{20}$Ca
(e) $^{43}_{20}$Ca

Q12B.2 How many protons are in each of the following nuclei? How many neutrons?
(a) $^{64}_{29}$Cu
(b) $^{23}_{12}$Mg
(c) $^{54}_{26}$Fe
(d) $^{56}_{26}$Fe
(e) $^{57}_{26}$Fe

Q12B.3 Compute the radii of the $^4_2$He and the $^{197}_{79}$Au nuclei. How many times bigger is the gold nucleus than the helium nucleus?

Q12B.4 Compute the radii of the $^{12}_6$C and the $^{238}_{92}$U nuclei. How many times bigger is the uranium nucleus than the carbon nucleus?

Q12B.5 The measured mass of a sodium $^{23}_{11}$Na atom is 22.989770 u. Compute its mass deficit in unified mass units (u) and its binding energy in megaelectronvolts (MeV).

Q12B.6 The measured mass of an iron $^{56}_{26}$Fe atom is 55.934939 u. Compute its mass deficit in unified mass units (u) and its binding energy in megaelectronvolts (MeV).

Q12B.7 The measured mass of a lead $^{208}_{82}$Pb atom is 207.976641 u. Compute its mass deficit in unified mass units (u) and its binding energy in megaelectronvolts (MeV).

Q12B.8 Using figure Q12.6, list the stable isotopes of $_{80}$Hg.

Q12B.9 Using figure Q12.6, list the stable isotopes of $_{20}$Ca.

Synthetic

Q12S.1 The total mass of a $^{12}_{6}$C atom is 12.00000 u by definition. What is the total mass of the $^{12}_{6}$C *nucleus*?

Q12S.2 What is the minimum kinetic energy that an alpha particle must have to get inside a gold nucleus?

Q12S.3 What is the minimum kinetic energy that an alpha particle must have to get inside a silver nucleus? (*Note:* $Z = 47$ and $A = 107$ for silver.)

Q12S.4 A plastic ball with charge $Q = +150$ nC and radius $R = 2.0$ cm touches a ball with charge $q = -75$ nC and a radius $r = 1.2$ cm.
(a) What is the binding energy of this system (in joules)?
(b) What is its mass deficit (in kilograms)?

Q12S.5 A speck of dust with charge $q = +0.020$ nC and negligible radius sits on the surface of a plastic ball of radius $R = 2.0$ cm and charge $Q = -120$ nC. What is the binding energy of this system (in joules)? What is its mass deficit (in kilograms)?

Q12S.6 Is it possible for a bound system to have negative binding energy? If not, why not? If so, qualitatively describe a system having negative binding energy.

Q12S.7 The earth is a bound system with a certain binding energy (mostly gravitational). Does the fact that the earth is rotating increase or decrease this binding energy, or is it irrelevant? Explain carefully, referring explicitly to the definition of binding energy. (*Hint:* When we say that a system is "at rest," we usually mean that we are using a frame where the system's *center of mass* is at rest. Consider the constituent parts of the system to be its atoms.)

Q12S.8 Imagine that you are trying to explain the concept of binding energy to your roommate. When you are finished, your roommate snorts in disgust, saying, "The whole idea makes no sense. You are telling me that when you take a bunch of initially separated objects and put them together in a bound system, the mass of the system is *less* than the total mass of the original parts? Ridiculous! Where does the mass go?" Answer your roommate in courteous language, explaining carefully where the "lost mass" goes.

Q12S.9 Consider the system consisting of the earth and the moon. Let us define the constituent parts of this system to be the earth and the moon.
(a) Compute the gravitational binding energy (in joules) and the mass deficit (in kilograms) of

this system. (The radius of the moon's orbit is 384,000 km, and its period is 27.3 days. Does it matter that the moon is *moving* in its orbit?)
(b) Also compute the ratio of the mass deficit to the mass of the moon. (For objects orbiting close to a black hole, this ratio can approach 1.)

Rich-Context

Q12R.1 At the end of a massive star's life, its core can undergo a catastrophic collapse that triggers a supernova explosion. The core remnant consists of about 1.4 solar masses' worth of pure neutrons and is called a *neutron star*.
(a) What is the approximate radius of such a neutron star, if neutrons are really like incompressible balls, as suggested by the text?
(b) The gravitational binding energy of any spherical object can only depend on its mass M, its radius R, and the universal gravitational constant G. Use dimensional analysis to find a formula for this binding energy that is valid up to some unknown dimensionless constant.
(c) Use the results of parts (a) and (b) to estimate the gravitational binding energy of a neutron star. (There will be an additional strong-interaction binding energy between the nucleons.)

Q12R.2 A certain 1.5-V battery (whose original mass is about 0.425 kg) has a capacity of 4.5 Ah. About how much lighter is this battery when it is discharged than when it is full? (*Hint:* It has the same number of electrons when "discharged" as it does when it is full.)

Advanced

Q12A.1 One can calculate the gravitational binding energy of a spherical object (such as the earth) as follows. Imagine disassembling the earth by removing a thin shell of material of thickness dr, then another shell of thickness dr, and so on. Assuming that the density of earth is a constant whose value is equal to $\rho = M/[(4/3)\pi R^3]$, where M is the mass of the earth and R is its radius, show that the energy required to move a shell of thickness dr away from the earth when its radius has already been pared down to r is

$$E = \frac{3GM^2r^4\,dr}{R^6} \tag{Q12.22}$$

Do an appropriate integral to find the gravitational binding energy of the sphere. Use this to find the earth's gravitational mass deficit (in kilograms).

ANSWERS TO EXERCISES

Q12X.1 In both cases, the alpha particle has an initial total energy of $E = K_0 + V(\infty) = K_0$. At r_{min} in the case shown in figure Q12.1b, the particle's kinetic energy is zero, so $E = K_0 = 0 + V_e(r_{min})$. But in the glancing interaction, the particle's kinetic energy $K_c \neq 0$ at $r = r_c$, so $E = K_0 = K_c + V_e(r_c)$. This implies that $V_e(r_{min}) > V_e(r_c)$, which (since $V_e \propto 1/r$) means that $r_{min} < r_c$.

Q12X.2 As we saw in the answer to exercise Q12X.1, $K_0 = V_e(r_{min})$ in the case of the collision shown in figure Q12.1b. Treating the nucleus and alpha particle as point particles, using the usual formula for the electrostatic potential energy and the fact that the charges of the nucleus and the alpha particle are Ze and $2e$, respectively, we get

$$K_0 = \frac{kq_1q_2}{r} = \frac{k(Ze)(2e)}{r_{min}} \qquad (Q12.23)$$

Solving this for r_{min} yields equation Q12.4.

Q12X.3 Plugging numbers into equation Q12.4, we get

$$r_{min} = \frac{2(79)(1.44 \text{ eV} \cdot \text{nm})}{7 \times 10^6 \text{ eV}} = 3.3 \times 10^{-5} \text{ nm} \qquad (Q12.24)$$

Since $1 \text{ fm} = 10^{-15} \text{ m} = 10^{-6} \text{ nm}$, $r_{min} \approx 33 \text{ fm}$.

Q12X.4 Since the proton and neutron have about the same mass, the mass of the nucleus is $m \approx Am_p$. The density of the nucleus is then (using equation Q12.5 to find r)

$$\rho = \frac{m}{V} = \frac{Am_p}{(4/3)\pi r^3} = \frac{Am_p}{(4/3)\pi r_0^3 A} = \frac{3m_p}{4\pi r_0^3} \qquad (Q12.25a)$$

$$= \frac{3(1.67 \times 10^{-27} \text{ kg})}{4\pi(1.2 \times 10^{-15} \text{ m})^3} = 2.3 \times 10^{17} \frac{\text{kg}}{\text{m}^3} \qquad (Q12.25b)$$

about 230 trillion times the density of water.

Q12X.5 The ratio of the magnitudes of the gravitational and electrostatic forces is

$$\frac{F_g}{F_e} = \frac{Gm_p^2/r^2}{ke^2/r^2} = \frac{Gm_p^2}{ke^2}$$

$$= \frac{(6.67 \times 10^{-11} \text{ J} \cdot \text{m/kg}^2)(1.67 \times 10^{-27} \text{ kg})^2}{(8.99 \times 10^9 \text{ J} \cdot \text{m/C}^2)(1.6 \times 10^{-19} \text{ C})^2}$$

$$= 8.1 \times 10^{-37} \approx 10^{-36} \qquad (Q12.26)$$

Q12X.6 The system's electrostatic potential energy is

$$V_{sys} = \frac{-kQ^2}{2R} + \frac{-kQ^2}{2R} + \frac{+kQ^2}{4R} = -\frac{3kQ^2}{4R}$$

$$= -\frac{3(9 \times 10^9 \text{ J} \cdot \text{m/C}^2)(10^{-7} \text{ C})^2}{2 \times 10^{-2} \text{ m}}$$

$$= -0.014 \text{ J} \qquad (Q12.27)$$

Presuming that all parts of the system are at rest, this is its total energy E_{sys} as well. When the three balls are dispersed to rest at infinity, the system's energy is $E_{parts} = 0$. Therefore, $E_b = E_{parts} - E_{sys} = +0.014 \text{ J}$.

Q12X.7 According to equation Q12.16, we have

$$\Delta m = 20(1.007825 \text{ u}) + 20(1.008665 \text{ u}) - 39.962591 \text{ u}$$

$$= 0.367209 \text{ u} \qquad (Q12.28)$$

Therefore

$$E_b = 0.367 \text{ u} \left(\frac{931.5 \text{ MeV}/c^2}{1 \text{ u}}\right) c^2$$

$$= 342 \text{ MeV} \qquad (Q12.29)$$

Q13 Stable and Unstable Nuclei

Chapter Overview

Introduction

In this chapter, we will explore some of the physical reasons for nuclear instablity and develop an approximate expression for nuclear binding energies. We will use this formula in chapter Q14 to make predictions about nuclear stability.

Section Q13.1:   The Weak Interaction

The **weak interaction** is the weakest of the four fundamental interactions except for gravity. It has the following characteristics:

1.   It affects both quarks and leptons (unlike the strong interaction).
2.   It affects both charged and uncharged quantons.
3.   It has a very short interaction range.
4.   It usually *transforms* one type of quanton to another.
5.   It often involves **neutrinos** or **antineutrinos** (unlike other interactions).

This interaction's main physical consequences are for nuclear structure. Important nuclear weak interaction processes include

$$\text{Neutron decay:} \qquad n \rightarrow e^- + p^+ + \bar{\nu} \qquad\qquad \text{(Q13.2)}$$

$$\text{Proton decay:} \qquad p^+ \rightarrow n + e^+ + \nu \qquad\qquad \text{(Q13.3)}$$

$$\text{Electron capture:} \qquad p^+ + e^- \rightarrow n + \nu \qquad\qquad \text{(Q13.4)}$$

These processes allow the conversion of protons to neutrons or vice versa.

Section Q13.2:   Why $N \approx Z$

The weak interaction explains why small nuclei have $Z \approx N$. To see why, model a nucleus as a collection of nuclei in a potential energy well. The Pauli exclusion principle implies that we can fit four nucleons into each of the well's energy levels (a pair of protons with opposing spins and a similar pair of neutrons). Let us also assume that the energy levels are roughly evenly spaced.

These assumptions mean that the lowest-energy configuration for a given number of nucleons occurs when the proton and neutron levels are equally filled. If $Z > N$, protons will have to occupy energy levels that are above open neutron energy levels. A nucleus can lower its overall energy by using a weak-interaction process to convert some protons to neutrons. Therefore a nucleus with an excess of protons will be unstable, eventually decaying to a nucleus where Z is closer to N. A similar argument implies that nuclei with $Z < N$ are also unstable.

Section Q13.3:   Why $N > Z$ for Large Nuclei

In larger nuclei, the electrostatic repulsion between protons significantly lifts proton energy levels relative to neutron levels. Nuclei become stable when the top proton level is about equal to the top neutron level; so, larger nuclei that are stable tend to have fewer protons than neutrons.

Section Q13.4: Classical Terms in the Binding Energy

Our goal in the remainder of this chapter is to create a nuclear model that will allow us to develop a formula that can predict the binding energy E_b of a nucleus given its values of A and Z. In this section we will consider terms in this formula that have a simple newtonian interpretation.

Because of the short range of the strong interaction, each nucleon essentially interacts with only its nearest neighbors. Any completely surrounded nucleon will therefore have a certain fixed number of bonds. If these bonds contribute a total interaction energy a_I to the binding energy for each such nucleon, the total contribution of A such nucleons will be $a_I A$.

But nucleons on the surface have fewer bonding partners. Since each surface nucleon occupies the same amount of surface area and a nucleus' surface area is $\propto$ $r^2 \propto (A^{1/3})^2$, the number of surface nucleons is proportional to $A^{2/3}$. Each of these nucleons will have the same deficit of bonds compared to an interior nucleon, so the real binding energy will be smaller than $a_I A$ by something proportional to $A^{2/3}$.

The protons will also repel each other electrostatically. A nucleus with Z protons has $\frac{1}{2}Z(Z-1) \approx \frac{1}{2}Z^2$ distinct proton pairs. These protons will be separated by an average distance that increases in direct proportion to the nuclear radius $r \propto A^{1/3}$, so the electrostatic potential energy of these protons will reduce the binding energy by something proportional to $Z^2 A^{-1/3}$.

Section Q13.5: The Asymmetry Term

The most important quantum effect is the one discussed in section Q13.2. If ΔE is the difference between adjacent nuclear energy levels, and we start with a hypothetical symmetric nucleus and convert n pairs of protons to neutrons to get the actual nucleus, then the total cost of shifting n pairs (ignoring the Coulomb repulsion effect already handled) turns out to be $2n^2 \Delta E$ (see the section). Since we need to shift $n = \frac{1}{4}(A - 2Z)$ pairs and $\Delta E \propto A^{-1}$, this effect reduces the binding energy by something proportional to $(A - 2Z)^2/A$.

Section Q13.6: Checking Against Reality

Putting all these terms together, we get the **semiempirical binding energy formula**

$$E_b(A, Z) = a_I A - a_S A^{2/3} - a_C Z^2 A^{-1/3} - a_A (A - 2Z)^2 A^{-1} \quad \text{(Q13.15)}$$

where

$$a_I = 15.56 \text{ MeV} \qquad \text{(Q13.16}a\text{)}$$

$$a_S = 17.23 \text{ MeV} \qquad \text{(Q13.16}b\text{)}$$

$$a_C = 0.697 \text{ MeV} \qquad \text{(Q13.16}c\text{)}$$

$$a_A = 23.285 \text{ MeV} \qquad \text{(Q13.16}d\text{)}$$

Purpose: This equation specifies a nucleus' binding energy E_b in terms of its mass number A and atomic number Z.

Symbols: a_I, a_S, a_C, and a_A are empirically determined constants of proportionality that we call the **interior bonding, surface correction, Coulomb repulsion,** and **asymmetry coefficients,** respectively.

Limitations: This is an *approximate* formula that only applies with any accuracy when $A > 20$ and even then is only correct to within about ±1%.

Note: The **binding energy per nucleon** $e_b \equiv E_b/A$.

This formula is reasonably accurate, and it helps us understand the contributing factors to E_b better than a more accurate but abstract formula would.

Q13.1   The Weak Interaction

If our discussion of the strong interaction in chapter Q12 were the whole story, it would seem that many more nuclei would be stable than are observed to be stable. Since the strong interaction between protons overpowers their electrostatic repulsion, there seems no reason why we could not have nuclei that contain a number of protons and no neutrons. Nuclei with lots of neutrons would seem even more likely: the neutrons attract each other through the strong interaction without even repelling each other electrostatically!

However, such nuclei do not seem to exist, so they *must* be unstable. The key to understanding *why* they are unstable is to recognize that in addition to the strong, electromagnetic, and gravitational interactions, there is yet another way that protons, neutrons, and electrons can interact. We call this fourth interaction the *weak interaction*.

The nature of this interaction is perhaps most easily illustrated by the collision of a neutrino and a neutron. A **neutrino** is a fundamental particle in the lepton family (the same family as the electron and muon) that has zero charge and a rest energy that is not precisely known but is negligibly small. Experiments have shown that an incoming neutrino (whose symbol is ν, the Greek letter nu) can react with the neutron to form a proton and an electron as follows:

$$\nu + n \rightarrow e^- + p^+ \qquad (Q13.1)$$

(This reaction is often used by physicists to detect neutrinos produced by the sun or other astrophysical sources.)

What kind of interaction could possibly cause this transformation of the neutron? It cannot be the strong interaction, because only quarks (not leptons) can participate in a strong interaction. Since the interaction involves electrically neutral particles, it cannot be an electromagnetic interaction. Since the neutrino has negligible mass, it cannot be a gravitational interaction (which is too weak to have significant effects on particle reactions anyway).

This reaction thus points to the existence of a fourth kind of fundamental interaction that can act between two quantons. We call this interaction the **weak interaction,** so named because under general circumstances its effects are much weaker than those of either the strong or the electromagnetic interactions (though it is still much stronger than gravity!). Experiments have shown that the weak interaction has the following characteristics:

1. The weak interaction (unlike the strong interaction) affects *both* quarks and leptons (unlike the electromagnetic interaction), can affect electrically neutral particles, and (unlike gravity) does *not* affect photons.
2. The effective range of the weak interaction is $\approx 10^{-18}$ m $= 10^{-3}$ fm, much shorter than any other kind of interaction.
3. Unlike *any* other kind of interaction, the weak interaction often has the effect of *transforming* one type of quanton to another (e.g., a neutron into a proton) rather than just transferring momentum.
4. Weak interactions characteristically involve the creation or absorption of neutrinos, which are not observed in any other context.

An important weak interaction process for nuclear physics is the **neutron decay** process

$$n \rightarrow e^- + p^+ + \bar{\nu} \qquad (Q13.2)$$

This process is much like the one shown in equation Q13.1 except that instead of absorbing a neutrino, the decaying neutron emits an **antineutrino,** the

Evidence for the existence of the weak interaction

Characteristics of the weak interaction

antiparticle corresponding to the neutrino (the bar over the ν indicates that it is an antiparticle). When neutrons are not attached to a nucleus, they spontaneously decay in this way with a characteristic half-life of about 10.25 min.

The other important weak processes for nuclear physics are the **proton decay** and **electron capture** processes

$$p^+ \rightarrow n + e^+ + \nu \tag{Q13.3}$$

$$p^+ + e^- \rightarrow n + \nu \tag{Q13.4}$$

where e^+ is the symbol for the **positron,** which is the antiparticle corresponding to the electron (having the same mass and spin as the electron but opposite charge). Since the combined rest energy of the proton and electron is roughly 0.8 MeV smaller than the rest energy of the neutron, proton decay is *not* possible for free protons, and the electron capture only works if the electron brings in more than 0.8 MeV of kinetic energy. But either of these processes can occur spontaneously in nuclei if the transformation causes a shift in nuclear structure that releases the required energy.

We will see that the reactions described by equations Q13.2, Q13.3, and Q13.4 are commonly observed in unstable nuclei and are part of the answer to why light nuclei have $Z \approx N$. The key thing to understand from this section is that *the weak interaction provides a means for converting a neutron to a proton or vice versa.*

Q13.2   Why $N \approx Z$

We are now in a position to show why nuclei in general have roughly equal numbers of protons and neutrons, and why they are especially stable when they have an even number of neutrons, protons, or both. The qualitative argument that follows depends only on the principle of energy quantization, the existence of the weak nuclear force, the fact that protons and neutrons have spin-$\frac{1}{2}$, and the Pauli exclusion principle, and *not* on the detailed workings of the strong interaction.

Consider, for the sake of argument, a $^{12}_{6}$C nucleus, and let's temporarily ignore the effects of the electrostatic repulsion between the protons. The six protons and six neutrons are tightly bound together in this nucleus by the strong nuclear force: it takes a large amount of energy to pull any one of these nucleons out of the nucleus. Let us pretend, then, that the nucleons essentially reside in some kind of potential energy *well* created by the strong nuclear interaction acting between them. Figure Q13.1 shows a schematic diagram of the well in this case, with its energy levels drawn as horizontal lines. I have drawn the levels as if they were equally spaced for simplicity: the real energy levels of quantons in a three-dimensional (spherical) well are complicated, but they turn out to be *roughly* equally spaced (unlike the one-dimensional well, where the separation between levels increases as n increases).

The exact characteristics of the strong-interaction potential energy curve turn out to be unimportant for this argument: all that we really need to know is that for every bound nucleon, there is a discrete series of possible quantized energy eigenstates. The well model is therefore a reasonable visualization for this qualitative argument.

The Pauli exclusion principle tells us that each of the well energy levels can be occupied by at *most* four nucleons: two protons (one with spin *up,* one with spin *down*) and two neutrons (one with spin *up,* one with spin *down*). I have schematically illustrated this by placing dots on the energy-level lines, black dots indicating protons and white dots indicating neutrons

A well model for the nucleus

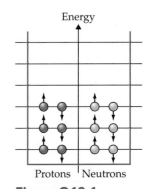

Figure Q13.1
The ground-state configuration for the nucleons in the $^{12}_{6}$C nucleus.

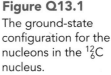

The Pauli exclusion principle permits two protons and two neutrons per level

Figure Q13.2
(a) The ground-state configuration for a hypothetical $^{15}_{6}$C nucleus. (b) This nucleus is unstable because it can lower its total energy by transforming the highest neutron into a proton, which allows it to drop to a lower energy level (this creates a $^{15}_{7}$N nucleus).

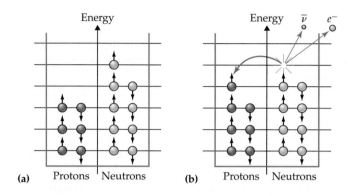

(a)   Protons Neutrons   (b)   Protons Neutrons

Unbalanced nuclei can lower energy by transforming nucleons

(the attached arrows show the spin orientation of the nucleon in each case). Figure Q13.1 shows the configuration of six protons and six neutrons having the lowest possible total energy: this is the *ground-state configuration* for the carbon-12 nucleus.

Now let's consider a case where the numbers of protons and neutrons are unbalanced. Figure Q13.2a shows the lowest-energy configuration for the $^{15}_{6}$C nucleus. Because of the Pauli exclusion principle, the last neutron has to go into an energy level that is higher than the lowest open proton energy level. Therefore, if that high neutron decays via the weak interaction into a proton, electron, and an antineutrino, the resulting proton can drop to one of the lower vacant proton energy levels, lowering the total energy of the nucleus. This kind of argument leads us to predict that $^{15}_{6}$C should be unstable and will transform itself to the lower-energy $^{15}_{7}$N nucleus by emitting an electron and an antineutrino. This prediction turns out to be true: $^{15}_{6}$C nuclei *have* been created in the laboratory and are indeed observed to decay (with a measured half-life of 2.45 s) to $^{15}_{7}$N, an electron, and an antineutrino.

Exercise Q13X.1

Now let's consider a nucleus with an overabundance of protons. $^{13}_{8}$O is observed to decay with a half-life of 8.9 ms to $^{13}_{7}$N, a positron, and a neutrino. Use the blank energy-level diagrams below to explain why.

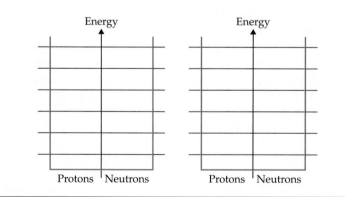

Protons Neutrons   Protons Neutrons

Exercise Q13X.2

Explain why a neutron in a $^{12}_{6}$C *cannot* decay to a proton or vice versa. (*Hint:* Consider conservation of energy. The difference between energy levels is on the order of magnitude of several megaelectronvolts at least.)

Exercise Q13X.3

$^{14}_{6}C$ decays to $^{14}_{7}N + e^{-} + \bar{\nu}$ with a half-life of 5730 y. How *can* it decay when the proton level below the highest neutron level is already full? (*Hint:* Why do free neutrons decay?)

We see from these examples that the Pauli exclusion principle and the quantization of energy imply that there is often a significant energy advantage to transforming excess neutrons to protons or excess protons to neutrons. The weak interaction provides a way that this can happen; and if a physical process that lowers a system's rest energy *can* take place, it eventually *will*. Thus any nucleus with very unbalanced numbers of protons and neutrons will transform itself to one where these numbers are more balanced ($N \approx Z$).

Conclusion: why $N \approx Z$

Q13.3   Why $N > Z$ for Large Nuclei

As Z increases, the mutual electrostatic repulsion between the protons becomes more and more important. This has the qualitative effect of raising the proton energy levels relative to the neutron energy levels (since the protons have a positive electrostatic potential energy that the neutrons do not).

A naïve analysis based on the ideas in section Q13.2 would suggest that $^{44}_{22}Ti$ (titanium) would be stable, since it has $N = Z$. But in this fairly large nucleus, the energy levels of the protons are raised so high by electrostatic repulsion that the top filled proton energy level is actually higher than the first empty neutron levels, making it *unstable:* as shown in figure Q13.3, it first decays to $^{44}_{21}Sc$ (scandium) by transforming one proton to a neutron, and then $^{44}_{21}Sc$ decays to $^{44}_{20}Ca$ (calcium), by transforming the other proton in the same way. The final stable nucleus has $N > Z$. The energy needed to create the

Electrostatic repulsion lifts proton levels relative to neutron levels

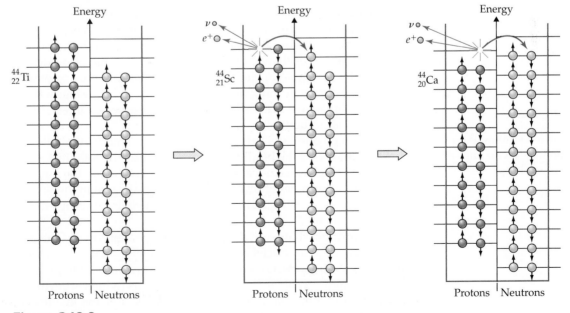

Figure Q13.3
The $^{44}_{22}Ti$ nucleus, even though it has $N = Z$, decays because electrostatic repulsion lifts the proton energy levels somewhat relative to the neutron levels.

neutron and positron in each decay (which together have more rest energy than the proton) comes from the energy released as the proton moves to the lower nuclear energy level.

The degree to which the proton energy levels get raised by electrostatic repulsion only gets larger as Z increases. Therefore, as Z gets larger, there is an increasing energy advantage to be obtained by transforming more and more protons (which are causing the trouble) to neutrons. This explains why $N > Z$ for large nuclei and why the discrepancy grows as Z increases.

Nuclei with odd Z

This shifting of the proton energy levels also explains why there are no stable odd-odd nuclei (where N and Z are odd) above $^{14}_{7}$N. In an odd-odd nucleus, there is one open space in the top proton level and one in the top neutron level, and since the proton level is not at the same energy as the neutron level, there is always an advantage to transforming a proton to a neutron or vice versa (this is what happens to the odd-odd nucleus $^{44}_{21}$Sc in figure Q13.3). The diagram implies that there will generally be *exactly one* stable isotope of a nucleus with large odd Z: the isotope whose uppermost occupied neutron level is full and just below the level with the odd proton (in the case of scandium, this isotope is $^{45}_{21}$Sc). If that neutron level is *not* full, the odd proton will transform to fill it, and if there are neutrons *above* the odd proton's level, one will transform to a proton to fill it.

Exercise Q13X.4

$^{42}_{20}$Ca, $^{43}_{20}$Ca, and $^{44}_{20}$Ca are all stable nuclei with $N > Z$. Using figure Q13.3, explain why these nuclei *cannot* decay.

Q13.4   Classical Terms in the Binding Energy

We have seen in the last two sections how the Pauli exclusion principle, energy quantization, and the weak interaction's ability to transform nucleons help us understand why $N \approx Z$ for small nuclei. When the effects of the electrostatic repulsion are taken into consideration, it becomes clear why N is somewhat larger than Z in larger nuclei, why large odd-odd nuclei don't exist, and why stable nuclei tend to have *even* numbers of protons and/or neutrons. When you think about it, this is pretty amazing: with only a few basic principles, we are able to answer a number of interesting questions about nuclear structure!

Our goal is to estimate the binding energy given only A and Z

We can go one step further by developing a model that allows us to predict quantitatively the binding energy (and thus the total rest energy) of any nucleus. Each possible nucleus is fully characterized by the number of nucleons A and the number of protons Z. Our ultimate goal in this section and section Q13.5 is to construct a formula for the nucleus' binding energy $E_b(A, Z)$ given only A and Z.

The formula that we will create will be geared toward the prediction of binding energies for nuclei with $A > 20$. In nuclei smaller than this, fairly significant quantum effects come into play that are difficult to predict without a fully quantum-mechanical model of the nucleus (which is beyond our means here). On the other hand, with large nuclei, with some justification we can make assumptions about nucleon behavior on the *average*, making it easier to arrive at a useful formula.

The model that we will construct will be a "semiclassical" model of the nucleus. While we will consider certain quantum effects on the binding energy in section Q13.5, in this section we will imagine that we can treat the nucleons in a nucleus as if they were tiny balls exerting classical forces on each other instead of being quantons occupying definite energy states. While this may not be very realistic, it does make the model much easier to understand and use. Moreover, the ultimate justification of any model in physics is its predictive power, and we will see that this model does pretty well, considering what it ignores.

We'll begin by considering the influence of the strong interaction in the binding energy. In chapter Q12, we discussed that while the strong interaction binds nucleons together very strongly (it binds protons, e.g., more strongly than the electrostatic interaction repels them), it also has a very short range. This range is on the order of 1 to 2 fm, which is also about the size of a nucleon. What this means is that a given nucleon inside a nucleus will create a strong-interaction bond with each of its *nearest* neighbors, but will not significantly interact with nucleons that don't actually touch it.

Imagine that a strong-interaction bond that a nucleon makes with a neighbor takes an energy ΔE_{bond} to break. Since this is part of the energy that has to be supplied to disassemble the nucleus, each strong-interaction bond will contribute ΔE_{bond} to the nucleus' binding energy. If each nucleon has n nearest neighbors, then the total strong-interaction contribution to the binding energy will be

Interior bonding term

$$\text{interior bonding contribution} = \left(\frac{\text{number of}}{\text{nucleons}} \right) \left(\frac{\text{bonds}}{\text{nucleon}} \right) \frac{\text{energy}}{\text{bond}}$$

$$= A \cdot n \cdot \Delta E_{bond} = a_I A \qquad (Q13.5)$$

where a_I is some constant (the **interior bonding coefficient**) that depends on the bond energy E_{bond} and the number of nearest neighbors that an interior nucleon might have: we will have to determine this constant experimentally.

This formula actually overestimates the total strong force binding energy because some nucleons lie on the surface of the nucleus and therefore have fewer neighbors to make bonds with, as illustrated in figure Q13.4. If a nucleon on the surface has j fewer bonds than one in the interior, then the amount by which we should *reduce* the binding energy calculated by equation Q13.5 to account for the nucleons on the surface should be

Surface correction term

$$\text{Surface correction} \approx (\text{number of surface nucleons}) \cdot j \cdot \Delta E_{bond} \quad (Q13.6)$$

Now, we discussed in chapter Q12 that the radius of a nucleus is empirically proportional to $A^{1/3}$: this means that the surface area of the nucleus will be proportional to $A^{2/3}$. The fact that $r \propto A^{1/3}$ (as we discussed) implies that nucleons always have roughly the same size; the number of nucleons on the surface of the nucleus will be proportional to the surface area and thus to $A^{2/3}$. Substituting this into equation Q13.6 and then subtracting the result from equation Q13.5, we find that the *total* strong-interaction contribution to the binding energy will be roughly

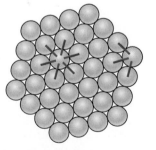

Figure Q13.4
The number of bonds of a nucleon with its nearest neighbors depends on whether it is at the nucleus' surface or is in its interior.

$$\text{strong-interaction contribution} = a_I A - a_S A^{2/3} \qquad (Q13.7)$$

where the **surface correction coefficient** a_S is another constant that we'll have to determine experimentally. Let me remind you that this surface correction is subtracted because the *lack* of bonds available to nucleons on the surface makes the nucleus less strongly bound than the first term would predict.

Finally, let's consider the effect of electrostatic repulsion on the binding energy. We can find the total electrostatic contribution to the binding energy by finding the electrostatic potential energy between each pair of protons and then summing. Now, the potential energy due to a given pair of protons

Coulomb repulsion term

A and B is ke^2/r_{AB}, so the *average* electrostatic energy per proton pair is thus

$$\frac{V_e}{\text{pair}} = ke^2 \left(\frac{1}{r}\right)_{\text{avg}} \tag{Q13.8}$$

where $(1/r)_{\text{avg}}$ is the value of $1/r$ between pairs averaged over all pairs. We have no idea what this average is, but this need not stop us. I claim that whatever $(1/r)_{\text{avg}}$ is, it is going to be proportional to the inverse radius of the nucleus: as you double the size of the nucleus, the average distance between nucleons will also double. The radius, in turn, is proportional to $A^{1/3}$, as we have seen. So the average electrostatic potential energy per pair can be written

$$\frac{V_e}{\text{pair}} = ke^2 \left(\frac{1}{r}\right)_{\text{avg}} \propto ke^2 \frac{1}{r_{\text{nucleus}}} \propto ke^2 \frac{1}{A^{1/3}} \propto \frac{1}{A^{1/3}} \tag{Q13.9a}$$

$$\Rightarrow \qquad V_e \propto (\text{number of proton pairs})(A^{-1/3}) \tag{Q13.9b}$$

Now, I claim that the number of proton pairs in a nucleus with Z protons is $\frac{1}{2}Z(Z-1)$ (see exercise Q13X.5), which (for large Z) is essentially equal to $\frac{1}{2}Z^2$. Therefore,

$$V_e \approx a_C Z^2 A^{1/3} \tag{Q13.10}$$

where the **Coulomb repulsion coefficient** a_C is another constant to be determined experimentally (a_C contains the factor of $\frac{1}{2}$ from the $\frac{1}{2}Z^2$ as well as the other constants of proportionality in equation Q13.9a). Equation Q13.10 means that the total binding energy due to the basic classical effects of the strong interaction and electrostatic repulsion can be written

$$\text{classical binding energy} \approx a_I A - a_S A^{2/3} - a_C Z^2 A^{1/3} \tag{Q13.11}$$

Note that this formula depends on A and Z (as desired) as well as some empirical constants a_I, a_S, and a_C that we hope will have the same values for all nuclei.

Exercise Q13X.5

Argue that $\frac{1}{2}Z(Z-1)$ does in fact give the number of distinct pairs of protons in a nucleus with Z protons. (*Hint:* Let us pick a proton to be the first member of the pair. How many choices do we have? Now let us pick the other member of the pair. How many choices do we have for this second member? Then think about why the factor of $\frac{1}{2}$ is there.) Also show (by directly counting the pairs) that this formula is correct for $Z = 1, 2, 3$, and 4.

Exercise Q13X.6

Explain why the electrostatic repulsion term is *subtracted* from the binding energy in equation Q13.11.

Q13.5　The Asymmetry Term

The most important *quantum* effect on the binding energy is the issue that we discussed qualitatively in sections Q13.3 and Q13.4. In this section we will try to estimate (other things being equal) how much having a different number of protons than neutrons reduces the nucleus' binding energy.

Let's *assume* that the uppermost proton and neutron energy levels have an approximately constant spacing ΔE. Let us start with a symmetric nucleus

Energy cost for converting protons to neutrons

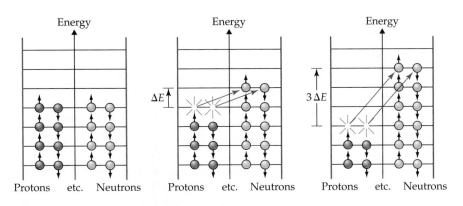

Figure Q13.5
The cost of transforming protons
to neutrons is ΔE for the first
pair, $3\,\Delta E$ for the second pair,
and so on.

where $N = Z$. We have already taken account of electrostatic repulsion in
equation Q13.11, so we will ignore it here and consider the proton levels to be
even with the neutron levels. Figure Q13.5 shows that (other things being
equal) if we convert one pair of protons to neutrons, it costs us ΔE per proton.
Converting the next pair of protons to neutrons costs us $3\,\Delta E$ per proton,
since we have to reach to the next-lower proton energy level and put them in
the next-higher neutron level. Converting the next pair costs us $5\,\Delta E$ per proton, and so on.

Now, there is an identity that says that $1 + 3 + 5 + \cdots + (2n - 1) = n^2$,
where n is the number of terms in the sum. So the cumulative cost of transforming n pairs of protons to neutrons is

$$\text{Cumulative energy cost} = 2 \cdot n^2 \cdot \Delta E \qquad \text{(Q13.12)}$$

Exercise Q13X.7

Check that $1 + 3 + 5 + \cdots + (2n - 1) = n^2$ and that n is the number of terms
in the sum for at least three small values of n.

Exercise Q13X.8

Argue that if N and Z are the numbers of neutrons and protons, respectively,
in our actual unbalanced nucleus and $N_0 = Z_0 = \frac{1}{2} A$ are the same for a hypothetical symmetric nucleus with the same A, the number of *pairs* of protons
we have to transform to get from the latter to the former is $n = \frac{1}{4}(N - Z)$.

Since $A - Z = N$, the result of exercise Q13X.8 means that the number of   Putting it all together
pairs of protons we would have to transform to get from a hypothetical com-
pletely symmetric nucleus to our actual nucleus with atomic number Z and
mass number A is $n = \frac{1}{4}(A - 2Z)$. So the total energy cost of having an unbalanced nucleus where $A \neq 2Z$ is

$$\text{Cumulative energy cost} = \tfrac{1}{8}(A - 2Z)^2\, \Delta E \qquad \text{(Q13.13)}$$

Now, one can show mathematically that the spacing ΔE between energy
levels in a three-dimensional well depends inversely on the volume of the well
(i.e., the nucleus) which is proportional to A. Therefore the energy cost associated with having an unbalanced nucleus above that of a balanced nucleus is

$$\text{Cumulative energy cost} = a_A (A - 2Z)^2 A^{-1} \qquad \text{(Q13.14)}$$

where a_A is an empirical constant called the **asymmetry coefficient**. Since as
this energy cost increases, the nucleus becomes less tightly bound, this term
contributes *negatively* to the binding energy.

Q13.6 Checking Against Reality

Combining this *quantum asymmetry* term with the classical *interior bonding, surface correction,* and *coulomb repulsion* terms, we predict that the total binding energy of a nucleus with atomic number Z and mass number A will be

The binding energy formula

$$E_b(A, Z) = a_I A - a_S A^{2/3} - a_C Z^2 A^{-1/3} - a_A (A - 2Z)^2 A^{-1} \quad \text{(Q13.15)}$$

Purpose: This equation specifies a nucleus's binding energy E_b in terms of its mass number A and atomic number Z.

Symbols: $a_I, a_S, a_C,$ and a_A are constant coefficients we call the *interior bonding, surface correction, Coulomb repulsion,* and *asymmetry coefficients,* respectively.

Limitations: This is an *approximate* formula that only applies with any accuracy when $A > 20$ and even then is only correct to within about ±1%.

Nuclear physicists call equation Q13.15 the **semiempirical binding energy formula**—*semiempirical* because although we determined the *form* of each term in the formula by using purely theoretical arguments, we have no way of finding the values of the coefficients a_I, a_S, a_C, and a_A from those arguments. The best that we can do is to find values that best fit the actually measured values of E_b. The formula fits observed binding energies *very* well if we choose

Empirical values for the various coefficients

$$a_I = 15.56 \text{ MeV} \quad \text{(Q13.16a)}$$

$$a_S = 17.23 \text{ MeV} \quad \text{(Q13.16b)}$$

$$a_C = 0.697 \text{ MeV} \quad \text{(Q13.16c)}$$

$$a_A = 23.285 \text{ MeV} \quad \text{(Q13.16d)}$$

In certain situations (particularly when we are considering nuclear reactions), it is more useful to consider the **binding energy per nucleon** $e_b \equiv E_b/A$:

The binding energy per nucleon

$$e_b(A, Z) = a_I - a_S A^{-1/3} - a_C Z^2 A^{-4/3} - a_A (A - 2Z)^2 A^{-2} \quad \text{(Q13.17)}$$

With the constants given in equations Q13.16, this formula correctly predicts the measured values of the binding energy per nucleon e_b for nuclei with $A > 20$ to within about ±1%. The comparison between predicted and actual values for e_b is illustrated for a variety of stable nuclei with odd A in figure Q13.6. (Note that higher points on the graph correspond to nuclei with *greater* binding energies per nucleon and thus to nuclei whose nucleons are more *tightly* bound.)

The fact that this formula works so well once appropriate values of the coefficients are chosen validates the theoretical arguments that we have made. This formula is really an intellectual *tour de force:* by simply *thinking* about the implications of some simple hypotheses we have created a model that predicts the binding energy of real nuclei surprisingly well. This is another example of the amazing ability of reason to illuminate the workings of the world.

Even so, the formula is not perfect. The formula fails badly for $A < 20$. Small but significant discrepancies show up even on the graph shown in figure Q13.6. What we are missing is a full and careful quantum-mechanical

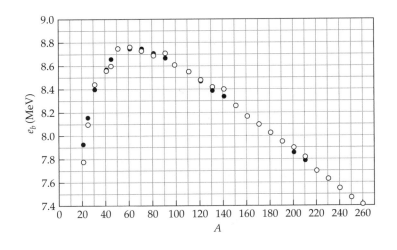

Figure Q13.6
A plot of binding energy per nucleon versus mass number A for every tenth odd-A nucleus with $A > 20$. The black dots show the values of e_b predicted by equations Q13.16 and Q13.17, while the white dots show the measured values of e_b. When only a white dot is seen, the discrepancy between predicted and actual values is less than can be indicated on the graph. (Adapted from W. S. C. Williams, *Nuclear and Particle Physics*, Oxford University Press, 1991.)

analysis of each possible nucleus. The complexity of the interactions involved make the problem so intractable that it has not yet been solved. *No theoretical formula exists that is able to predict perfectly the binding energy of a given nucleus.*

Part of the beauty of the line of reasoning leading to the binding energy formula is that it deftly sidesteps this complexity, and instead leads us directly to the most important physical determinants of the binding energy! Solving the full quantum-mechanical problem (if we could do it at all) might give more accurate results, but would probably not help us *understand* what is going on with nuclei nearly as well. Part of the art of doing physics is to find expressions of reality that not only work (in the sense of yielding correct results) but also help one understand what is going on in terms of simpler physical ideas. Equation Q13.17 is a beautiful example of this, as example Q13.1 illustrates.

This formula helps us understand and interpret the physics of nuclei

Example Q13.1   Destabilizing Effects in Small Nuclei

Problem   The curve displayed in figure Q13.6 has a maximum at $A \approx 60$. What is the chief reason why even *stable* nuclei with $A < 60$ become less tightly bound as A decreases? (*Hint:* $Z \approx \frac{1}{2}A$ for smaller nuclei.)

Solution   If we set $Z \approx \frac{1}{2}A$, then the formula for e_b becomes

$$e_b = a_I - a_S A^{-1/3} - \frac{1}{4}\frac{a_C A^2}{A^{-4/3}} - \frac{a_A(A - A)^2}{A^2}$$

$$= a_I - a_S A^{-1/3} - \frac{1}{4}a_C A^{2/3} - 0 \tag{Q13.18}$$

Of these terms, only the surface correction term actually gets larger as A gets smaller. (Even if A is different from $2Z$ by 1 or 2, the asymmetry term will still be much smaller than the surface term for $A > 20$.) This tells us that the most important reason that low-mass stable nuclei are less tightly bound than their more massive counterparts is that surface term. We might have guessed this qualitatively: as the size of the nucleus decreases, a greater proportion of the nucleons are on the surface, and since nucleons on the surface have fewer bonds, this contributes to a general decrease in the binding energy per nucleon.

Example Q13.2   The Binding Energy per Nucleon of Zinc

Problem   What is the binding energy per nucleon for $^{64}_{30}$Zn?

Solution   For this nucleus $Z = 30$ and $A = 64$, so $A^{1/3} = 4$ (convenient!). So

$$a_I = +15.56 \text{ MeV}$$

$$-a_S A^{-1/3} = -\frac{17.23 \text{ MeV}}{4} = -4.31 \text{ MeV}$$

$$-a_C Z^2 A^{-4/3} = \frac{-(0.697 \text{ MeV})(30)^2}{4^2} = -2.45 \text{ MeV}$$

$$\frac{-a_A(A - 2Z)^2}{A^2} = \frac{(23.29 \text{ MeV})(64 - 60)^2}{64^2} = -0.09 \text{ MeV}$$

The total is 8.71 MeV.

Evaluation   The measured value of e_b for $^{64}_{30}$Zn is 8.74 MeV, about a 0.3% difference.

Exercise Q13X.9

What is the chief reason that stable nuclei *above* $A = 60$ become less tightly bound as A increases? (*Hint:* $Z \approx \frac{2}{5} A$ for large nuclei.)

Exercise Q13X.10

The measured binding energy E_b for $^{40}_{20}$Ca is 342 MeV (see exercise Q12X.7). What is the value of E_b predicted by equation Q13.15?

TWO-MINUTE PROBLEMS

Q13T.1   One detector for solar neutrinos in current use utilizes a huge vat of carbon tetrachloride to absorb neutrinos. If a $^{37}_{17}$Cl atom absorbs a neutrino, what will it become? (Is the product chemically different from chlorine?)

A. $^{38}_{17}$Cl
B. $^{37}_{16}$S
D. $^{37}_{18}$Ar
E. $^{36}_{18}$Ar
F.   Other (specify)

Q13T.2   A series of possible particle processes are listed next. Which (if any) are probably *not* processes mediated by the weak interaction? (γ = photon, μ = muon = heavy electron. Don't worry about conservation of energy.)

A. $n + e^+ \rightarrow p^+ + \bar{\nu}$
B. $\nu + n \rightarrow p^+ + \mu^-$
C. $\bar{\nu} + p^+ \rightarrow n + e^+$
D. $e^- + e^+ \rightarrow \gamma + \gamma$
E. $\nu + e^+ \rightarrow \bar{n} + p^+$
F.   All might be weak processes

Q13T.3   The $^6_2$He nucleus is unstable. What will it decay to?

A. $^6_3$Li $+ e^- + \bar{\nu}$
B. $^5_3$Li $+ e^- + \bar{\nu}$
C. $^6_3$Li $+ e^+ + \nu$
D. $^5_3$Li $+ e^-$
E. $^5_1$H $+ e^+ + \nu$
F.   Other (specify)

Q13T.4 The $^{26}_{13}$Al nucleus is unstable. What will it decay to? (*Hint:* Electrostatic repulsion is significant here.)
A. $^{26}_{14}$Si $+ e^- + \bar{\nu}$
B. $^{25}_{14}$Si $+ e^- + \bar{\nu}$
C. $^{26}_{12}$Mg $+ e^- + \bar{\nu}$
D. $^{26}_{12}$Mg $+ e^+ + \nu$
E. $^{27}_{12}$Mg $+ e^+ + \nu$
F. Other (specify)

Q13T.5 The measured binding energy per nucleon for $^{15}_{7}$N is $e_b = 7.700$ MeV, while that for $^{15}_{8}$O is 7.464 MeV. Why is the average nucleon in the oxygen nucleus less tightly bound than in the nitrogen nucleus?
A. It has fewer nucleons overall.
B. It has more nucleons on the surface.
C. Electrostatic repulsion is greater.
D. The nucleus is more asymmetric.
E. Other (specify).

Q13T.6 The actual binding energy per nucleon for $^{20}_{10}$Ne is $e_b = 8.032$ MeV, while that for $^{21}_{10}$Ne is 7.972 MeV. Why is the average nucleon in the larger nucleus *less* tightly bound than in the smaller nucleus?
A. It has more nucleons on the surface.
B. Electrostatic repulsion is greater.
C. The nucleus is more asymmetric.
D. Other (specify).

Q13T.7 What is the binding energy per nucleon for $^{27}_{13}$Al, according to our semiempirical formula? (*Hint:* $27^{1/3} = 3$.)
A. 8.33 MeV
B. 22.79 MeV
C. 225 MeV
D. 8.36 MeV
E. 7.50 MeV
F. Other (specify)

HOMEWORK PROBLEMS

Basic Skills

Q13B.1 The $^{12}_{7}$N nucleus is unstable. Draw an energy-level diagram like those in figure Q13.2 and predict how it will decay. (Assume that electrostatic repulsion is *not* significant.)

Q13B.2 The $^{12}_{5}$B nucleus is unstable. Draw an energy-level diagram like those in figure Q13.2 and predict how it will decay. (Assume that electrostatic repulsion is *not* significant.)

Q13B.3 The $^{18}_{9}$F nucleus is unstable. Draw an energy-level diagram like those in figure Q13.2 and predict how it will decay. (Electrostatic repulsion *is* significant here, but it moves the proton levels up by less than the approximate spacing of the energy levels.)

Q13B.4 The $^{32}_{15}$P nucleus is unstable. Draw an energy-level diagram like those in figure Q13.2 and predict how it will decay. (Electrostatic repulsion *is* significant here, but it moves the proton levels up by less than the approximate spacing of the energy levels.)

Q13B.5 The $^{41}_{20}$Ca nucleus is unstable. Use figure Q13.3 to predict how this nucleus will decay.

Q13B.6 What are the binding energy and the binding energy per nucleon predicted by equations Q13.15 and Q13.17 for $^{125}_{52}$Te? (Te stands for *tellurium*.)

Q13B.7 What are the binding energy and the binding energy per nucleon predicted by equations Q13.15 and Q13.17 for $^{216}_{84}$Po? (Po stands for *polonium*.)

Synthetic

Q13S.1 One way to remember whether it is a neutrino or an antineutrino (or an electron or a positron) that is involved in a weak-interaction process is to use conservation laws. The weak interaction, in addition to conserving energy, momentum, and angular momentum, conserves charge, a quantity called the *lepton number* (electrons and neutrinos have lepton number +1, positrons and antineutrinos have lepton number −1, and protons and neutrons have lepton number 0), and a quantity called the *baryon number* (protons and neutrons have baryon number +1 and electrons, positrons, and neutrinos have baryon number 0).
(a) Show that the processes described by equations Q13.1 through Q13.4 conserve charge, lepton number, and baryon number.
(b) If we hit a proton with particle x having sufficient energy, it turns into a neutron and emits particle y. If x is either a neutrino or an antineutrino, which is it? What is particle y?

Q13S.2 The measured binding energy per nucleon for $^{39}_{18}$Ar is $e_b = 8.562$ MeV. The same for $^{39}_{19}$K is 8.557 MeV. Explain why an average nucleon in the argon nucleus is a little more tightly bound than in the potassium nucleus.

Q13S.3 The measured binding energy per nucleon for $^{61}_{28}$Ni is $e_b = 8.765$ MeV. The same for $^{60}_{28}$Ni is 8.781 MeV. Give at least one reason why an average nucleon in the more massive nucleus is a bit *less* tightly bound than one in the less massive nucleus, even though one might think that adding a neutron would add more strong-interaction "glue" to the nucleus.

Q13S.4 Imagine a nucleus with $Z = N$ initially. Imagine that we then add neutrons to the nucleus without changing Z. As the number of neutrons increases, there are *two* effects that lead to an average nucleon being more tightly bound and one effect that leads to it being less tightly bound. Describe these effects.

Q13S.5 Imagine a nucleus with $Z = N$ initially. Imagine that we then add protons to the nucleus without changing N. As the number of protons increases, there are two effects that lead to an average nucleon being *less* tightly bound and one effect that leads to it being *more* tightly bound. Describe these effects.

Q13S.6 In figure Q13.3, I drew the proton energy levels slightly lower in the rightmost diagram than in the leftmost diagram. Why?

Q13S.7 The experimentally measured mass of the $^{24}_{12}\text{Mg}$ atom is 23.985045 u. Find the binding energy per nucleon predicted by equation Q13.17, and compare with the *actual* value of e_b for this nucleus.

Q13S.8 The experimentally measured mass of the $^{35}_{17}\text{Cl}$ atom is 34.968853 u. Find the binding energy per nucleon predicted by equation Q13.17, and compare with the *actual* value of e_b for this nucleus.

Q13S.9 The experimentally measured mass of the $^{107}_{47}\text{Ag}$ atom is 106.905095 u. Find the binding energy per nucleon predicted by equation Q13.17, and compare with the *actual* value of e_b for this nucleus.

Q13S.10 The experimentally measured mass of the $^{200}_{80}\text{Hg}$ atom is 199.968316 u. Find the binding energy per nucleon predicted by equation Q13.17, and compare with the *actual* value of e_b for this nucleus.

Q13S.11 Compute the binding energy per nucleon for ^3_1H (whose atom has a measured mass of 3.016050 u), ^3_2He (3.016029 u), ^4_2He (4.002603 u), ^6_3Li (6.015123 u) and ^7_3Li (7.016004 u). Note how extraordinarily strongly bound the average nucleon in ^4_2He is

compared to its neighboring nuclei. Describe some of the advantages that ^4_2He has over its neighbors that might explain its exceptionally tight binding.

Rich-Context

Q13R.1 $^{24}_{12}\text{Mg}$, $^{25}_{12}\text{Mg}$, and $^{26}_{12}\text{Mg}$ are all stable. Draw energy-level diagrams (like those in figure Q13.3) for these nuclei, and use the diagrams to explain how it is possible that all *three* of these nuclei are stable. Can you say for certain that electrostatic repulsion is significant here?

Q13R.2 A neutrino must pass within about 10^{-18} m of a quark to have any chance of interacting with it. Thus the maximum target area that a quark presents to a neutrino is about 10^{-36} m^2. However, a neutrino moves so quickly and the weak interaction is so weak that experimentally only about 1 in 10^{12} typical neutrinos passing a quark within this radius actually interacts significantly with it. Something on the order of 10^{14} neutrinos per second from the sun fall on 1 m^2 of the earth's surface.
   (a) What is the (approximate) average time between neutrino interactions with your body when the sun is directly overhead?
   (b) What is the same at midnight when the entire bulk of the earth is between you and the sun?

Advanced

Q13A.1 When we found the asymmetry term in the binding energy, we found the cost for going from a hypothetical, completely symmetric nucleus to our actual asymmetric nucleus with the same A. But what about the extra energy that the nucleons in the *symmetric* nucleus have above what we'd expect classically because they are pushed above the ground level by the Pauli exclusion principle? Using ideas in section Q13.6, show that the contribution of this effect to the total binding energy should be proportional to A, which means that we can absorb it into the first term of the binding energy formula.

ANSWERS TO EXERCISES

Q13X.1 There are two protons in the $^{13}_8\text{O}$ in a level above a neutron level with one open space. One of the protons will thus decay (emitting a positron and a neutrino) to a neutron and drop into that open space. After the transition, the nucleus is an $^{13}_7\text{N}$.

Q13X.2 In order for a proton to decay, it would have to jump *up* a level to get to the unoccupied neutron level. The same thing is true if a neutron were to decay. In the first case, there is no possible source for the energy required to boost the proton. In the second, the energy required to jump up a level is greater than

the 0.8 MeV that is gained from the difference between the rest energy of the neutron and the sum of the rest energies of the proton and electron.

Q13X.3 In this case, even though the proton level below the top neutron level is full, the nucleus can *still* lower its total energy by converting a neutron to an electron and a proton, moving the proton over to the proton level of the *same* height as the neutron level. This is so because the rest energy of the neutron is slightly more than the rest energy of a proton and an electron. (This doesn't always work, though, because electrostatic repulsion lifts the proton energy levels slightly relative to the neutron levels, as we will see in the next section. If the proton level is lifted too far, there will be no energy to gain from this process.)

Q13X.4 In the case of $^{42}_{20}$Ca, the neutron level below the highest proton level is full, so a proton decaying to a neutron would have to move to a *higher* energy level: such a decay therefore cannot conserve energy. In the case of $^{43}_{20}$Ca and $^{44}_{20}$Ca, the proton level below the highest neutron level is full, so a neutron decaying to a proton would have to move to the next-higher energy level. Again this is unlikely to be consistent with conservation of energy.

Q13X.5 In a nucleus containing Z protons, imagine that we go through each of the Z protons and count up all the other protons that this proton interacts with. For each proton, their are $Z - 1$ other protons it could interact with, so the total number of pairs at first glance seems to be $Z(Z - 1)$. But this is *not* the number of distinct pairs, since this counting process counts each pair twice. (Think about it: the pair consisting of protons A and B gets counted once when A gets chosen first and again when B gets chosen first.) So the total number of distinct pairs is $\frac{1}{2}Z(Z - 1)$. Checking this, we find that when $Z = 1$, number of pairs = 0 (correct); when $Z = 2$, number of pairs = $\frac{1}{2}2(1) = 1$ (correct); when $Z = 3$, number of pairs = $\frac{1}{2}3(2) = 3$ (correct); and when $Z = 4$, the number of pairs is $\frac{1}{2}4(3) = 6$ (correct, see figure Q12.5).

Q13X.6 The Coulomb repulsion term is subtracted because an increase in repulsion tends to push nucleons apart, making the nucleus *less* tightly bound and thus *decreasing* the binding energy. Therefore, this term should appear as a negative number in the binding energy formula.

Q13X.7 For $n = 1, 2n - 1 = 1$. There is, therefore, *one* term in the sum (1) which does indeed equal $n^2 = 1^2 = 1$. If

$n = 3$, then $2n - 1 = 5$, so the sum becomes $1 + 3 + 5 = 9$, which does have three terms and which is indeed equal to n^2. If $n = 6$, then $2n - 1 = 11$, and the sum becomes $1 + 3 + 5 + 7 + 9 + 11 = 36$. We see that there are six terms in the sum and that the result is again equal to n^2.

Q13X.8 Each pair of protons that we transform increases N by 2 and decreases Z by 2, for a total change in $N - Z$ of 4. Thus if we want to change our nucleus from one where $N_0 - Z_0 = 0$ to one where $N - Z \neq 0$, we must transform $(N - Z)/4$ pairs.

Q13X.9 If we set $Z \approx \frac{2}{5}A$, then the formula for the binding energy per nucleon becomes

$$e_b \approx a_I - a_S A^{-1/3} - \frac{1}{4}\frac{a_C A^2}{A^{-4/3}} - \frac{a_A\left(\frac{1}{5}A\right)^2}{A^2}$$

$$= a_I - a_S A^{-1/3} - \tfrac{1}{4}a_C A^{2/3} - \tfrac{1}{25}a_A \qquad \text{(Q13.19)}$$

Of these terms, only the electrostatic repulsion term actually gets larger as A gets larger. This tells us that the most important reason that high-mass stable nuclei are less tightly bound than nuclei around $A = 60$ is an increasing amount of electrostatic repulsion. We might have guessed this qualitatively: as the number of protons increases, the number of repelling proton pairs increases sharply.

Q13X.10 For the $^{40}_{20}$Ca nucleus, $A = 40$ and $Z = 20$, so

$$a_I A = (15.56 \text{ MeV})(40) = +622.4 \text{ MeV}$$

$$-a_S A^{2/3} = (17.23 \text{ MeV})(40^{2/3}) = -201.5 \text{ MeV}$$

$$-a_C Z^2 A^{-1/3} = (0.697 \text{ MeV})(20)^2 40^{-1/3} = -81.5 \text{ MeV}$$

$$\frac{-a_A(A - 2Z)^2}{A} = \frac{(23.29 \text{ MeV})(0)^2}{40} = 0.0 \text{ MeV}$$

The total is 339.4 MeV, a difference of 2.6 MeV $\approx$ 0.8% from the measured value of 342 MeV.

Q14 Radioactivity

Chapter Overview

Introduction

In this chapter, we will use the binding energy formula developed in chapter Q13 to predict when nuclei will become unstable against beta or alpha decay. We will also see why the activity of a sample of unstable nuclei decreases exponentially with time.

Section Q14.1:   Beta Decay

We have seen that nuclei with either too many protons or too many neutrons are unstable. **Beta decay** processes are weak-interaction processes that adjust the N/Z ratio without changing A. The three fundamental beta decay processes are

Neutron decay (β^-):	${}_Z^A\text{Xi} \to {}_{Z+1}^A\text{Xf} + e^- + \bar{\nu}$	(lowers N/Z)	(Q14.1a)
Proton decay (β^+):	${}_Z^A\text{Xi} \to {}_{Z-1}^A\text{Xf} + e^+ + \nu$	(raises N/Z)	(Q14.1b)
Electron capture (EC):	$e^- + {}_Z^A\text{Xi} \to {}_{Z-1}^A\text{Xf} + \nu$	(raises N/Z)	(Q14.1c)

Conservation of energy and a little algebra imply that both the β^- and EC processes are possible if and only if $m_i > m_f$, where m_i and m_f are the *atomic* masses for the initial and final nuclei, respectively. The proton decay (β^+) process is only possible if $m_i > m_f + 2m_e$. We see therefore that *the nucleus with the smallest atomic mass for a given A is stable against beta decay.*

The optimal value for Z will minimize the atomic mass of a nucleus with a given A. If we write out a formula for the atomic mass, using the semiempirical binding energy formula from chapter Q13, and minimize with respect to Z, we find that

$$Z = \left(\frac{1 + [a_M/4a_A]}{2 + [a_C/2a_A]A^{2/3}} \right) A = \left(\frac{1.0084}{2 + 0.015A^{2/3}} \right) A \qquad \text{(Q14.9)}$$

Purpose: This equation estimates the optimal atomic number Z for a nucleus with a given mass number A.

Symbols: $a_M = 0.782$ MeV, $a_C = 0.697$ MeV, and $a_A = 23.285$ MeV.

Limitations: This result is based on the semiempirical binding energy formula and so shares its limitations: this result might be wrong in certain cases (because the binding energy formula is only approximate) and certainly is not accurate below $A = 20$.

Note: The optimal value of Z is within $\pm 3\%$ of $\frac{2}{5}A$ for A from 165 to 250.

This formula fairly accurately determines the optimal Z for nuclei with odd A. But since a filled energy level has a slightly lower energy than an unfilled one, it turns out that when A is even, we usually get *two* stable nuclei with even Z flanking the predicted optimal value of Z.

Section Q14.2:   Alpha Decay

Very massive nuclei shed ${}_2^4\text{He}$ nuclei to reduce electrostatic repulsion: we call this **alpha decay.** Conservation of energy (and a bit of algebra) implies that such a process is possible if and only if $m_i > m_f + m_{\text{He}}$, where m_i, m_f, and m_{He} are the *atomic* masses for the initial, final, and helium atoms, respectively. Now, if we plug equation Q14.9 into the binding energy formula, we get the binding energy $E_b(A)$ as a function of A alone for nuclei with optimal Z. A bit of algebra shows that the condition $m_i > m_f + m_{He}$ implies that

$$\frac{1}{4}E_{He} \geq \frac{dE_b(A)}{dA} \qquad (Q14.18)$$

(This turns out to be correct for all possible ejected nuclear fragments if we replace the left side with the fragment's binding energy per nucleon.) Comparing this criterion against a graph of $dE_b(A)/dA$, we find that only nuclei below about $A \approx 150$ are stable against alpha decay. This is (reasonably) consistent with the experimental evidence, which suggests that *some* nuclei with $146 \leq A \leq 209$ and *all* nuclei with $A > 209$ are unstable against alpha decay. Nuclei have a hard time shedding smaller fragments because such fragments have much lower binding energies per nucleon, meaning that A has to be much higher for equation Q14.18 to be satisfied. The ejection of larger conglomerates is simply statistically unlikely.

Because alpha decay processes can only change A by 4 (and beta decay processes do not change A), the value of A for decaying nuclei will remain within one of four sets of values where $A = 4n, 4n + 1, 4n + 2$, or $4n + 3$ (where n is an integer), respectively.

Section Q14.3: Gamma Decay

A nucleon in an excited state dropping to a lower nuclear energy level (without changing type) will emit a photon with an energy in the megaelectronvolt range. We call such processes **gamma decay** processes.

Section Q14.4: A Review of Exponentials and Logarithms

This section reviews the important properties of exponentials and natural logarithms. Read this section if you are rusty with regard to these important functions.

Section Q14.5: Decay Rates

A simple model that seems to apply to all quantum decay processes is that the decay probability during a given time interval is *fixed* for a given type of decay process for a given nucleus. This model implies that we can define a **decay constant** λ (not to be confused with wavelength) for a given process that is the process's characteristic probability of decay during an interval dt divided by that dt (in the limit that $dt \rightarrow 0$). A sample's **activity** is then λN: this is the instantaneous rate at which atoms in the sample decay in counts per second (1 **becquerel** $= 1$ Bq $\equiv 1$ decay per second; 1 **curie** $= 1$ Ci $= 3.7 \times 10^{10}$ Bq). The number of undecayed atoms N remaining therefore satisfies $dN/dt = -\lambda N$, whose solution is

$$N(t) = N_0 e^{-\lambda t} \qquad (Q14.29)$$

Purpose: This equation specifies the number of nuclei $N(t)$ remaining in a sample at time t if we start with N_0 nuclei at time $t = 0$.
Symbols: λ is the decay constant.
Limitations: This equation works very well as long as $N \gg 1$.

A decay's **half-life** $t_{1/2}$ is the time it takes N to decrease by one-half. Equation Q14.29 implies that

$$\lambda = \frac{\ln 2}{t_{1/2}} \qquad (Q14.31)$$

Q14.1   Beta Decay

General reasons for nuclear instability

A given nucleus will be unstable if a physical process exists that enables its nucleons to settle into a more tightly bound quantum state so that energy can be released during the process. If such a lower-energy state *and* a means for getting to that state both exist, then a nucleus will (sooner or later) decay.

Lower-energy states generally exist for a nucleus if (1) it doesn't have the optimal mix of protons and neutrons, (2) it has roughly the right mix of protons and neutrons but is so large that excessive electrostatic repulsion between the protons makes it energetically favorable for the nucleus to fragment into smaller pieces, or (3) it has a nucleon that (for some reason) is in an energy level higher than the lowest level available to that nucleon. These sources of instability give rise to *beta, alpha,* and *gamma decay* processes, respectively.

Beta decay processes adjust the N/Z ratio

Originally, the term **beta decay process** meant a radioactive decay process that produced a beta particle (electron), but physicists now use the term to describe any weak-interaction process that adjusts the ratio of neutrons to protons (N/Z) in a nucleus without changing the total number of nucleons A. The three most common beta decay processes (and their conventional shorthand symbols) are

Neutron decay (β^-): $^A_Z Xi \rightarrow\ ^A_{Z+1} Xf + e^- + \bar{\nu}$ (lowers N/Z)   (Q14.1a)

Proton decay (β^+): $^A_Z Xi \rightarrow\ ^A_{Z-1} Xf + e^+ + \nu$ (raises N/Z)   (Q14.1b)

Electron capture (EC): $e^- +\ ^A_Z Xi \rightarrow\ ^A_{Z-1} Xf + \nu$ (raises N/Z)   (Q14.1c)

where Xi and Xf should be replaced by the chemical symbols for the initial and final nuclei, respectively. Note that in the electron capture process, the electron captured by the nucleus is one of the initial atom's own inner electrons.

What is the condition for stability against beta decay?

For any given value of A, there will be a nucleus with the ideal N/Z ratio that represents the most stable nucleus for that A. What is the criterion for stability? Is the stable nucleus the one with the smallest nuclear rest mass, the one with the largest binding energy, or what?

A nucleus is stable if it cannot undergo any kind of decay process. A decay process releases energy by definition. With this in mind, let us consider the neutron decay process shown in equation Q14.1a. Let m_i^{nuc} and m_f^{nuc} be the rest masses of the initial and final nuclei, respectively, and let m_e be the mass of an electron (the rest energy of an antineutrino is zero). For this β^- process to occur, conservation of energy requires that

$$m_i^{nuc} c^2 = m_f^{nuc} c^2 + m_e c^2 + \Delta E$$

$$\Rightarrow\quad m_i^{nuc} > m_f^{nuc} + m_e \quad \text{(for } \beta^- \text{ decay)} \qquad (Q14.2)$$

since ΔE must be positive if the process releases energy. If we add the mass of Z electrons to both sides, this becomes

$$m_i^{nuc} + Zm_e > m_f^{nuc} + (Z+1)m_e \quad \text{(for } \beta^- \text{ decay)} \qquad (Q14.3a)$$

But since the original *atom* in this case had Z electrons and the final atom has $Z + 1$ electrons, this criterion boils down to

$$m_i > m_f \quad \text{(for } \beta^- \text{ decay)} \qquad (Q14.3b)$$

where m_i and m_f are the initial and final *atomic* masses, respectively, of the atoms involved. Therefore, this process is possible if the decaying atom's *atomic* mass exceeds the final atom's *atomic* mass.

In a similar way, *you* can show that the *same* criterion applies to EC decays:

$$m_i > m_f \quad \text{(for EC decays)} \qquad (Q14.4)$$

Exercise Q14X.1

(Important!) Verify that equation Q14.4 is correct.

Now consider the β^+ process described in equation Q14.1b. As in the β^- process, conservation of energy requires that

$$m_i^{\text{nuc}} > m_f^{\text{nuc}} + m_e \qquad \text{(for } \beta^+ \text{ decay)} \qquad \text{(Q14.5)}$$

If we add the mass of Z electrons to both sides in this case, we get

$$m_i^{\text{nuc}} + Zm_e > m_f^{\text{nuc}} + (Z+1)m_e \qquad \text{(for } \beta^+ \text{ decay)} \qquad \text{(Q14.6a)}$$

The left side is the initial atom's *atomic* mass, as before. But the atom appearing on the right of equation Q14.1b has only $Z - 1$ electrons, so the criterion in this case becomes

$$m_i > m_f + 2m_e \qquad \text{(for } \beta^+ \text{ decay)} \qquad \text{(Q14.6b)}$$

Therefore, this decay process is possible only if the mass of the initial *atom* exceeds that of the final *atom* by more than two electron masses.

We can see that a nucleus in the set of nuclei with a given A will be stable if and only if it is the one in that set with the *smallest atomic mass*. If it has the smallest atomic mass, then equation Q14.3a implies that it cannot decay by electron emission, equation Q14.4 implies that it cannot decay by electron capture, and equation Q14.6b implies that it cannot decay by positron emission.

The nucleus with the smallest *atomic* mass for a given A is stable

---

Exercise Q14X.2

The atomic mass of $^{30}_{15}\text{P}$ is 29.978307 u, while the atomic mass of $^{30}_{14}\text{Si}$ is 29.973770 u. Which decays to which, and by what process?

We can use the binding energy formula to predict the most stable nucleus for a given A as follows. What we want to do is to find the minimum value of the nucleus's atomic rest energy mc^2, which is given by

How to use the binding energy formula to *predict* the stable nucleus

$$mc^2 = Zm_{\text{H}}c^2 + (A - Z)m_nc^2 - E_b = Am_nc^2 - Z(m_n - m_{\text{H}})c^2 - E_b$$

$$= Am_nc^2 - Za_M - a_I A + a_S A^{2/3} + a_C Z^2 A^{-1/3} + \frac{a_A(A - 2Z)^2}{A}$$

$$= Am_nc^2 - Za_M - a_I A + a_S A^{2/3} + \frac{a_C Z^2}{A^{1/3}} + \frac{a_A(A^2 - 4AZ + 4Z^2)}{A}$$

$$\text{(Q14.7a)}$$

where

$$a_M \equiv (m_n - m_{\text{H}})c^2 = 0.782 \text{ MeV} \qquad \text{(Q14.7b)}$$

Note that this function has the form $mc^2 = b_1 + b_2 Z + b_3 Z^2$, where $b_1, b_2,$ and b_3 are constant for a given A, so this will be some kind of parabola. To find the value of Z that minimizes this for a given A, we take the derivative of equation Q14.7a with respect to Z and set it equal to zero:

$$0 = 0 - a_M - 0 + 0 + \frac{2a_C Z}{A^{1/3}} + \frac{a_A(0 - 4A + 8Z)}{A}$$

$$= \frac{2a_C Z}{A^{1/3}} + \frac{8a_A Z}{A} - a_M - 4a_A \qquad \text{(Q14.8)}$$

Solving this for Z, we get

Formula for the optimal Z for a given A

$$Z = \left(\frac{1 + [a_M/4a_A]}{2 + [a_C/2a_A]A^{2/3}} \right) A = \left(\frac{1.0084}{2 + 0.015 A^{2/3}} \right) A \qquad (Q14.9)$$

Purpose: This equation estimates the optimal atomic number Z for a nucleus with a given mass number A.
Symbols: $a_M = 0.782$ MeV, $a_C = 0.697$ MeV, and $a_A = 23.285$ MeV.
Limitations: This result is based on the semiempirical binding energy formula and so shares its limitations: this result might be wrong in certain cases (because the binding energy formula is only approximate) and certainly is not accurate below $A = 20$.

Note that this formula makes some intuitive sense: when A is small, $Z \approx \frac{1}{2}A$, and as A gets larger, Z becomes a smaller and smaller fraction of A.

Exercise Q14X.3

Verify equation Q14.9, and check the values of the constants.

Exercise Q14X.4

Show that the optimal value of Z is $Z = \frac{2}{5}A$ to better than $\pm 3\%$ from about $A = 165$ to $A = 250$.

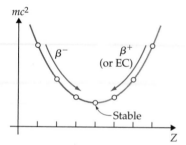

Figure Q14.1
The atomic rest energies of nuclei with A but differing Z.

A nucleus whose initial value of Z is *not* optimal for its value of A will undergo beta decay processes to get closer to the minimum atomic mass. If we plot atomic mass versus Z (as in figure Q14.1), nuclei on the *left* slope of the minimum (too few protons) will undergo β^- (neutron decay) processes to get closer to the minimum, and nuclei on the right (too many protons) will undergo β^+ if possible (EC processes otherwise) to get closer to the minimum.

Example Q14.1   The Stable Nucleus for $A = 65$

Problem   Find the stable value of Z for $A = 65$, and predict how nuclei near that value of Z will decay.

Solution   According to equation Q14.9, the value of Z that yields the minimum atomic mass is

$$Z = \frac{65(1.0084)}{2 + 0.015(65)^{2/3}} = 29.2 \qquad (Q14.10)$$

This is nearest to 29, so we predict that $^{65}_{29}$Cu is stable, and that $^{65}_{27}$Co and $^{65}_{28}$Ni will be unstable and decay via β^- processes (which emit electrons), while $^{65}_{30}$Zn and $^{65}_{31}$Ga will decay by β^+ processes (which emit positrons), as shown in figure Q14.2a. This is exactly what is observed.

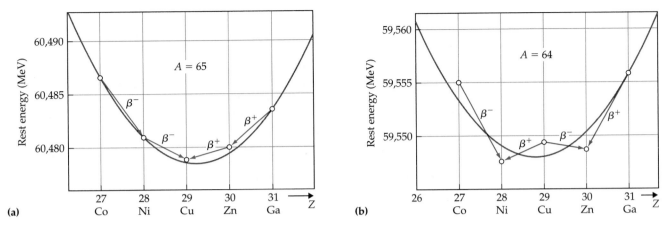

Figure Q14.2

(a) The atomic rest energies and predicted decays for nuclei having $A = 65$. White dots are measured values of the rest energy, and the gray curve is the prediction of equation Q14.7a. (b) The same for nuclei with $A = 64$. Note that the difference in binding between even-even and odd-odd nuclei causes the actual values of the adjusted binding energy to oscillate above and below the predicted curve.

Equation Q14.9 fairly successfully predicts the stable values of Z for *odd* values for A above 20. For even values of A, however, there is a complicating factor. If the predicted value of Z is an *odd* number for an *even* value of A, then this means that N would be odd, too. Such odd-odd nuclei are less stable than the binding energy formula would predict, because a filled proton or neutron energy level actually has a slightly lower energy value than one with an open spot. Nuclei with odd A always have either Z or N odd, so either the top proton level or the top neutron level has an open spot so equation Q14.9 works well. On the other hand, nuclei with *even* A either have both N and Z even (extra stable) or both odd (extra unstable). Thus if equation Q14.9 yields an odd value of Z for an even A, what usually happens is that *both* (extra stable) even values of Z on either side of this odd Z are stable, each representing a *local* minimum in the atomic mass. (Since the nucleus can only decay to its nearest neighbors, it will be stable if it is at a *local* atomic mass minimum). See figure Q14.2b for an example involving $A = 64$ (where the predicted Z is 28.8).

The pairing effect in nuclei with even A

Exercise Q14X.5

Considering this effect, what would you predict the stable nucleus or nuclei having $A = 100$ to be?

Q14.2   Alpha Decay

As the number of protons in a nucleus increases, even when the mix of protons and neutrons is optimal, the nucleus becomes more loosely bound because of the growing electrostatic repulsion between its protons. This repulsion eventually becomes so strong that all nuclei with $Z > 83$ (polonium and above) actually become unstable. These nuclei shed the excess protons two at a time by emitting an alpha particle ($^4_2$He nucleus). This process is called **alpha decay.**

Conservation of energy during this process requires that

$$m_i^{nuc}c^2 = m_f^{nuc}c^2 + m_{He}^{nuc}c^2 + \Delta E \qquad (Q14.11)$$

where ΔE is the energy released (which essentially goes entirely to kinetic energy in the ejected alpha particle). If the initial nucleus has Z protons, the final nucleus has $Z - 2$ protons. If we add Zm_ec^2 to both sides, apportioning $Z - 2$ of the electron masses added on the right to the final nucleus and two to the helium, we find that

$$m_ic^2 = m_fc^2 - m_{He}c^2 + \Delta E \qquad (Q14.12)$$

where m_i, m_f, and m_{He} are now *atomic* masses. Since ΔE has to be greater than zero, we see that *this process will occur spontaneously if and only if the initial atomic mass is greater than the final atomic mass plus the atomic mass of helium.*

Exercise Q14X.6

$^{226}_{88}$Ra (radium), whose atomic mass is 226.025402 u, decays to $^{222}_{86}$Rn (radon), whose atomic mass is 222.017570 u. The atomic mass of helium is 4.002603 u. What is the energy of the ejected alpha particle?

How big does a nucleus have to be for alpha decay to happen? We can again use the binding energy formula to find out! Using the definition of the binding energy $mc^2 = Zm_Hc^2 + (A - Z)m_nc^2 - E_b$, we see that equation Q14.12 becomes

$$\left(\sum m_{parts,i}\right)c^2 - E_{bi} = \left(\sum m_{parts,f}\right) - E_{bf} + \left(\sum m_{parts,He}\right) - E_{He} + \Delta E$$
$$(Q14.13)$$

where E_{He} is the binding energy of helium and $\sum m_{parts}$ is the sum of the mass of the protons, neutrons, and electrons for each atom. Since (*unlike* in the case of beta decay) we are not *changing* any nucleons (merely rearranging them), we can cancel these masses from both sides of the equation, leaving

$$-E_{bi} = -E_{bf} - E_{He} + \Delta E \quad \Rightarrow \quad E_{bf} + E_{He} = E_{bi} + \Delta E \quad (Q14.14)$$

In other words, the alpha decay process can take place only if the total final binding energy of the two products is greater than the initial binding energy:

$$E_{bf} + E_{He} \geq E_{bi} \quad \text{or} \quad E_{He} \geq E_{bi} - E_{bf} \qquad (Q14.15)$$

How can we find out when this condition is satisfied? First, note that if we plug equation Q14.9 into the binding energy formula, we can eliminate Z, yielding E_b as a function of A alone:

$$E_b(A) = a_I A - a_S A^{2/3} - a_C A^{5/3}\left(\frac{1.0084}{2 + 0.015 A^{2/3}}\right)^2 - a_A A\left[1 - \frac{2(1.0084)}{2 + 0.015 A^{2/3}}\right]$$
$$(Q14.16)$$

This formula specifies the binding energy for those nuclei with optimal values of Z at the given value of A, which are the only nuclei that can possibly be stable anyway. Now, E_{bi} and E_{bf} correspond to this function evaluated at A and $A - 4$, respectively, so we can approximate the difference in equation Q14.15 by the derivative

$$E_{bi} - E_{bf} = 4\frac{E_b(A) - E_b(A - 4)}{4} \approx 4\frac{dE_b(A)}{dA} \qquad (Q14.17)$$

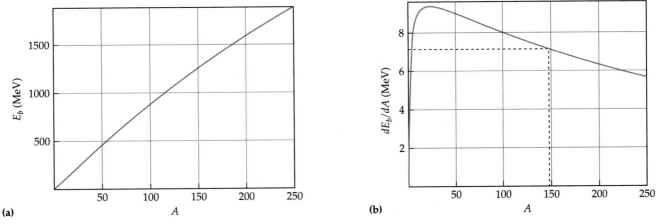

Figure Q14.3

(a) A graph of the curve of binding energy $E_b(A)$ as a function of A for nuclei with optimal values of Z. (b) A graph of dE_A/dA.

Therefore, the condition given by equation Q14.15 becomes

$$\frac{1}{4}E_{\text{He}} \geq \frac{dE_b}{dA} \qquad \text{(Q14.18)}$$

The value of $\frac{1}{4}E_{\text{He}} = 7.074$ MeV, so nuclei with optimal values of Z will be unstable against alpha decay at approximately the value of A where $dE_b(A)/dA$ drops below 7.074 MeV.

Figure Q14.3 displays graphs of $E_b(A)$ and $dE_b(A)/dA$ for values of A ranging up to about 250. We see dE_b/dA drops below 7.074 MeV at approximately $A = 150$. Therefore, we would predict on the basis of the binding energy formula that nuclei stable against beta decay would become unstable against alpha decay above about $A = 150$.

Of course, this is an *approximate* result based on our semiempirical binding energy formula. Empirically, the nucleus $^{146}_{62}\text{Sm}$ is very close to having an optimal N/Z, and yet is observed to undergo alpha decay: this is one of the lowest-A nuclei with nearly optimal N/Z known to do so. A number of other such alpha-unstable nuclei with $146 < A < 209$ exist with nearly optimal N/Z, intermixed with others that are stable. *All* nuclei with $A > 209$ are unstable; those with optimal N/Z ratios undergo alpha decay. So our estimate is not all that far off.

Why do nuclei having too many protons emit alpha particles instead of just ejecting protons, ^2_1H nuclei, ^3_2He nuclei, or a myriad of other possibilities? The answer lies in the exceptionally large binding energy of the ^4_2He nucleus. Note that the quantity $\frac{1}{4}E_{\text{He}}$ that appears on the left side of equation Q14.18 is the *binding energy per nucleon* of the ^4_2He nucleus. In fact, equation Q14.18 applies to *any* possible ejected combination of nucleons, as long as we replace $\frac{1}{4}E_{\text{He}}$ with the binding energy per nucleon of whatever combination is ejected (see problem Q14S.9). A graph of the binding energy per nucleon of various low-A nuclei is shown in figure Q14.4. Note how the binding energy per nucleon for ^4_2He is unusually high compared to its neighbors. This is mostly due to the strong pairing effect that favors even-even nuclei (especially in small nuclei).

If we look back at equation Q14.18, we see that as the binding energy per nucleon for the ejected particle gets smaller, the minimum A where ejection of such a particle becomes energetically possible becomes larger. The minimum value of A for ^2_1H ejection, for example, is above 800 (see problem Q14B.10)!

Why not emit other kinds of fragments?

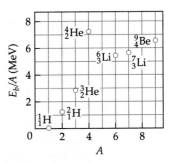

Figure Q14.4

Binding energies per nucleon for various low-mass nuclei.

But there are a host of nuclei (starting with $^{12}_6C$) that have binding energies per nucleon larger than that for the alpha particle. Why don't massive nuclei eject larger fragments such as these?

Ejection of a larger nuclear fragment is called **spontaneous fission,** and it is observed in artificially-created nuclei with extremely large masses. But experimental evidence suggests that the probability of ejecting a fragment rapidly decreases as the size of the ejected fragment increases. This might be "explained" (in terms of newtonian concepts) as follows: as the nucleons randomly move around in the nucleus, it might be fairly likely for two protons and two neutrons to randomly end up close to each other and traveling with the same velocity, so that they might escape together. But it would be much *less* probable for six protons and six neutrons to randomly end up close together and with the same velocity to eject a nucleus, and this becomes increasingly unlikely as the required conglomeration gets larger. So it is not that spontaneous fission is *energetically* impossible, it is just very *improbable* if the fragment is much bigger than an alpha particle.

Alpha decay sequences

A succession of alpha decays tends to leave a massive nucleus with too many neutrons, since the optimal ratio of neutrons to protons in a massive nucleus is about 3 to 2, but alpha decay removes protons and neutrons in equal numbers. Therefore, massive radioactive nuclei decay to a stable endpoint by both alpha decay (to trim down) and β^- decay (to increase the ratio of protons to neutrons).

Note that alpha decays reduce a nucleus' value of A by 4, while β^- decays do not change A. Therefore, a nucleus with $A = 4n$ (where n is an integer) always decays to other nuclei satisfying $A = 4n$ as well (where n may be a different integer). Similarly, a nucleus with $A = 4n + 2$ always decays to nuclei also satisfying $A = 4n + 2$. Thus, most of the alpha emitters found in nature (which are all near the optimal N/Z ratio) lie somewhere along one of four decay sequences near the optimum mix line, one sequence for nuclei

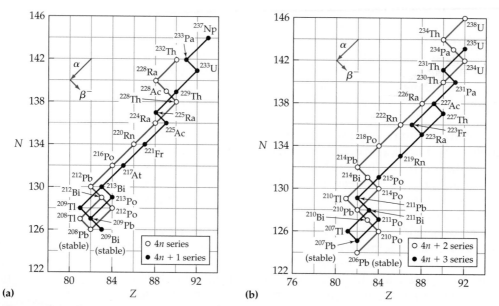

Figure Q14.5
The four alpha decay sequences for nearly stable nuclei just above the top stable value of Z. The top nucleus in each sequence is the member with the longest half-life, while the bottom nucleus in each is stable. (Adapted from Beiser, *Concepts of Modern Physics,* 4th ed., McGraw-Hill, New York 1987.)

having $A = 4n$, $A = 4n + 1$, $A = 4n + 2$, and $A = 4n + 3$, respectively, where n is some integer. (Even nuclei that are far from having the optimal N/Z will typically decay rapidly to a member of one of these sequences.) These sequences are shown in figure Q14.5. The top nucleus in each sequence shown happens to decay *very* slowly compared to the other unstable nuclei in the sequence. The top nuclei in the $4n$, $4n + 2$, and $4n + 3$ sequences, in fact, have lifetimes long enough to have survived since the earth was formed and thus are found in nature: the other unstable nuclei in these chains exist in nature only because they are decay products of the top nucleus.

Q14.3   Gamma Decay

Often an alpha or beta decay process leaves the nucleus in an "excited state" where at least one nucleon is sitting in an energy level above the lowest one available to it. After a certain amount of time, the nucleon will spontaneously drop to the lower level, emitting a photon that carries off the energy difference between the two levels. This process, which is called **gamma decay,** is illustrated in figure Q14.6.

This situation is completely analogous to an electron in an atom undergoing a spontaneous transition from a higher to a lower energy level. The main difference is that the *energies* of photons radiated by transitions between electron energy levels in atoms are typically on the order of a few electronvolts. Because of the strength of the strong nuclear force and the tiny size of the nucleus, the energy difference between adjacent levels in a nucleus is typically a *million* times larger, and so gamma decay photons typically have energies on the order of a few megaelectronvolts!

The effective wavelength $\lambda = hc/E$ of such photons is thus about a million times shorter than the wavelength of visible light (≈ 500 fm $\approx$ 5×10^{-4} nm instead of ≈ 500 nm). This is even quite a bit shorter than the wavelength of low-energy *electrons* (which is about 0.1 nm). The wave nature of such *gamma ray photons* is thus hard to observe: for all practical purposes they behave as high-energy electrically neutral particles do.

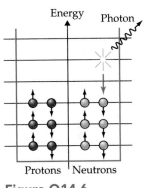

Figure Q14.6
An illustration of gamma decay.

Q14.4   A Review of Exponentials and Logarithms

Section Q14.5 requires that you be very familiar with the exponential and natural logarithm functions. If you are already comfortable with these functions, you can skip ahead to section Q14.5. If not, this section will help bring you up to speed.

The **exponential** function e^x is a function $f(x)$ mathematically defined to be that function whose derivative is itself:

$$\frac{de^x}{dx} \equiv e^x \qquad \text{(Q14.19)}$$

Mathematically, this turns out to be equivalent to saying that e^x is equivalent to taking the irrational number $e = 2.71828 \cdots$ to the x power:

$$e^x = (2.71828 \cdots)^x \qquad \text{(Q14.20)}$$

Since $a^{n+m} = a^n a^m$ and $a^0 \equiv 1$ for any a, n, and m (you can most easily check this for integer powers), the exponential function has the following properties:

$$e^{x+y} = e^x e^y \qquad \text{(Q14.21a)}$$

$$e^{ax} = (e^x)^a \qquad \text{(Q14.21b)}$$

$$e^0 = 1 \qquad \text{(Q14.21c)}$$

$$e^{-x} = \frac{1}{e^x} \qquad \text{(Q14.21d)}$$

where the last follows from the fact that $e^x e^{-x} = e^{x-x} = e^0 = 1$.

Equation Q14.19 implies that if we plot e^x starting at $x = 0$, we find that the curve starts at a value of 1 and slope of 1. Since the slope is positive, the curve will go to larger values as x increases. But as e^x increases with x, so does the slope de^x/dx. Therefore, both the value and the slope of e^x increase together without limit as x becomes large. On the other hand, when x is negative, we have $e^x = e^{-|x|} = 1/e^{|x|}$, so e^x must decrease asymptotically to zero as x becomes increasingly negative. Figure Q14.7a shows a graph of e^x.

The **natural logarithm** is defined to be the inverse function of the exponential: what the exponential does to an x, the logarithm undoes, and vice versa:

$$\ln e^x = x \qquad e^{\ln x} = x \qquad \text{(Q14.22)}$$

Note that this definition implies the following properties for the logarithm

$$\ln 1 = \ln e^0 = 0 \qquad \text{(Q14.23a)}$$

$$\ln \frac{1}{x} = \ln \left(\frac{1}{e^{\ln x}} \right) = \ln(e^{-\ln x}) = -\ln x \qquad \text{(Q14.23b)}$$

$$\ln xy = \ln(e^{\ln x} e^{\ln y}) = \ln e^{\ln x + \ln y} = \ln x + \ln y \qquad \text{(Q14.23c)}$$

$$\ln \frac{x}{y} = \ln x + \ln \frac{1}{y} = \ln x - \ln y \qquad \text{(Q14.23d)}$$

The definition also implies that (if we define $y \equiv e^x$)

$$1 = \frac{dx}{dx} = \frac{d\ln e^x}{dx} = \frac{d\ln y}{dy}\frac{dy}{dx} = \frac{d\ln y}{dy}y \quad \Rightarrow \quad \frac{d\ln y}{dy} = \frac{1}{y} \quad \text{(Q14.24)}$$

for any y (since $dy/dx = de^x/dx = e^x = y$). Figure Q14.7b shows a graph of $\ln x$.

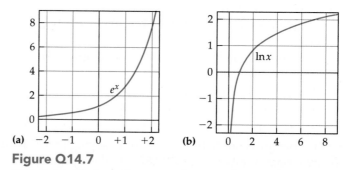

Figure Q14.7

(a) A graph of e^x as a function of x. Note that the slope and value of this function both increase together as x increases.
(b) A graph of $\ln x$. Since $\ln x$ is the inverse of e^x, its graph is the same as that for e^x with the horizontal and vertical axes exchanged. Note that $\ln x$ is undefined for $x < 0$, since e^x is always positive.

Q14.5   Decay Rates

We have seen in this chapter that the way that an unstable nucleus decays depends on why it is unstable. If the nucleus has more than the optimal number of neutrons, it will decay by electron emission (β^- decay) processes, which convert neutrons to protons. If it has more than the optimal number of protons, it will decay by positron emission (β^+ decay) or electron capture (EC) processes, which convert protons to neutrons. Some nuclei with $A > 146$ (and all nuclei with $A > 209$) decay by emitting alpha particles, even when they have the optimal value of Z: this relieves the stress caused by the electrostatic repulsion of all those protons. If a nucleon ends up in an energy level above the lowest possible energy level available to that nucleon, it will eventually decay to that lower level by emitting a gamma ray photon.

The thing that all these processes have in common is that they are all spontaneous transition processes that take a nucleus from one quantum state to a lower-energy quantum state. Quantum field theory (supported by empirical evidence) implies that for any given spontaneous transition process, there is always a *fixed probability* that it will occur during a given time interval Δt. (This applies even to ordinary atomic transitions that emit photons!)

Quantum transitions have constant probability of occurring in a given interval

For example, imagine that a given nucleus has a 50% probability of decaying during a 1.0-s time interval. If the nucleus survives the first second, it will then have a 50% probability of surviving the next second, and so on. It is as if the nucleus flipped a coin every second to see whether it should decay. When the coin finally comes up heads, the nucleus decays. Such a nucleus might decay during the first second, or it might survive for many seconds (just as a coin might come up heads the first time or come up tails for many sequential throws). The best that we can say about an individual such nucleus is that it will most *probably* decay within a handful of seconds.

The behavior of a statistically large number N of such nuclei is more predictable. Since every nucleus has a fixed probability of decaying in a given time interval, for every nucleus we can define a **decay constant** λ as follows

$$\lambda \equiv \frac{\text{decay probability during } dt}{dt} \equiv \frac{(\text{decays in } N \text{ nuclei during } dt)/N}{dt}$$

Definition of the *decay constant* λ

(Q14.25)

where dt is an "infinitesimal" time interval, that is, an interval sufficiently short that the value of N doesn't change much during dt. (Please note that this λ has nothing to do with wavelength.) This means that

$$\frac{\text{Number of decays in } N \text{ nuclei during } dt}{dt} = \lambda N \qquad (Q14.26)$$

Definition and units for the activity of a sample

We call λN the **activity** or **decay rate** of the sample. The SI unit for the activity of a radioactive sample is the **becquerel** (Bq), where 1 Bq $\equiv$ 1 decay per second. An older unit is the **curie** (Ci), where 1 Ci $\equiv$ the activity of 1.0 g of $^{226}_{88}$Ra (radium) $\approx 3.7 \times 10^{10}$ Bq.

The activity λN of a sample *would* express the number of nuclei that decay per second *if* N were to remain constant during that second. However, as these nuclei decay, the number of atoms N that are left to decay becomes smaller:

$$\frac{dN}{dt} = -\lambda N \qquad (Q14.27)$$

The number of undecayed nuclei decreases exponentially with time

where dN is the *change* in the number of undecayed nuclei N during the infinitesimal time interval dt (dN is *negative* because N is *decreasing*). If λ is so small that $N \approx$ constant during a 1-s interval, then λN *does* fairly accurately give the number of nuclei that decay per second in a sample of N nuclei.

We can determine how N depends on time as follows. If we multiply both sides of equation Q14.22 by dt/N and take the integral of both sides, we get

$$\frac{dN}{N} = -\lambda\,dt \quad \Rightarrow \quad \int_{N_0}^{N(t)} \frac{dN}{N} = -\lambda \int_0^t dt = -\lambda t$$

$$\ln N(t) - \ln N_0 = -\lambda t \quad \Rightarrow \quad \ln \frac{N(t)}{N_0} = -\lambda t \qquad \text{(Q14.28)}$$

since $\int y^{-1}\,dy = \ln y$ (see equation Q14.24). Taking the exponential of both sides of this equation, we get

$$\frac{N(t)}{N_0} = e^{-\lambda t} \quad \Rightarrow \quad N(t) = N_0 e^{-\lambda t} \qquad \text{(Q14.29)}$$

Purpose: This equation specifies the number of nuclei $N(t)$ remaining in a sample at time t if we start with N_0 nuclei at time $t = 0$.
Symbols: λ is the decay constant.
Limitations: This equation works very well as long as $N \gg 1$.

We see that if the probability of a nucleus decaying in a given time interval is a constant, the number of undecayed nuclei will decrease exponentially with time.

Exercise Q14X.7

Check that the result in equation Q14.29 is consistent with equation Q14.27 by evaluating the time derivative of $N(t)$ and showing that it is indeed equal to $-\lambda N(t)$.

Exercise Q14X.8

Argue that the *activity* of a sample of radioactive material also decreases exponentially in time.

Definition of the *half-life*

We often describe how fast a given sample of radioactive material decays in terms of its **half-life** $t_{1/2}$, which is defined to be the time that it takes the number of undecayed nuclei to decrease to one-half of its original value. According to equation Q14.29, this means that

$$\frac{1}{2} = \frac{N(t_{1/2})}{N_0} = e^{-\lambda t_{1/2}} \qquad \text{(Q14.30)}$$

By taking the inverse and then the natural logarithm of both sides of this equation, you can show that

$$t_{1/2} = \frac{\ln 2}{\lambda} \qquad \text{(Q14.31)}$$

Exercise Q14X.11

Check this.

Thus the half-life is a fixed constant for a given nucleus that is simply a different way of talking about the decay constant λ. Note in particular that the half-life of a sample does *not* depend on our choice of $t = 0$: if we look at a sample of N_0 nuclei now and find that one-half ($\frac{1}{2}N_0$) decay in 1 y, if we come back 30 y later and look at the (much smaller) sample of N_0' nuclei that remain, we will *still* find that $\frac{1}{2}N_0'$ will decay during the next year. This is a basic consequence of the idea that the probability that a given nucleus will decay in a given time is constant, irrespective of how long it has already lasted.

Example Q14.2   Safety Issues in Storing Tritium

Problem   Imagine that we have a bottle containing 4.5 g of $^3_1$H (tritium), which decays by a β^- process to $^3_2$He with a half-life of 12.26 y. **(a)** What is the activity of this sample (in becquerels and curies)? **(b)** About how long would it take the number of undecayed atoms (and thus the activity of this sample) to decrease by a factor of 1 billion?

(a) *Model*   Since Avogadro's number of protons or neutrons has a mass of about 1 g, Avogadro's number of tritium atoms would have a mass of 3 g. Thus our sample contains about $1.5 \times$ Avogadro's number $\approx 9 \times 10^{23}$ nuclei. According to equation Q14.31, the decay constant for the decay of this nucleus is

$$\lambda = \frac{\ln 2}{t_{1/2}} = \frac{\ln 2}{12.26 \text{ y}}\left(\frac{1 \text{ y}}{3.16 \times 10^7 \text{ s}}\right) = 1.8 \times 10^{-9} \text{ / s} \qquad \text{(Q14.32)}$$

Solution   The activity of this sample is thus

$$\lambda N = (1.8 \times 10^{-9} \text{ / s})(9.0 \times 10^{23})\left(\frac{1 \text{ Bq}}{1 \text{ / s}}\right) = 1.6 \times 10^{15} \text{ Bq}$$

$$= (1.6 \times 10^{15} \text{ Bq})\left(\frac{1 \text{ Ci}}{3.7 \times 10^{10} \text{ Bq}}\right) = 44{,}000 \text{ Ci} \qquad \text{(Q14.33)}$$

Evaluation   Such a sample would be *very* hazardous, as we will see in chapter Q15.

(b) *Model*   We want to find the time T such that

$$10^{-9} = \frac{N(T)}{N_0} = e^{-\lambda T} \qquad \Rightarrow \qquad 10^9 = e^{+\lambda T}$$

Solution   Solving for T, we get

$$T = \frac{\ln 10^9}{\lambda} = \frac{\ln 10^9}{1.8 \times 10^{-9} \text{ / s}}\left(\frac{1 \text{ y}}{3.16 \times 10^7 \text{ s}}\right) = 360 \text{ y} \qquad \text{(Q14.34)}$$

Evaluation   The units are right, and the answer seems reasonable. The activity of the sample at this point would be 44 μC, which is still pretty radioactive (for comparison, the sources that can be used for education without a special license must be smaller than 1 μC). Therefore, we would need to store our tritium sample securely for at *least* this long before it became safe.

One of the most common and instructive uses of radioactive materials is in radioactive dating of artifacts, rocks, and so on. In general, the technique involves somehow inferring the ratio of the number of atoms of a radioactive isotope N to the number of atoms N_0 that existed at some time $t = 0$. According to equation Q14.29, $N/N_0 = e^{-\lambda t}$, where t is the elapsed time since $t = 0$. If we know λ for the isotope, we can find t. This method generally yields pretty accurate results as long as t is of the same order of magnitude as the half life. (A more technical way to state the requirement is that we should have $0.2 < \lambda t < 5$ for best results.)

A simple example is the use of $^{40}_{19}K$ (potassium) to date rocks. As in the case of $^{64}_{29}Cu$ (see figure Q14.2b), this odd-odd nucleus decays with a half-life of 1.25×10^9 y to either of *two* stable nuclei $^{40}_{18}Ar$ or $^{40}_{20}Ca$ in a ratio of 1 argon atom for every 8 calcium atoms. While argon is found in the atmosphere, it is all but excluded from rocks as they form because of its chemical inactivity. So essentially all the argon in a rock is there as a result of decays of $^{40}_{19}K$ since the rock solidified (solidification prevents the argon from moving out).

Example Q14.3   Potassium Dating

Problem Imagine that we find 1.1×10^{11} $^{40}_{18}Ar$ atoms in a sample of rock that also contains 3.2×10^{11} $^{40}_{19}K$ atoms. When was this rock formed?

Model The existence of 1.1×10^{11} $^{40}_{18}Ar$ atoms implies that there must have been 9 times this much, or 1.0×10^{12} atoms, of $^{40}_{19}K$ in the rock when it solidified (eight-ninths of which decayed to calcium, which cannot be distinguished from natural calcium in the rock). Combining this with the number of $^{40}_{19}K$ atoms still remaining in the sample, we find the total number N_0 of $^{40}_{19}K$ atoms that were in the rock at the time of its formation to be 1.32×10^{12}. The number currently remaining is $N = 3.2 \times 10^{11}$. The equation $\lambda = (\ln 2)/t_{1/2}$ allows us to solve for λ, and $N/N_0 = e^{-\lambda t}$ for the elapsed time t since the rock's formation.

Solution Doing the algebra, we get

$$\frac{N}{N_0} = e^{-\lambda t} \quad \Rightarrow \quad \frac{N_0}{N} = e^{+\lambda t} \quad \Rightarrow \quad t = \frac{\ln(N_0/N)}{\lambda} = t_{1/2}\frac{\ln(N_0/N)}{\ln 2}$$

(Q14.35a)

$$t = \frac{(1.25 \times 10^9 \text{ y})}{\ln 2} \ln\left(\frac{1.32 \times 10^{12}}{0.32 \times 10^{12}}\right) = 2.6 \times 10^9 \text{ y} \qquad (Q14.35b)$$

Evaluation This result has the right units and a reasonable magnitude.

Carbon-14 dating is based on a somewhat different principle. Cosmic rays continually produce $^{14}_{6}C$ in the upper atmosphere. This means that even though this isotope has a half-life of only 5730 y, there is always a constant amount of $^{14}_{6}C$ in the environment (about one carbon atom in 7.7×10^{11} is $^{14}_{6}C$). This means that a *living* creature, which is constantly shifting carbon in and out of its body, will maintain about the same ratio of $^{14}_{6}C$ to $^{12}_{6}C$ as is found in the environment. If that creature dies and is buried, however, it no longer is exchanging carbon with the environment. The number of $^{12}_{6}C$ atoms in the creature's body remains constant, while the number of $^{14}_{6}C$ atoms decreases exponentially. The ratio of $^{14}_{6}C$ to $^{12}_{6}C$ in an organic sample can be used to determine how long it has been since that sample was part of a living creature.

Example Q14.4   Carbon-14 Dating

Problem   Imagine a sample of leather from a tomb is burned to obtain 0.12 g of carbon. The measured activity of the sample due to $^{14}_{6}C$ decay is 0.012 Bq (about 43 decays per hour). How old is the sample? For $^{14}_{6}C$, $t_{1/2} = 5730$ y.)

Model   Note that $\lambda = (\ln 2)/t_{1/2} = (\ln 2)/(5730 \times 3.16 \times 10^7 \text{ s}) = 3.8 \times 10^{-12}$ per second for $^{14}_{6}C$ decay. Now, 12 g of carbon corresponds to 1 mol of carbon, so the sample has 0.01 mol $= 6 \times 10^{21}$ carbon atoms. While the leather was still part of a living cow, it would have had $N_0 = (6 \times 10^{21})/(7.7 \times 10^{11}) = 7.8 \times 10^9$ $^{14}_{6}C$ atoms and thus an *original* activity of

$$A_0 = \lambda N_0 = (3.8 \times 10^{-12} / \text{s})(7.8 \times 10^9) = 0.030 \text{ Bq} \quad \text{(Q14.36a)}$$

This is related to the current activity by $A/A_0 = \lambda N/\lambda N_0 = N/N_0 = e^{-\lambda t}$ so $A_0/A = e^{+\lambda t}$, where t is the time since the cow died.

Solution   Solving for t, we get

$$t = \frac{\ln(A_0/A)}{\lambda} = \frac{\ln(0.030/0.012)}{(3.8 \times 10^{-12} / \text{s})}\left(\frac{1 \text{ y}}{3.16 \times 10^7 \text{ s}}\right) = 7600 \text{ y} \quad \text{(Q14.36b)}$$

Evaluation   Again, the units are right, and the answer seems reasonable.

Scientists applying potassium dating to meteorites (which were presumably formed when the solar system was formed) get a consistent value of 4.5 Gy. Similarly, potassium dating puts the time of formation of lunar rocks returned by the Apollo astronauts at various times between 3.7 Gy and 4.6 Gy, but no samples are older than 4.6 Gy. These are but two of the many radioactive dating results implying that the age of both the earth and the solar system is about 4.6 Gy.

TWO-MINUTE PROBLEMS

Q14T.1 $^{60}_{27}Co$ has an atomic mass of 59.933819 u, and $^{60}_{28}Ni$ has an atomic mass of 59.930788 u. Which is the correct decay process?

A. $^{60}Ni \to {}^{60}Co$
B. $^{60}Ni \to {}^{60}Co + e^- + \bar{\nu}$
C. $^{60}Ni \to {}^{60}Co + e^+ + \nu$
D. $^{60}Co \to {}^{60}Ni + e^- + \bar{\nu}$
E. $^{60}Co \to {}^{60}Ni + e^+ + \nu$
F. Some other decay process (specify)

Q14T.2 $^{79}_{35}Br$ has an atomic mass of 78.918336 u, and $^{79}_{36}Kr$ has an atomic mass of 78.920084 u. Which is the correct decay process?

A. $^{79}Kr \to {}^{79}Br$
B. $^{79}Kr \to {}^{79}Br + e^- + \bar{\nu}$
C. $^{79}Kr \to {}^{79}Br + e^+ + \nu$
D. $^{79}Br \to {}^{79}Kr + e^- + \bar{\nu}$
E. $^{79}Br \to {}^{79}Kr + e^+ + \nu$
F. Some other process (specify)

Q14T.3 $^{55}_{26}$Fe, which has an atomic mass of 54.938296 u, decays with a half-life of 2.7 y to $^{55}_{25}$Mn, which has an atomic mass of 54.938047 u. Does this likely decay by a β^+ process, or *must* it decay by electron capture?
A.   A β^+ process
B.   An EC process
C.   What do you mean? It decays by a β^- process!
D.   What do you mean? $^{55}_{25}$Mn must decay to $^{55}_{26}$Fe!

Q14T.4 The likely stable value(s) for Z when A = 88 is/are
A.   38
B.   39
C.   38 and 39
D.   38 and 40
E.   Some other value or values (specify)

Q14T.5 The main reason that massive nuclei simply don't shed protons to relieve the instability caused by too much electrostatic repulsion is that
A.   Energy would not be conserved for typical values of A.
B.   This would lead to openings in proton energy levels.
C.   Protons aren't repelled as strongly as alpha particles.
D.   Isolated protons can't exist outside an atom.
E.   Other (specify).

Q14T.6 $^{235}_{92}$U (uranium) could decay to $^{206}_{82}$Pb (lead) by appropriate combination of alpha and beta decay processes, true (T) or false (F)?

Q14T.7 Imagine that we have 6.4×10^{20} undecayed radioactive nuclei at $t = 0$. After 1 h, we have 3.2×10^{20}. How many will we have left after 3 h more?
A.   They will be all gone after 1 h more.
B. 8×10^{19}
C. 4×10^{19}
D. 2×10^{19}
E.   Less than 1×10^{19}

Q14T.8 The amount of $^{40}_{18}$Ar in a rock is one-ninth the amount of $^{40}_{19}$K. If the half-life of the latter is T, what is the age of the rock?
A. T
B. $T/9$
C. $9T$
D. $\ln 9 T$
E. $T/\ln 9$
F.   Other (specify)

Q14T.9 The amount of $^{14}_{6}$C in a bone is about one-fourth of what one would expect based on the relative abundance of $^{14}_{6}$C and $^{12}_{6}$C in living tissue. Since the half-life of $^{14}_{6}$C is a bit less than 6000 y, the age of the bone is closest to
A.   36,000,000 y
B.   24,000 y
C.   12,000 y
D.   4000 y
E.   1500 y

HOMEWORK PROBLEMS

Basic Skills

Q14B.1 $^{19}_{8}$O, which has an atomic mass of 19.003577 u, decays via β^- decay to $^{19}_{9}$F, which has an atomic mass of 18.998403 u. How much energy is released in this decay process? (This energy goes to the kinetic energy of the electron and the antineutrino.)

Q14B.2 $^{13}_{7}$N, which has an atomic mass of 13.005738 u, decays via β^+ decay to $^{13}_{6}$C, which has an atomic mass of 13.003355 u. How much energy is released in this decay process? (This energy goes to the kinetic energy of the positron and the neutrino.)

Q14B.3 $^{37}_{18}$Ar, which has an atomic mass of 36.966776 u, decays via electron capture to $^{37}_{17}$Cl, which has an atomic mass of 36.965903 u. Predict the energy of the emitted neutrino (assuming that the captured electron does not bring in much energy other than its rest energy).

Q14B.4 Predict the optimal value of Z for A = 109.

Q14B.5 Predict the optimal value of Z for A = 209.

Q14B.6 $^{238}_{92}$U (atomic mass 238.050784 u) decays via alpha decay to $^{234}_{90}$Th (atomic mass 234.043593 u). What is the kinetic energy of the emitted alpha particle?

Q14B.7 $^{239}_{94}$Pu (atomic mass 239.052157 u) decays via alpha decay to $^{235}_{92}$U (atomic mass 235.043924 u). What is the kinetic energy of the emitted alpha particle?

Q14B.8 A sample containing 6×10^{16} undecayed nuclei of a certain type decays at an instantaneous rate of 2×10^9 Bq. What is λ for this nucleus? What is its half-life?

Q14B.9 A sample containing 5.4×10^{18} undecayed nuclei of a certain type is measured to decay at an instantaneous rate of 8×10^{11} Bq. What is λ for this nucleus? What is its half-life?

Q14B.10 Figure Q14.8 shows a graph of dE_b/dA for A up to 1100. Assume that equation Q14.18 applies to any small fragment being ejected from a massive nucleus, as long as the last term is the binding energy per nucleon of the ejected fragment. Estimate the

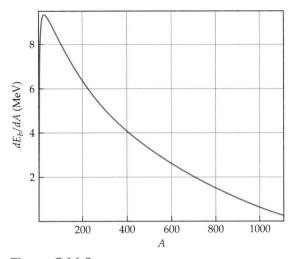

Figure Q14.8
A graph of dE_A/dA for A up to 1100.

minimum value of A required for ejection of a $_1^2$H fragment (you can use figure Q14.3 to estimate the binding energy for $_1^2$H).

Synthetic

Q14S.1 Why is it hard to observe the wavelike behavior of gamma decay photons? Why not just set up two slits and display the interference pattern, as we do for visible light? Explain your response carefully.

Q14S.2 A given nucleus has a 0.05% probability of decaying in a 1-s time interval. What would be the half-life of a sample of this isotope?

Q14S.3 Prove that $_{18}^{37}$Ar *cannot* decay to $_{17}^{37}$Cl by a β^+ process. (See problem Q14B.3 for necessary data.)

Q14S.4 $_{20}^{41}$Ca (atomic mass 40.962278 u) decays to $_{19}^{41}$K (atomic mass 40.961825 u). Can this decay be a β^+ decay, or must it go by electron capture?

Q14S.5 Use the formula for the optimal value of Z to predict how $_{80}^{197}$Hg will decay.

Q14S.6 Use the formula for the optimal value of Z to predict how $_{47}^{111}$Ag will decay.

Q14S.7 The atomic masses of $_{18}^{39}$Ar, $_{19}^{39}$K, and $_{20}^{39}$Ca are 38.964315 u, 38.963999 u, and 38.970718 u, respectively. Which one is stable? How will the others decay? Will any of these nuclei decay by electron capture but not β^+ decay?

Q14S.8 The atomic masses of $_{34}^{81}$Se, $_{35}^{81}$Br, $_{36}^{81}$Kr, and $_{37}^{81}$Rb are 80.917990 u, 80.916289 u, 80.916589 u, and 80.918990 u, respectively. Which one is stable?

How will the others decay? Will any of these nuclei decay by electron capture but not β^+ decay?

Q14S.9 Prove that equation Q14.18 applies to any reasonably small fragment being ejected from a massive nucleus, as long as the left side is the binding energy per nucleon of the ejected fragment.

Q14S.10 $_6^{14}$C has a half-life of 5370 y. A certain sample originally contained 0.10 g of this isotope 47,200 y ago. How much is left now?

Q14S.11 $_{13}^{28}$Al has a half-life of 2.24 min. What is the value of λ for this nucleus? What percentage of a given sample of this isotope would remain after 1.0 h?

Q14S.12 A certain radioactive sample has an activity of 2.25 MBq and a half-life of 9.2 h. What will be the activity of this sample after exactly 1 week?

Q14S.13 Imagine that a given radioactive nucleus has a 50% probability of decaying within a 1.0-s time interval. What is the probability that it will survive for 5.0 s? (*Hint:* $\lambda \neq 0.5$ per second in this case.)

Q14S.14 A given nucleus has a 20% probability of decaying in a 1.0-s time interval. What would be the half-life of a sample of this nucleus? (*Hint:* $\lambda \neq 0.20$ per second in this case, though it will not be too far from this value.)

Rich-Context

Q14R.1 Imagine that you are one of the first astronauts to visit Mars. Your crew stumbles on evidence of ruins created by some alien race an unknown amount of time ago. You personally discover some kind of device that proves to have spots that are faintly radioactive. When you run an analysis of the spots, you find that the spots are about 0.012% radioactive $_{43}^{99}$Tc (which does not exist in nature), mixed with about 28% $_{79}^{197}$Au, 8% $_{41}^{93}$Nb, 2.2% $_{44}^{99}$Ru, and about 1.1% $_{26}^{57}$Fe, insignificant traces of other nuclei, and the remainder $_{26}^{56}$Fe (the percentages are by numbers of atoms, and all nuclei except the technicium are stable). The computer tells you that the only *stable* isotopes of gold and niobium are the ones you found; the relative abundances of the iron isotopes are consistent with their natural abundances; and about 13% of natural ruthenium is $_{44}^{99}$Ru. The computer has no records on technicium (since the mission planners did not expect to have anyone find any), but your measurements indicate that a sample of purified technicium with a mass of 1.98 μg reads about 1200 Bq of activity. Assuming that the technicium in the device was artificial, how long ago was the device made? (*Hint:* You have all the information you need. It may help to remember that 1 g is about equal to the mass of Avogadro's number of nucleons.)

Advanced

Q14A.1 Imagine that a nucleus of type A decays with decay constant λ_A to a nucleus of type B, which itself decays with decay constant λ_B to some third nucleus. Assume that we start with a sample of pure type A nuclei.

(a) Find the differential equation analogous to equation Q14.27 that describes the rate of change of the number $N_B(t)$ of B nuclei.

(b) Show that the following is a solution to the differential equation you found in part (a):

$$N_B(t) = \frac{N_0 \lambda_A}{\lambda_B - \lambda_A}(e^{-\lambda_A t} - e^{-\lambda_B t}) \qquad (Q14.37)$$

(c) Find an expression for the time when $N_B(t)$ is maximum. [*Hint:* Define $R \equiv \lambda_B/\lambda_A$. Note that $e^{-\lambda_B t} = (e^{-\lambda_A t})^{\lambda_B/\lambda_A} = (e^{-\lambda_A t})^R$.]

ANSWERS TO EXERCISES

Q14X.1 Energy conservation for the EC decay process implies that

$$m_e c^2 + m_i^{\text{nuc}} c^2 = m_f^{\text{nuc}} c^2 + \Delta E$$

$$\Rightarrow \quad m_e + m_i^{\text{nuc}} > m_f^{\text{nuc}} \qquad (Q14.38)$$

In this case, if the initial nucleus has Z protons, the final nucleus has $Z - 1$ protons. If we add $(Z - 1)m_e$ to both sides, the left side becomes the atomic mass of the initial nucleus with Z protons and the right side becomes the atomic mass of a nucleus with $Z - 1$ protons. The equation therefore reduces to

$$m_i > m_f \qquad (Q14.39)$$

as claimed.

Q14X.2 $^{30}_{14}\text{Si}$ is slightly less massive, so it will be the stable nucleus. $^{30}_{15}\text{P}$ will have to decay to it by either a β^+ or EC process (since we are reducing Z by 1). If it is a β^+ process, the energy released is

$$\Delta E = m_P c^2 - m_{\text{Si}} c^2 - 2m_e c^2$$

$$= (29.978307\ \text{u} - 29.973770\ \text{u})c^2 - 2(0.511\ \text{MeV})$$

$$= (0.004537\ \text{u}c^2)(931.5\ \text{MeV}/\text{u}c^2) - 1.02\ \text{MeV}$$

$$= 3.2\ \text{MeV} \qquad (Q14.40)$$

Since this is greater than zero, it is allowed by conservation of energy and thus is more likely than the EC process.

Q14X.3 Starting with equation Q14.8, we have

$$0 = \left(\frac{2a_C}{A^{1/3}} + \frac{8a_A}{A}\right)Z - 4a_A - a_M$$

$$\Rightarrow \quad \left(\frac{2a_C}{A^{1/3}} + \frac{8a_A}{A}\right)Z = 4a_A + a_M \qquad (Q14.41)$$

Multiplying both sides by $A/4a_A$, we get

$$\left(\frac{a_C A^{2/3}}{2a_A} + 2\right)Z = A\left(1 + \frac{a_M}{4a_A}\right) \qquad (Q14.42)$$

which, upon division by $2 + a_C A^{2/3}/2a_A$ becomes equation Q14.9. Note that

$$\frac{a_C}{2a_A} = \frac{0.697\ \text{MeV}}{2(23.29\ \text{MeV})} = 0.015 \qquad (Q14.43)$$

and

$$\frac{a_M}{4a_A} = \frac{0.782\ \text{MeV}}{4(23.29\ \text{MeV})} = 0.0084 \qquad (Q14.44)$$

Q14X.4 The optimal value of Z/A for $A = 165$ is

$$\frac{Z}{A} = \frac{1.0084}{2 + 0.015(165)^{2/3}} = 0.411 \qquad (Q14.45a)$$

which is just under 3% larger than $\frac{2}{5} = 0.400$. The optimal value of Z/A for $A = 250$ is

$$\frac{Z}{A} = \frac{1.0084}{2 + 0.015(250)^{2/3}} = 0.389 \qquad (Q14.45b)$$

which is just over 3% smaller than $\frac{2}{5}$.

Q14X.5 The optimal value of Z for $A = 100$ is

$$Z = \frac{1.0084(100)}{2 + 0.015(100)^{2/3}} = 43.4 \qquad (Q14.46)$$

If it weren't for the pairing effect, $^{100}_{43}\text{Tc}$ (technicium) would be the stable nucleus for $A = 100$, but since this is an odd-odd nucleus, it is not as stable as the $^{100}_{42}\text{Mo}$ (molybdenum) and $^{100}_{44}\text{Ru}$ (ruthenium) nuclei on either side of it. (This prediction turns out to be correct.)

Q14X.6 According to equation Q14.12, we have

$$\Delta E = m_{\text{Ra}} c^2 - m_{\text{Rd}} c^2 - m_{\text{He}} c^2$$

$$= (226.025402\ \text{u} - 222.017570\ \text{u} - 4.002603\ \text{u})c^2$$

$$= (0.005229\ \text{u}c^2)(931.5\ \text{MeV}/\text{u}c^2)$$

$$= 4.87\ \text{MeV} \qquad (Q14.47)$$

Q14X.7   Taking the derivative of $N(t)$, we get

$$\frac{dN}{dt} = N_0 \frac{d}{dt}e^{-\lambda t} = N_0 e^{-\lambda t}\frac{d}{dt}(-\lambda t) \qquad \text{(Q14.48)}$$

according to equation Q14.29 and the chain rule. Completing the derivative, we get

$$\frac{dN}{dt} = -\lambda(N_0 e^{-\lambda t}) = -\lambda N \qquad \text{since } N = N_0 e^{-\lambda t}$$
$$\text{(Q14.49)}$$

Thus the derivative is equal to $-\lambda N$, as claimed in equation Q14.27.

Q14X.8   The activity is defined to be $\lambda N = \lambda N_0 e^{-\lambda t}$, so the activity decreases by the same exponential factor $e^{-\lambda t}$ as the number of undecayed nuclei.

Q14X.9   Taking the inverse and then the logarithm of both sides of equation Q14.30, we get

$$2 = e^{+\lambda t_{1/2}} \qquad \Rightarrow \qquad \ln 2 = +\lambda t_{1/2} \qquad \text{(Q14.50)}$$

Dividing both sides of the last equation by λ, we get equation Q14.31.

Q15 Nuclear Technology

Chapter Overview

Introduction

This closing chapter of the subdivision on nuclear physics examines some of the technological applications and societal implications of nuclear physics.

Section Q15.1: The Penetrating Ability of Radiation

A high-energy quanton will ionize atoms as it passes. A slow-moving quanton with lots of charge does this best, because it has both the means and the time to interact with atomic electrons as it passes. This process drains energy from the quanton, eventually stopping it. This model implies that (1) alpha particles are very effective ionizers and are easiest to stop, (2) electrons are less effective and travel farther, (3) uncharged gamma photons are much more penetrating, and (4) neutrinos (which hardly interact with anything) are by far the most penetrating kind of radiation (neutrinos move through the earth with ease).

Section Q15.2: The Biological Effects of Radiation

This ionization disrupts cellular activities. If it is intense enough, it can kill a cell outright, but even a single ionization event can modify DNA to create cancer (though the probability is small). Alpha and beta emitters are most deadly when ingested, because once inside the body, they intensely radiate a relatively small region of tissue, greatly increasing the risk of cancer. Neutron radiation is deadly outside the body, because it easily penetrates most inorganic materials that might be used for shielding easily but interacts very effectively with hydrogen.

Units of radiation include the **roentgen** (the amount of radiation that produces 3×10^{-10} C of ionization in 1 cm^3 of air), the **rad** (the amount of radiation that deposits 10^{-2} J of energy per kilogram of exposed material $\approx$1 roentgen), and the **gray** (= 100 rad). The most important biological unit of radiation is the **rem,** which is defined to be 1 rad times a **relative biological effectiveness (RBE) factor** that specifies the effectiveness of radiation damage in various kinds of living tissue. A whole-body dose > 600 rem kills within weeks, > 5000 rem within days. Low-level radiation causes roughly 300×10^{-6} cancer death per rem per person exposed. The typical natural background exposure is about 130 mrad/y.

Section Q15.3: Uses of Radioactive Substances

In addition to radioactive dating (see section Q14.5), radioactive substances are used as tracers for biological activity, in treatments for cancer, and to create nuclear power sources for deep space missions.

Section Q15.4: Introduction to Nuclear Energy

Because the binding energy curve has a maximum at $A \approx 60$, **fusion** of small nuclei can increase the total binding energy of a given collection of nucleons with $A < 60$, and **fission** of large nuclei to fragments with $A > 60$ can also increase the total binding energy of a given collection of nucleons.

Section Q15.5:   Fission

Certain nuclei (such as $^{235}_{92}$U) undergo fission when struck by a neutron. Since fission produces more free neutrons, one can set up a chain reaction in which earlier fission reactions initialize other reactions. The reaction dies out if the fuel is **subcritical** (meaning that loss of neutrons implies that less than one future reaction is caused by any given reaction) but grows geometrically if the fuel is **supercritical.** The reaction is exactly self-sustaining if the fuel is **critical.**

To create a nuclear bomb, the problem is to assemble a supercritical mass and hold it long enough for the reaction to take place. This is much easier with pure $^{235}_{92}$U, but this is hard to separate from the more common $^{238}_{92}$U, which does not undergo fission. While $^{239}_{94}$Pu is easy to produce, very complicated bomb designs are required to create a supercritical mass.

In a fission reactor, the fuel itself is deliberately kept quite subcritical by itself. The fuel assembly is then brought closer to criticality by embedding the fuel in a **moderator** such as graphite or heavy water that increases the effectiveness of neutrons in the chain reaction by slowing them down. Neutron-absorbing control rods keep the reaction from becoming supercritical. The main dangers with fission reactors are the remote possibility of a **meltdown** (if systems cooling the reactor fail) and the long-term safe disposal of radioactive waste.

Section Q15.6:   Fusion

To create a nuclear fusion reaction, hydrogen must be heated to 10^8 K and confined long enough for the reaction to take place. In a **thermonuclear bomb** (hydrogen bomb), a fission bomb provides the high temperature and confinement. For controlled fusion reactors, the three most promising choices for confinement mechanisms are **magnetic confinement,** where cleverly designed magnetic fields confine a hot plasma; **inertial confinement,** where a pellet of fuel is suddenly imploded by a symmetric array of lasers; and **gravitational confinement,** where the fuel is compressed and maintained at high temperatures by gravity. The latter, the method used successfully by stars such as the sun, is the only method of controlled fusion known to work, but requires stellar masses to do the job. The sun uses the **proton-proton chain** of reactions that effectively converts four hydrogen atoms to one helium atom (this very clean fusion reaction requires densities too high for practical artificial fusion reactors).

Q15.1   The Penetrating Ability of Radiation

In chapter Q12, we saw that alpha particles are readily absorbed by even a thin sheet of paper, while beta particles can penetrate even thin sheets of steel and gamma rays can penetrate thick blocks of lead. All these particles have kinetic energies in the range of several megaelectronvolts. Why should these three kinds of particles be so different in their ability to go through a barrier?

Radiation quanta moving through a medium lose energy by ionizing atoms

When a high-energy quanton moves through any kind of medium, it often collides with atomic electrons in the medium. The impact of these collisions is generally enough to **ionize** the atom (i.e., knock an atomic electron completely free). The process of liberating an electron from an atom requires the high-energy quanton to lose an amount of energy roughly equal to the binding energy of the electron involved (which may range from a fraction of an electronvolt to thousands of electronvolts). Each collision thus drains some of the quanton's energy, and after a sufficient number of collisions, the quanton is stopped and essentially absorbed into the medium.

Charged particles such as alpha particles, electrons, and positrons interact with atomic electrons even at a distance through electrostatic attraction or repulsion. On the other hand, a gamma ray photon is electrically neutral and thus will only interact with an atomic electron if it directly hits it. As a result, gamma ray photons do not interact with a medium much, so they are able to travel a long way before losing significant amounts of energy. The most effective shielding for gamma ray photons is some kind of very dense material (such as lead), so that the largest possible number of possible targets can be presented along the path of a gamma photon.

Implications for the penetrating abilities of types of radiation

Alpha particles are more easily stopped in a medium than either electrons or positrons because (1) they are twice as strongly charged and (2) they move much more *slowly* than either an electron or a positron of comparable energy, which gives them more *time* to transfer significant amounts of energy to atomic electrons along the way. This means that an alpha particle will lose more energy per collision than an electron would and thus will be absorbed sooner.

Exercise Q15X.1

Roughly how many times more slowly does an alpha particle with kinetic energy of 2.0 MeV move than an electron with the same kinetic energy? (Note that an electron with this energy will be highly relativistic, since a kinetic energy of 2.0 MeV is about 4 times as large as its rest energy of 0.511 MeV. Therefore, an electron moves with speed $\approx c$.)

Neutrinos are the most penetrating radiation

We haven't talked much about neutrinos, which are produced in every beta decay process. Neutrinos are by far the most penetrating kind of radiation, because (since they are electrically neutral) they can only interact with matter via the weak interaction, which has such an extraordinarily tiny range that the probability that a neutrino will happen to pass close enough to an atomic electron or quark in a nucleus to interact with it is very small. While it may take a meter or two of lead to have an even chance of stopping a gamma photon, it takes thousands of *light-years* of lead to have the same chance of stopping a neutrino. Neutrinos that are produced in prodigious quantities by the nuclear reactions in the sun flow through the earth more easily than light goes through a pane of glass. At midnight, neutrinos from the sun flow up through your sleeping body on their way to interstellar space without even *noticing* the mere planet between you and the sun.

Why doesn't anyone mention the hazards associated with neutrino exposure? This is because there essentially aren't any! As we will see in section Q15.2, other kinds of radiation are hazardous precisely *because* they interact with living tissue. Neutrinos are so penetrating precisely because they have an extremely low probability of interacting with *anything*.

Q15.2   The Biological Effects of Radiation

At the most basic level, radiation does damage to living cells because it ionizes atoms in the cell through which it passes. This can disrupt chemical reactions taking place in a cell, and sufficiently high levels of ionizing radiation can kill the cell outright by disrupting too many of its internal operations all at once.

A single passing high-energy particle can also scramble a cell's genetic information by breaking one or more chemical bonds in that cell's DNA, causing it to rearrange itself in an inappropriate manner, or by creating free radicals in the cell that damage the DNA. Sometimes, the damaged DNA is part of the genetic information that limits a cell's reproductive abilities. In such a case, the cell might begin reproducing without limits, which causes cancer. Even though this will only happen *very* rarely, this is by far the most serious health consequence of relatively low doses of radiation because while losing a handful of cells by direct disruption is not going to hurt one very much, once a cancer gets started, it too often kills.

Because alpha particles will be stopped by your skin, and electrons and positrons by simple shielding (even a few meters of air), the most dangerous kinds of radiation from sources *outside* your body are gamma ray photons and neutrons, since they are able to go through a limited amount of shielding and penetrate deep within the body. Fortunately (as in the case of neutrinos), the same reason that gamma rays are able to go through thick shielding implies that they will also often go right through your body without doing *much* damage. Still, sufficiently intense gamma rays can kill, and even in the case of weak radiation, it only takes one ionization in the wrong place on a DNA molecule to cause cancer.

As we will see, fission and fusion reactions produce large numbers of neutrons, which are more dangerous than gamma photons. High-energy neutrons can penetrate simple shielding without difficulty (because they, like gamma photons, are neutral), but are very effectively absorbed by *living* tissue, because they collide and lose energy effectively in collisions with *hydrogen* atoms (which are plentiful in living tissue). A neutron colliding with a proton may knock the proton clear out of the hydrogen atom, breaking any bond to which the hydrogen atom contributes. A *neutron bomb* is designed to produce prodigious amounts of neutrons without releasing much actual explosion energy, and so it can kill people fairly directly and painfully (by exposure to the neutron radiation) without harming property (in my opinion a macabre reversal of the most basic moral principles).

The very property that makes alpha and beta emitters relatively benign *outside* the body makes them especially deadly when ingested. For example, a tiny sample of an alpha emitter such as plutonium lodged in a person's lung will ionize atoms in a relatively small region of tissue (since the range of alpha particles is very small). That small region of tissue, however, will be intensely irradiated, as all the energy of the alpha particles will be absorbed

Ionization disrupts cell processes

Main danger of low-level radiation is cancer

Radiated neutrons are dangerous

Ingestion is dangerous

there. The probability that the endless random ionizations in this small region will lead to a cancer-causing mutation becomes greater with time. It has been estimated that a dust particle containing 1 μg of plutonium is sufficient to cause lung cancer with virtual certainty within a 30-y period.

Some radioactive isotopes get concentrated in the food chain as a result of their chemical similarity to elements needed by living things. For example, $^{90}_{38}$Sr (a common by-product of fission reactions) is chemically similar to calcium, and so $^{90}_{38}$Sr ingested by a vertebrate is concentrated in its bones, where it irradiates the marrow that produces blood cells. The result in the long run is typically leukemia.

Units to describe radiation absorbed by tissue

Four different kinds of units are used to express the amount of radiation absorbed by tissue. The original unit was the **roentgen** (R), defined to be the amount of ionizing radiation that produces an ionization charge of 3×10^{-10} C in 1 cm^3 of air under normal conditions. This unit was later replaced by the **rad** (standing for *radiation absorbed dose*), defined to be the amount of radiation that would deposit 10^{-2} J/kg in any material. This unit was chosen to be roughly equivalent to the roentgen. The standard SI unit for an absorbed dose of radiation is the **gray,** defined to be 1 J/kg = 100 rad.

The amount of actual biological damage done by radiation depends not only on the energy deposited but also on the kind of radiation (e.g., radiation that produces a lot of ionization in a small volume, such as alpha radiation, is more effective in causing damage than radiation that doesn't, such as gamma radiation). The **rem,** standing for *roentgen equivalent in man* (sorry, scientists in the 1950s were not yet concerned about sexist language), is defined to be the amount of radiation that has the same biological effect as 1 rad of beta or gamma radiation. Mathematically, the rem is defined as follows:

$$1 \text{ rem} = 1 \text{ rad} \times \text{RBE} \qquad\qquad (\text{Q15.1})$$

where the RBE (**relative biological effectiveness factor**) can be determined for a given type of radiation at a certain energy by using tables. For example, the RBE for gamma photons at typical energies is 1, but the RBE for high-energy alpha particles may be as high as 20.

Whole-body doses of over 600 rem in a short time generally cause death in humans within a few weeks: over 5000 rem causes death within a few hours. For doses under 50 rem, the main danger is cancer. The average of a variety of recent studies suggests that the rate at which a single whole-body exposure event induces cancer within 30 y is about 300 deaths per rem per million people.

Cancer production seems entirely random

The best available evidence suggests that whether an ionizing particle causes cancer or not is a totally random thing: exposure to even a single ionizing particle *might* cause cancer (even if the probability is vanishingly small). This means there is no safe amount of exposure to radiation, and if possible, exposure to radiation should be avoided. The flip side is that a person can get killed crossing the street, too, and many consider that risk worth taking, considering the benefits obtained. Moreover, there is no way to live a life free from radiation, as many natural sources of radiation are unavoidable (see table Q15.1). There are no easy answers to questions about radiation safety.

One of the sad aspects of the history of research into radiation is that its real hazards took so long to be discovered. It became known almost immediately that intense sources of radiation would produce injuries similar to burns (Becquerel accidentally burned himself by carrying radioactive samples in his pocket, and subsequently Pierre Curie deliberately burned himself with a radioactive sample to explore the consequences). Unfortunately, the most serious biological consequences of exposure to radiation generally show themselves only years afterward. In the 1930s people began to realize

Table Q15.1 Typical sources of human exposure to radiation

(The key word to note in the artificial category is *averaged:* someone who happens to be near a nuclear power plant having an accident or downwind from a testing range will get a much higher dose.)

Natural Sources	mrad/y
Cosmic rays	45
External terrestrial	60
Ingested isotopes	25
	130

Artificial Sources (Averaged)	mrad/y
Diagnostic X-rays	70
Weapons testing	<1
Nuclear power production	<1
Occupational	<1

Adapted from S. C. Bushong, *The Physics Teacher*, vol. 15, 3, 1977.

that exposure to radiation had serious *long-term* health effects (and began to protect themselves), but a real understanding of the health effects was not available until the 1950s. As a result, many nuclear physicists died early from the effects of their research (e.g., Marie Curie died in 1934 at the age of 67 from radiation-induced leukemia). This is all the more tragic considering that people knew that *X-rays* produced cancer as early as 1904 (after one of Thomas Edison's assistants became suddenly and dramatically ill with cancer while working on an X-ray lamp).

Q15.3   Uses of Radioactive Substances

Because radioactive substances are easily detected by the radiation they produce, low-level radioactive materials are often used as *tracers* to explore how certain molecules are transported in living creatures or in the environment. For example in *positron emission tomography* (PET), a sample of glucose is prepared where some normal atoms are replaced by a positron emitter with a short half-life. This glucose is injected in a volunteer or patient, and the places where radioactive glucose ends up in the person's body can be imaged by detecting the gamma rays produced when the emitted positrons annihilate electrons in the person's body. This technique has been particularly helpful in understanding brain function: the places in the brain where glucose (a food energy source) accumulates indicate the parts of the brain that are being used. This enables researchers to determine which parts of the brain are used when the volunteer is asked to perform a certain kind of task. (The short half-life helps ensure that the person's exposure to radiation is limited.) Figure Q15.1 shows a PET scan of a human brain during a certain kind of behavior.

Radioactive tracers

Focused radiation is also often used as a treatment for cancer as both a supplement and an alternative to surgery. Because cancer cells are constantly dividing, they are more susceptible to disruption from ionizing radiation.

Treating cancer

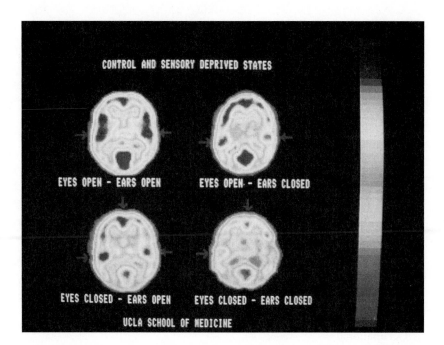

Figure Q15.1
A PET image of a human brain. The areas marked are areas of increased uptake of some radioactive marker due to brain activity.

The treatment can also be made more effective by focusing the radiation on the area where cancer cells are likely to be found, minimizing the damage to other tissues.

Nuclear "batteries"

Finally, the electric power for deep space probes like *Cassini* or *Galileo* is generated by a kind of "nuclear battery." Such a battery consists of a substantial sample of a fairly radioactive alpha emitter. The energy liberated by the radioactive decay keeps the sample thermally hot, and a thermoelectric generator is used to convert this heat directly to electricity. Since the alpha particles have very limited range, the radioactivity can be fairly well confined to the sample, and thus does not affect the probe's sensitive electronic circuits.

Q15.4   Introduction to Nuclear Energy

Basic principles of fission and fusion reactions

If we plot the binding energy per nucleon e_b as a function of A (see figure Q15.2), we find that e_b *increases* from low A to $A \approx 60$ (due mostly to the increase in the ratio of internal to surface nucleons as A increases) and then *decreases* as A continues to increase (due mostly to the increased electrostatic repulsion between protons). This in general means that if you combine a pair of nuclei whose combined $A < 60$ (a process we call nuclear **fusion**), you will get a final nucleus with a greater binding energy per nucleon than you had originally, and thus energy will be released. Similarly, if you take a nucleus having an $A > 120$ and pull it into two pieces (a process we call nuclear **fission**), the resulting smaller nuclei will both have greater binding energy per nucleon than the original nucleus, and therefore energy will be released.

Like alpha decay processes, these processes do not change the number or type of nucleons involved, they simply rearrange them. This means that the rest energies of all the parts of these nuclei remain the same, and can be canceled out of the energy conservation equation (see the argument leading up to equation Q14.14). This leaves

$$-E_{bi} = -E_{bf} + \Delta E \qquad (Q15.2)$$

if we interpret E_{bi} as the total binding energy going into a reaction and E_{bf} as the total final binding energy after the reaction is over. Therefore, any energy

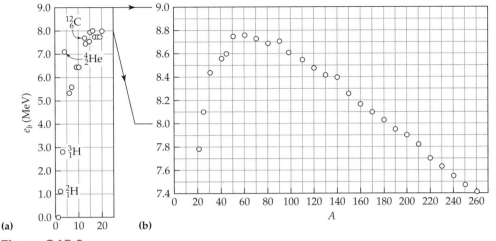

Figure Q15.2
(a) Graph of binding energy per nucleon for all stable nuclei (as well as $^3_1$H) with $A < 20$.
(b) A graph of the binding energy per nucleon for selected stable nuclei with $A > 20$, using an expanded vertical scale and a compressed horizontal scale.

that is released by such a reaction must be entirely due to an *increase* in the system's binding energy:[†]

$$\Delta E = E_{bf} - E_{bi} \qquad\qquad (Q15.3)$$

Purpose: We can use this equation to determine the energy ΔE released in a nuclear reaction.
Symbols: E_{bi} is the total binding energy of the initial nucleus or nuclei, and E_{bf} is the same for the final nuclei.
Limitations: This equation only applies to nuclear reactions that merely rearrange nucleons, not to weak-interaction processes that change nucleons.

Using this equation, we can use the binding energy graphs in figure Q15.2 to estimate the kind of energy that we can expect to produce.

A particularly productive fusion reaction that people are studying for possible use in power generation is

$$^2_1\text{H} + {}^3_1\text{H} \rightarrow {}^4_2\text{He} + n \qquad\qquad (Q15.4)$$

According to figure Q15.2, it looks as if $^2_1$H has a binding energy per nucleon of about 1.1 MeV, $^3_1$H has a binding energy per nucleon of about 2.8 MeV, and helium has a binding energy of about 7.1 MeV per nucleon. The neutron has no binding energy at all. Thus the energy released in this reaction is about

$$\Delta E = E_{bf} - E_{bi} = 4(7.1 \text{ MeV}) - 2(1.1 \text{ MeV}) - 3(2.8 \text{ MeV}) \approx 17.8 \text{ MeV}$$
$$(Q15.5)$$

(the precise result is about 17.6 MeV).

[†]Remember that binding energy is the energy that is *released* when a system forms from parts originally dispersed at infinity. A process that increase a system's binding energy will therefore release an amount of energy equal to the change in the system's binding energy.

A typical fission reaction, on the other hand, might start with $^{235}_{92}$U and produce roughly equal-size fragments ($A \approx 116$) and two or three free neutrons. So the energy produced in this reaction is

$$\Delta E = E_{bf} - E_{bi} = 232(8.49 \text{ MeV}) + 0 - 235(7.62 \text{ MeV}) \approx 180 \text{ MeV}$$

$$(Q15.6)$$

(The free neutrons do not have any binding energy.)

Example E15.1   Energy Density in Uranium

Problem   Calculate (in joules) the amount of energy that would be released by the fission of 1 g of $^{235}_{92}$U atoms. Compare to the energy released by burning 1 g of gasoline (44 kJ).

Solution   235 g of $^{235}_{92}$U would correspond to 1 mol, so 1 g corresponds to

$$\frac{1 \text{ mol}}{235} \left(\frac{6.0 \times 10^{23} \text{ atoms}}{1 \text{ mol}} \right) \left(\frac{180 \times 10^6 \text{ eV}}{1 \text{ atom}} \right) \left(\frac{1.6 \times 10^{-19} \text{ J}}{1 \text{ eV}} \right) = 7.4 \text{ GJ}$$

$$(Q15.7)$$

Evaluation   This is more than 150,000 times the energy produced by 1 g of gasoline.

Exercise Q15X.2

How much energy could we get from the complete fusion of 1 g of an equal mixture of $^2_1$H and $^3_1$H?

Q15.5   Fission

The discovery of fission

At the end of 1938 the German radiochemists Otto Hahn and Fritz Straussmann (who had been working in collaboration with the physicist Lise Meitner before she fled Germany in July of that year) obtained conclusive evidence that a radioactive isotope of barium was created in a sample of uranium that had been bombarded with neutrons. Since barium is a much lighter nucleus than uranium, Meitner concluded (after receiving an account from Hahn) that the neutrons were inducing some of the uranium nuclei to undergo fission.

Later study suggested that when a massive nucleus absorbs a bombarding neutron, it will sometimes oscillate or "ring." The oscillations of $^{235}_{92}$U turn out to be violent enough to tear the nucleus apart (see figure Q15.3). Because the fission fragments tend to be too rich in neutrons (remember that massive nuclei have proportionally more neutrons than lighter nuclei), on average about 2.5 neutrons become detached from the fragments in the fission process.

Creating a chain reaction

Because the fission process produces neutrons, a **chain reaction** becomes possible. Imagine a sphere of pure $^{235}_{92}$U. An initial neutron (from a cosmic ray or a spontaneous fission) causes one nucleus to split. This releases two or three other neutrons, which in turn induce two or three neighboring nuclei to

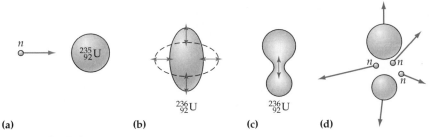

Figure Q15.3

(a) When a $^{235}_{92}$U nucleus absorbs a neutron, (b) it is transformed into an excited oscillating state of $^{236}_{92}$U. (c) Most other excited nuclei would simply emit a photon and settle down to the ground state, but this nucleus is unstable and (d) comes apart into two small fragments plus two to three neutrons.

split, which release more neutrons which induce more nuclei to split, and so on. The number of nuclei participating in this chain reaction would increase geometrically in time, leading to an explosive release of energy.

The ratio of the number of neutrons produced in a step of the chain reaction to the number produced in the previous step is called the **multiplication factor.** If every neutron produced in a given step were to induce a new fission, this factor would be simply the average number of neutrons produced by each fission reaction (i.e., about 2.5 in the case of $^{235}_{92}$U). However, in any given sample of fissionable material (the more technical term is **fissile** material), some neutrons will either be absorbed without producing a fission or simply escape from the sample, reducing the *multiplication factor*. We describe a sample of fissile material in a configuration such that the multiplication factor is exactly 1 (the desired situation in a nuclear reactor) as being **critical.** A configuration such that the multiplication factor is less than 1 is **subcritical,** and the chain reaction will not sustain itself. A sample in such a configuration that the multiplication factor is greater than 1 is **supercritical,** and any fission process initiated in such a sample will grow geometrically.

In a sample of pure $^{235}_{92}$U, the main problem in maintaining a chain reaction is that neutrons escape from the surface of a sample. The larger the sample is, the longer a neutron has to travel through the material before it can escape, and so the more likely it is to be absorbed. We call the smallest mass of a spherical sample of a fissionable material for which the multiplication factor is above 1 the **critical mass** of that isotope: for $^{235}_{92}$U this turns out to be 53 kg (which is contained in a sphere about 18 cm in diameter).

To create a nuclear bomb, all that you need to do is to suddenly create a supercritical mass from one or more subcritical masses of some fissionable isotope. The simplest design for such a bomb is the gun barrel design shown in figure Q15.4. In this design, one subcritical mass of $^{235}_{92}$U is driven by an explosive charge into another to create a supercritical mass. This was the design used in the bomb that destroyed Hiroshima on August 6, 1945, killing more than 100,000 people. (This design was never tested before that date, partly because it was considered so foolproof and partly because a test would have seriously depleted the United States' entire supply of $^{235}_{92}$U at the time.)

The bomb design tested July 15, 1945, at Alamogordo, New Mexico (see figure Q15.5), was a much more ambitious design that used $^{239}_{94}$Pu as the fissile material. $^{239}_{94}$Pu was available in larger amounts than $^{235}_{92}$U at the time, but a chain reaction in supercritical $^{239}_{94}$Pu escalates much more rapidly than it does in $^{235}_{92}$U, and in a simple gun barrel design, the energy released would blow the bomb apart before the critical mass had time to assemble and completely

Atomic bombs

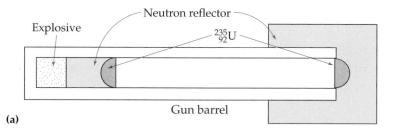

(a)   Gun barrel

Figure Q15.4

(a) Basic design for a gun barrel atomic bomb. When the bomb is detonated, the explosive drives one subcritical mass of uranium into the other, creating a supercritical mass (the neutron reflectors help keep neutrons from escaping the mass, thus hastening the reaction).
(b) Photograph of the "Little Boy" atomic bomb dropped on Hiroshima that used this design.

(b)

Figure Q15.5

A photograph of the Alamogordo explosion produced by an imploding plutonium shell design.

react. A bomb using $^{239}_{94}$Pu can only be created by surrounding a slightly sub-critical sphere with explosives. When the explosives are detonated, the shock wave crushes the plutonium sphere to a high enough density that it becomes critical. The inward motion of the plutonium prevents it from flying apart be-fore it completely reacts.

What is it that prevents every nation in the world from having nuclear weapons? There are two answers. It is relatively easy to *design* a bomb based on $^{235}_{92}$U, but it is extremely difficult to produce enough sufficiently pure $^{235}_{92}$U to create a critical mass. $^{235}_{92}$U is chemically identical to the far more common nonfissile isotope $^{238}_{92}$U. The easiest way to separate the isotopes is to put a uranium compound in solution and use a centrifuge to stratify the solution in terms of mass. This is a difficult, time-consuming, and expensive process, and few nations possess the knowledge, patience, and money to do it.

On the other hand, $^{239}_{94}$Pu is comparatively easy to obtain: it is produced automatically in fission reactors from neutron bombardment of $^{238}_{92}$U and can be chemically separated quite easily from the uranium reactor fuel. The bomb design, however, is *much* more difficult, because creating an explosion that will *symmetrically* compress a sphere is very difficult.

The key to preventing proliferation of nuclear weapons is to keep people from getting *any* fissile material of sufficient purity to use in a bomb (*especially* $^{235}_{92}$U), to keep successful *plutonium* bomb designs from becoming public, to monitor nations using nuclear reactors to make sure that they are not extracting and purifying plutonium, and to ban the testing of weapon designs. The Nuclear Nonproliferation Treaty and the Comprehensive Test Ban treaties are designed to keep participating nations from doing these things and to internationally isolate those nations that will not sign.

In a fission reactor, the idea is to arrange and control an assembly of fissile material so that its multiplication factor is exactly 1. If the fissile material is even a little subcritical, the chain reaction will die out; if it becomes supercritical, the chain reaction will grow uncontrollably.

Fission reactors do not use weapons-grade $^{235}_{92}$U as a fuel: most use enriched uranium that is mostly normal $^{238}_{92}$U but contains several percent $^{235}_{92}$U (instead of the 0.7% that occurs in nature). This is done for two reasons. First, "enriched" uranium is much cheaper than weapons-grade uranium, since the uranium does not have to be completely purified. Second, enriched uranium can never become supercritical on its own, as the $^{238}_{92}$U absorbs too many of the neutrons produced by the fission of $^{235}_{92}$U. This means that a well-designed reactor can never explode due to a runaway chain reaction.

The trick, then, is to *increase* the effectiveness of neutrons in inducing fission so that the fuel becomes just barely critical. It turns out that the probability that a neutron will induce fission is roughly inversely proportional to its speed, essentially because a slow-moving neutron has more time to interact with a nucleus. Typical neutrons emitted by a fission reaction move much too fast to effectively induce more fissions. The process can be made *much* more efficient if the energy of the neutrons is significantly reduced.

In nuclear reactors this is done with the help of a **moderator,** which is typically ordinary water. When a neutron collides with a hydrogen nucleus in the water molecule, it can transfer much of its kinetic energy to the proton. After a sufficient number of collisions, the kinetic energy of the neutron is reduced to about 0.1 eV. Such a neutron is thousands of times more effective at inducing fission than before. This makes it possible to sustain a chain reaction in an otherwise subcritical assembly of enriched uranium.

In a typical power reactor, the enriched uranium is encased in metal rods (called fuel rods) and immersed in water as shown in figure Q15.6. The fuel rods are separated by control rods, which are made of a material that absorbs neutrons (such as cadmium). When these control rods are lifted from between the fuel rods, neutrons produced by fission reactions are allowed to radiate into the water moderator, where their speed is reduced by collisions with hydrogen nuclei. The neutrons then diffuse back into the fuel rods, where they cause more fission reactions. If the reaction begins to go too fast, lowering the control rods absorbs neutrons, quenching the reaction.

One of the advantages of this design is that it is fairly stable. If the reaction starts to run away, the temperature of the water moderator increases and its density decreases, making it less effective as a moderator, which in turn slows the reaction. Also, if the temperature gets too high, the control rod

Fission power reactors

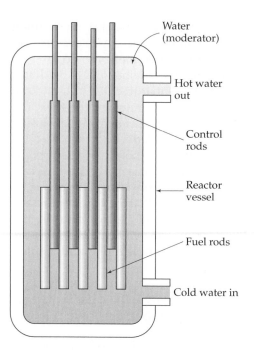

Water
(moderator)

Hot water
out

Control
rods

Reactor
vessel

Fuel rods

Cold water in

Figure Q15.6
The design of a typical water-
moderated nuclear power reactor.

Things that can go wrong

linkages are designed to break, dropping the control rods by gravity and
quenching the reaction.

This doesn't mean that things cannot go catastrophically wrong. Even
when the chain reaction is completely shut down, the fuel rods are so radioactive that they generate a large amount of heat just sitting there. Under
normal circumstances, this heat is carried away by the circulating water; but
if for some reason the water does not circulate, the heat generated by radioactive decay in the fuel rods is sufficient to boil the water away and melt
the fuel rod assembly into a hot molten puddle at the bottom of the reactor
vessel. This puddle will continue to generate heat and will eventually eat its
way through the bottom of the vessel and into the ground, releasing large
amounts of violently radioactive substances into the environment. This kind
of catastrophe is called a **meltdown.**

The disaster at Three Mile Island in the late 1970s barely avoided a meltdown: by the time that water circulation was restored, the reactor core was
already melted out of shape. At Chernobyl (in what was at that time the So-
viet Union), a runaway reaction caused a steam explosion that ruptured the
containment vessel. The graphite that was used as a moderator in this archaic
design then caught fire, spreading radioactive material far and wide in the
smoke.

Second- and third-generation reactor designs being developed by various
companies make the chances of such disasters much less likely than first-
generation designs. However, the power reactors currently operating in the
United States are first-generation designs.

The problem of disposing
radioactive waste

The long-term normal operation of *any* type of fission power plants also
presents significant environmental problems. When the fuel is spent, it must
be removed from the reactor, carefully reprocessed to remove any fissile materials that remain, and then safely disposed of. The problem is that the fuel
is intensely radioactive and remains dangerous for tens of thousands of
years. To date, there is no consensus about how one might safely transport
and dispose of such waste (at present, such waste is "temporarily" stored at
reactor sites, which is ultimately very unsafe).

Although, in general, an object placed near a radioactive substance does *not* itself become radioactive, bombardment by *neutrons* can transmute non-radioactive materials into radioactive isotopes, because nuclei in the initially nonradioactive material can absorb neutrons and become too neutron-rich to be stable. During the active life of a nuclear reactor, the reactor containment vessel is constantly bombarded with neutrons, and as a consequence it becomes fairly radioactive. When the reactor has reached the end of its useful life (about 30 y), the containment vessel has to be carefully disassembled and treated as fairly high-level radioactive waste as well.

Q15.6   Fusion

Another approach to producing nuclear energy is fusion. Promising fusion reactions that are known to be possible are

Basic fusion processes

$$\ce{_1^2H} + \ce{_1^2H} \rightarrow \ce{_1^3H} + \ce{_1^1H} + 4.0 \text{ MeV} \tag{Q15.8a}$$

$$\ce{_1^2H} + \ce{_1^2H} \rightarrow \ce{_2^3He} + n + 3.3 \text{ MeV} \tag{Q15.8b}$$

$$\ce{_1^2H} + \ce{_1^3H} \rightarrow \ce{_2^4He} + n + 17.6 \text{ MeV} \tag{Q15.8c}$$

The advantage of the first two reactions is that $\ce{_1^2H}$ is in good supply: even though this isotope of hydrogen comprises only 0.015% of all hydrogen atoms, the oceans contain roughly 10^{15} tons of this isotope. The third reaction produces much more energy than the other two, but $\ce{_1^3H}$ is a radioactive isotope with a relatively short half-life (12.3 y) that has to be manufactured.

The trick with any of these reactions is to get the hydrogen atoms close enough together to react. The problem is that these nuclei strongly repel each other because of their positive electric charge. Yet the effective range of the strong nuclear force that acts to fuse these nuclei together is only 1 or 2 fm (1 fm = 10^{-15} m), so the nuclei have to be brought quite close together to react.

The problem of getting nuclei to react

Exercise Q15X.3

Imagine two $\ce{_1^2H}$ nuclei approaching head on with equal speeds. Show that each nucleus must have a minimum initial kinetic energy of about 360 keV if they are to approach within 2 fm.

The only known way to create a self-sustaining fusion reaction is to confine gas within a small volume and heat it to incredibly high temperatures (higher than 10^8 K). At sufficiently high temperatures, the violent random motion of the atoms will lead to reasonably frequent collisions of sufficient energy to cause fusion to take place. The energy released in the fusion reaction will increase the temperature of the gas, causing more fusion reactions to take place, and so on.

Creating an uncontrolled fusion reaction is not that difficult. **Thermonuclear bombs** are constructed by enclosing a normal fission bomb with a sheath of $\ce{_1^2H}$ and $\ce{_1^3H}$. The exploding fission bomb creates the high temperatures and extreme pressures necessary to ignite the fusion reaction shown in equation Q15.8c, which in turn greatly increases the amount of energy released.

Uncontrolled reactions

Achieving controlled fusion under laboratory conditions is *much* more difficult. To get the reaction to work, the gas must be compressed to a fairly high density (so that collisions between atoms will be frequent), heated to a very high temperature (so that collisions will be sufficiently energetic), and

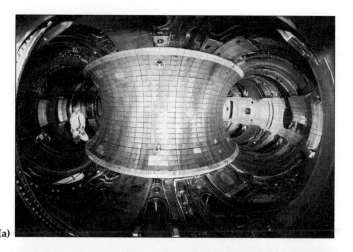

Figure Q15.7
(a) A magnetic confinement fusion reactor. (b) The solar system's only continuously operating fusion reactor.

(a)

(b)

then confined for a long enough time to allow the collisions to happen. It is not easy to keep an extremely hot gas confined for a sufficient amount of time. Several techniques for doing this are currently being explored.

Magnetic confinement

Hydrogen gas at 10^8 K is ionized: that is, the atoms are converted to bare nuclei and free electrons. The fact that the particles in such a gas (which is called a **plasma**) are electrically charged makes it *possible* (not easy!) to confine and compress the gas by using cleverly designed magnetic fields. One of the most significant achievements to date was an experiment in November 1991 in which European physicists used **magnetic confinement** to contain a fusion reaction that lasted nearly 2 s and released nearly 2 MJ of energy. Unfortunately, this was still far short of the energy needed to maintain the reaction. Figure Q15.7a shows a magnetic confinement fusion reactor.

Inertial confinement

In a reactor using **inertial confinement,** a small pellet of frozen $^2_1$H is injected into a chamber where it is suddenly irradiated by laser light from all directions. In theory, the intense laser light will compress the pellet by a factor of 10,000 and heat it to well over 10^8 K so suddenly that a large number of fusion reactions (releasing about 1 MJ) take place before the forces produced by the fusion reactions have time to accelerate the pellet material outward. Essentially the aim is to use each pellet to produce a tiny thermonuclear explosion. While a variety of inertial confinement schemes are under development, none as yet has proved commercially promising.

Blessings and problems with fusion power

If and when fusion reactors become available, they should be able to deliver relatively large amounts of energy at a relatively low environmental cost. "Mining" seawater for $^2_1$H should not be very destructive environmentally.

The products of fusion reactions (particularly the reaction shown in equation Q15.8b) are relatively benign compared to highly radioactive fission fragments. The very difficulty of sustaining a fusion reaction makes it nearly impossible for a fusion reactor to go out of control, so fusion reactors should be comparatively safe.

However, even fusion reactors will present some problems and hazards. The main radioactive waste that would be produced in a fusion reactor is ^3_1H, whose 12.3-y half-life is short enough that it will not need to be stored for millennia (like fission waste) but long enough that storage will be a problem. Moreover, hydrogen atoms (because of their small size) tend to leak out of even the best containment vessels. Since ^3_1H behaves chemically as ordinary hydrogen does, any that escapes into the environment has a good chance of being taken up in the food chain, where it can do real harm. Furthermore, fusion reactions will still bombard the reaction chamber with neutrons, making the chamber radioactive in time (creating a disposal problem when the reactor is dismantled).

The most natural means of generating fusion power is the method employed by the sun (and all other stars). The tremendous weight of the outer layers of the sun confines hydrogen plasma in its core, compressing it to densities exceeding 100 g/cm^3 and to temperatures on the order of 1.5×10^7 K.

Fusion in the stars

Under these conditions of constant high density and temperature, more complicated chains of thermonuclear reactions can take place than are practical for artificial fusion reactors. In the sun, hydrogen is converted into helium in a four-step process, called the **proton-proton chain.**

$$^1_1\text{H} + {}^1_1\text{H} \rightarrow {}^2_1\text{H} + e^+ + \nu \quad (0.42 \text{ MeV}) \tag{Q15.9a}$$

$$^1_1\text{H} + {}^2_1\text{H} \rightarrow {}^3_2\text{He} + \gamma \quad (5.49 \text{ MeV}) \tag{Q15.9b}$$

$$^3_2\text{He} + {}^3_2\text{He} \rightarrow {}^4_2\text{He} + {}^1_1\text{H} + {}^1_1\text{H} + \gamma \quad (12.85 \text{ MeV}) \tag{Q15.9c}$$

$$e^- + e^+ \rightarrow 2\gamma \quad (1.02 \text{ MeV}) \tag{Q15.9d}$$

where the γ's represent gamma ray photons. The net effect of this chain of reactions is to convert four protons to a ^4_2He nucleus. Two each of the first, second, and fourth reactions and one of the third are required to produce the single ^4_2He nucleus, releasing a total of 26.72 MeV, of which roughly 0.26 MeV is carried directly out of the sun by the two neutrinos (which hardly react with anything and are thus able to stream out of the sun's core without being absorbed). The gamma ray photons' energy and the kinetic energy of the nuclei are deposited in the solar core near to the reaction site. The net result is that about 6.6 MeV of energy is deposited in the solar core for every hydrogen atom fused.

More complicated fusion chains using ^4_2He and/or $^{12}_6\text{C}$ as catalysts are possible in larger stars, where the higher temperatures make it possible for hydrogen nuclei to get close enough to react with these more highly charged nuclei.

The sun is the ultimate energy source of all life.[†] Fusion reactions make it possible for the sun to shine at an essentially constant rate for long enough for life to evolve and flourish on the earth's surface. Even though the sun produces energy at the rate of 3.9×10^{26} W, there is enough hydrogen in the sun to keep it shining for a total of roughly 10 billion y.

[†]One of my professors, when asked whether he thought that magnetic confinement or inertial confinement was the better way to generate fusion power, replied that he thought that **gravitational confinement** was by far the best method. He went on to say that 93 million mi is about as close as he wanted to be to a nuclear reactor anyway.

TWO-MINUTE PROBLEMS

Q15T.1 The most important reason that an alpha particle can be stopped by a single sheet of paper while an electron of the same energy might not be stopped by hundreds of sheets of paper is that the alpha particle
A. Is a nucleus, while the electron is not
B. Moves more slowly than the electron
C. Participates in stronger interactions than the electron does
D. Has greater mass than the electron
E. Has more charge than the electron

Q15T.2 The *main* reason that gamma photons are more penetrating than either alpha particles or electrons is that
A. They are uncharged.
B. They have no rest mass.
C. They travel at speed c.
D. They have more energy than electrons.
E. They don't ionize atoms as electrons do.

Q15T.3 A positron will travel about as far through a given material as an electron with the same energy, true (T) or false (F)? (Make a guess.)

Q15T.4 A terrorist organization seeking to construct a nuclear bomb would likely be more interested in obtaining $^{235}_{92}$U than $^{239}_{94}$Pu, T or F?

Q15T.5 Exposure to radiation of any kind for long enough will tend to make an object radioactive, T or F?

Q15T.6 The *moderator* in a fission reactor
A. Slows down neutrons to make them more effective
B. Absorbs neutrons to help control the chain reaction
C. Scatters neutrons evenly throughout the fuel
D. Keeps the chain reaction progressing slowly
E. Has some other purpose (specify)

Q15T.7 Which of the following events is most likely to *directly* cause a meltdown in a nuclear fission reactor?
A. The fuel rods become old and brittle.
B. The reactor is operated at too high a power.
C. The chain reaction slowly starts to get out of control.
D. Pumps that circulate water in the reactor fail.
E. Either B or C.

HOMEWORK PROBLEMS

Basic Skills

Q15B.1 Determine the energy released by the particular fission reaction

$$n + {}^{239}_{94}\text{Pu} \rightarrow {}^{90}_{38}\text{Sr} + {}^{146}_{56}\text{Ba} + 4n \qquad (Q15.10)$$

and compare with the estimate that we made in section Q15.4. (The measured atomic masses of $^{239}_{94}$Pu, $^{90}_{38}$Sr, and $^{146}_{56}$Ba are 239.052158 u, 89.907738 u, and 145.930113 u, respectively.)

Q15B.2 The Hiroshima bomb released about 8×10^{13} J of energy. About how many kilograms of $^{235}_{92}$U actually underwent fission in that explosion?

Q15B.3 Verify that the energy released by the reaction shown in equation Q15.8a is as stated in that equation. (The mass of $^2_1$H is 2.014102 u, that of $^3_1$H is 3.016049 u, and that of $^1_1$H is 1.007825 u.)

Q15B.4 In exercise Q15X.3, we found that two $^2_1$H nuclei approaching each other with a kinetic energy of 360 keV might get close enough to each other to undergo fusion. The average kinetic energy of an atom in a gas at an absolute temperature T is about $(3/2)k_B T$ (where k_B = Boltzmann's constant = 1.38×10^{-23} J/K). Estimate the temperature that the $^2_1$H gas would need to have to undergo fusion readily.

Q15B.5 Estimate the kinetic energy that two $^{12}_6$C nuclei would have to have to get close enough ($\approx$ 3-fm center-to-center separation) to undergo fusion. The average kinetic energy of an atom in a gas at an absolute temperature T is about $(3/2)k_B T$ (where k_B = Boltzmann's constant = 1.38×10^{-23} J/K). Estimate the temperature that the $^{12}_6$C gas would need to have to undergo fusion readily.

Q15B.6 Explain in each case exactly why mere exposure to even intense radiation consisting of alpha particles, electrons, positrons, and/or gamma particles will (under ordinary circumstances) *not* make something radioactive. Explain how exposure to *neutrons*, on the other hand, might very well make something radioactive.

Synthetic

Q15S.1 From the data given in section Q15.2, estimate the number of cancer deaths per year caused by exposure to background radiation. (*Hint:* Treat each year as a new "exposure event" that will cause cancers down the line. Assume an RBE of about 1, which will yield a minimal estimate.)

Q15S.2 Use the graph in figure Q15.2 to estimate the energy released by the fusion of $^4_2$He and $^{12}_6$C to form $^{16}_8$O. (This reaction takes place in some very massive stars.)

Q15S.3 Use the graph in figure Q15.2 to estimate the energy released by the fusion of three $^4_2$He to form $^{12}_6$C.

Q15S.4 (Adapted from H. Ohanian, *Modern Physics*, Prentice-Hall, Englewood Cliffs, N.J., 1987.) One can estimate the damage caused by nuclear weapons by using *scaling laws* that describe how the radii of a given severity of damage scale with the energy released by the weapon. The scaling law for the blast wave produced by a nuclear weapon is

$$\frac{R}{R_0} = \left(\frac{E}{E_0} \right)^{1/3} \qquad \text{(Q15.11)}$$

where R is the radius of damage caused by a weapon that releases energy E and R_0 and E_0 are the same for a reference weapon. Similarly, the radius of damage for thermal effects is given by

$$\frac{R}{R_0} = \left(\frac{E}{E_0} \right)^{1/2} \qquad \text{(Q15.12)}$$

The Hiroshima bomb released about 15 kT of explosive energy (where 1 T is the energy released by 1 ton of dynamite). The blast from this bomb completely destroyed buildings within a radius of about 1.0 mi and ignited wooden objects a distance of 1.25 mi from ground zero.
(a) What would be the corresponding distances for a 100-kT bomb (the yield of a typical bomb in the United States arsenal)?
(b) What would be the corresponding distances for a 10-MT bomb (roughly the yield of the largest nuclear bombs in the United States arsenal)?
(c) Is it the blast or thermal effects that create the greatest damage in large bombs?

Q15S.5 The sun has a mass of about 2.0×10^{30} kg and radiates energy at a rate of roughly 3.9×10^{26} W. Imagine that the sun was a big ball of burning gasoline suspended in an oxygen atmosphere. About how long would the sun be able to shine? (1 g of gasoline releases about 44,000 J of energy when it burns. The discrepancy between the sun's age estimated this way and other evidence of the earth's great age was quite disconcerting to physicists until the process of fusion was discovered.)

Q15S.6 The sun has a mass of about 2.0×10^{30} kg and radiates energy at a rate of roughly 3.9×10^{26} W. When the sun was first formed, it consisted virtually entirely of hydrogen ($^1_1$H). What fraction of its hydrogen has been consumed in the 5 billion y since?

Q15S.7 Using information in section Q15.6 and the facts that the sun radiates energy at the rate of 3.9×10^{26} W and its distance from the earth is 1.5×10^{11} m, estimate the number of neutrinos that pass through 1 m^2 of surface on the earth that directly faces the sun. (Roughly this many neutrinos go through your body every second!)

Rich-Context

Q15R.1 Imagine that you are trying to design a nuclear "battery" for a deep-space probe. The design specifications call for the battery to still be able to produce at least 35 W of power after 12 y of operation. The thermoelectric generator that you have available converts thermal energy to electric energy at an efficiency of about 12%. You want to use an alpha emitter for the radioactive core of the battery since it is easy to shield the electronics from alpha particles. Imagine that you can obtain sufficient amounts of either $^{208}_{84}$Po or $^{209}_{84}$Po, which are both alpha emitters with half-lives of 2.9 y and 105 y, respectively. The former decays to $^{204}_{82}$Pb, which is stable, while the latter decays to $^{205}_{82}$Pb, which in turn decays by electron capture to $^{205}_{81}$Tl which is stable. Which of these two isotopes would be better for the battery, and how much of the better one do you need? (The atomic masses of $^{208}_{84}$Po, $^{209}_{84}$Po, $^{204}_{82}$Pb, $^{205}_{82}$Pb, and $^{205}_{81}$Tl are given by 207.981222 u, 208.982404 u, 203.973020 u, 204.974458 u, and 204.974401 u, respectively.)

Advanced

Q15A.1 Imagine that an atomic bomb causes about 15 kg of $^{235}_{92}$U to undergo fission. About how long did the chain reaction last? We'll answer the question in stages.
(a) Assume that the multiplication factor is equal to 2. This means that a stray neutron induces one fission, which causes two fissions, which cause four fissions and so on. About how many such doublings will it take before the entire 15 kg is consumed?
(b) Argue that 99.9% of the energy is released in the last 10 doublings.
(c) Estimate the speed of a neutron having 1 MeV of kinetic energy, which is about the average energy that a fission neutron will have. (It is OK to assume that the neutron is nonrelativistic, since its kinetic energy is lots less than its rest energy, which equals 939 MeV.)
(d) About how long will it take such a neutron to travel the width of the uranium sphere, which will be about 10 cm across? This will roughly be the time between successive doublings. Explain why.
(e) How long does the *entire* chain reaction take? Over what time period is 99.9% of the energy released?

ANSWERS TO EXERCISES

Q15X.1   The 2.0-MeV kinetic energy of an alpha particle is a tiny fraction of its rest energy of about 4(940 MeV) (where 940 MeV is the approximate rest energy of a nucleon). Therefore, the alpha particle will be *non-relativistic*, so its kinetic energy will be $K \approx \frac{1}{2}mv^2$, which implies that $v = \sqrt{2K/m}$, or

$$\frac{v}{c} = \sqrt{\frac{2K}{mc^2}} = \sqrt{\frac{2(2.0 \text{ MeV})}{4(940 \text{ MeV})}} = 0.033 \qquad \text{(Q15.13)}$$

The electron, on the other hand, with a kinetic energy of nearly 4 times its rest energy, will be highly relativistic and move with nearly the speed of light. We see, therefore, that an alpha particle moves about 30 times more slowly than the electron of the same energy.

Q15X.2   If we had 1 mol each of $_1^2\text{H}$ and $_1^3\text{H}$, the mixture would have a mass of 5 g. Therefore, a 1-g mixture of equal proportions of $_1^2\text{H}$ and $_1^3\text{H}$ will contain about 0.20 mol of each $= 0.20(6 \times 10^{23} \text{ atoms}) = 1.2 \times 10^{23}$ atoms of each type. If we allow these nuclei to react, each reaction produces 17.6 MeV. Therefore, the total energy produced by 1.20×10^{23} reactions will be

$$(1.2 \times 10^{23} \text{ reactions}) \left(\frac{17.6 \times 10^6 \text{ eV}}{\text{reaction}} \right)$$

$$\times \left(\frac{1.6 \times 10^{-19} \text{ J}}{1 \text{ eV}} \right) = 3.4 \times 10^{11} \text{ J}$$

$$= 340 \text{ GJ} \qquad \text{(Q15.14)}$$

Note that this is quite a bit more energy than even 1 g of uranium can release (and far more than 1 g of gasoline can release).

Q15X.3   According to the problem description, the $_1^2\text{H}$ atoms approach each other with the same speed, so they must have the same initial kinetic energy K_i. According to conservation of energy,

$$K_i + K_i + 0 = 0 + 0 + \frac{ke^2}{r_{\text{min}}} \qquad \text{(Q15.15)}$$

since the initial potential energy of the nuclei at essentially infinite separation is zero, and their final kinetic energy is zero at the instant of closest approach. So

$$K_i = \frac{ke^2}{2r_{\text{min}}} = \frac{1.44 \text{ eV} \cdot \text{nm}}{2(2 \text{ fm})} \left(\frac{1 \text{ fm}}{10^{-6} \text{ nm}} \right)$$

$$= 360,000 \text{ eV} = 360 \text{ keV} \qquad \text{(Q15.16)}$$

as predicted.

Glossary

absolute square (of a number): equal to $a^2 + b^2$ if the number is a complex number of the form $a + ib$, and simply equal to the ordinary square if the number is real. (Section Q5.6)

absorption lines: dark lines that appear in the spectrum of white light after that light has gone through some transparent material (usually a gas). These lines correspond to frequencies of light that resonate with certain transitions in the material's atoms and thus are effectively absorbed. Absorption lines for a given kind of quantum system correspond to the same photon energies and wavelengths as *emission* lines for the same kind of system (the photons emitted or absorbed have energies corresponding to the possible differences between the system's energy levels). (Section Q8.4)

absorption spectroscopy: the process of identifying a substance from its pattern of absorption lines. (Section Q8.4)

activity or **decay rate:** defined to be λN. This gives the instantaneous rate that nuclei in a sample containing N nuclei decay. (Section Q14.5)

alpha decay: a decay process by which a nucleus with too many protons (and thus too much internal electrostatic repulsion) trims down by emitting a helium nucleus. (Section Q14.2)

alpha particle: another name for a completely ionized helium nucleus (the name became conventional before it was discovered that the alpha particle is a helium nucleus). (Sections Q12.2, Q12.6)

anode: a metal plate that receives electrons. (Section Q3.2)

antineutrino $\bar{\nu}$: the antiparticle to a neutrino. (Section Q13.1)

antinode: a position on a standing wave where the medium oscillates more violently than anywhere else. (Section Q1.5)

asymmetry term and **coefficient** a_A: a term in the binding energy that expresses the cost of having nuclei with $N \neq Z$ and an empirical constant appearing in this term. (Section Q13.5)

atomic number Z: the number of protons in a nucleus. (Section Q12.1)

barrier: an alternative term for a **classically forbidden region** in the context of a bound quantum system. (Section Q7.1)

becquerel: the SI unit of activity: 1 Bq = 1 decay per second. (Section Q14.5)

beta decay: any of a number of decay processes that adjust the N/Z ratio of a nucleus without changing its mass number A. The three basic beta decay processes are β^- decay $(n \to p^+ + e^- + \bar{\nu})$, β^+ decay $(p^+ \to n + e^+ + \nu)$, and *electron capture* or EC decay $(e^- + p^+ \to n + \nu)$. (Section Q14.1)

beta particles: an earlier name for electrons emitted during radioactive decay. (Section Q12.6)

binding energy E_b: the energy required to disperse a system at rest into its constituent parts, where each part is ultimately at rest at an essentially infinite distance from each other part. The binding energy E_b is almost always positive and gets larger as the system becomes more tightly bound. (Section Q12.4)

binding energy per nucleon e_b: quantity defined to be $e_b \equiv E_b/A$, where E_b is the nucleus's total binding energy. (Section Q13.6)

Bohr model: a greatly simplified model of the hydrogen atom that allows us to predict its allowed energies without doing a full three-dimensional analysis. (Section Q7.5)

Bohr radius a_0: radius $a_0 \equiv h^2/4\pi^2 mke^2 = \hbar^2/mke^2$. This is the smallest possible radius that an electron can have in the Bohr model. (Section Q7.5)

boson: a quanton whose spin quantum number s is an integer (such as 0, 1, 2, . . .). Photons are bosons, as are quantum systems whose internal angular momenta cancel. (Section Q8.6)

boundary: a place where the characteristics of the medium through which a wave travels change significantly. (Section Q1.4)

bound system: a system of quantons whose interactions keep them from separating beyond a certain point. (Section Q7.1)

box: a potential energy function $V(x)$ that is defined to be zero within a certain region ("inside the box") and infinity outside that region ("outside the box"). (Section Q7.3)

bubble chamber: a device used mostly in the 1960s that temporarily records the trajectories of energetic particles in high-energy particle experiments so that they can be photographed. (Section Q4.1)

cathode: a metal plate that releases electrons. (Section Q3.2)

chain reaction: a continuing fission reaction in which neutrons released by one fission reaction go on to induce more fission reactions. (Section Q15.5)

circular wave: a two-dimensional wave whose crests at any instant of time mark out concentric circles. (Section Q2.1)

classically allowed region: a region of space where a classical particle with energy E moving in response to a potential energy $V(x)$ has a positive (and thus a possible) kinetic energy $K = E - V(x)$. (Section Q7.1)

classically forbidden region: a region of space where a classical particle with energy E moving in response to a potential energy $V(x)$ would have a negative kinetic energy $K = E - V(x)$. Since a negative kinetic energy is impossible in classical mechanics, the particle cannot be found in such a region. (Section Q7.1)

classical: an adjective describing a model that fundamentally thinks of matter in terms of particles and interactions in terms of continuous fields. Newtonian mechanics, special relativity, and Maxwell's approach to electrodynamics are all classical theories. The antonym is "quantum." (Section Q5.3)

classical wave: a disturbance that moves through a medium and causes some kind of physical displacement of the medium. (Section Q1.2)

coherent (light): light whose photons have *identical* wavefunctions (i.e., the wavefunctions all have the same wavelength and are in step with each other). (Section Q9.4)

Collapse rule: the rule in quantum mechanics that states that when the value of a quanton's observable is determined, the quanton's state collapses to the eigenvector corresponding to that value. (Section Q6.2)

complex conjugate: a quantity defined to be $c^* = a - ib$ given a complex number $c \equiv a + ib$. (Section Q5.6)

complex number: a number of the form $a + ib$, where a and b are ordinary numbers and $i \equiv \sqrt{-1}$. The value of a quanton's wavefunction at a given position in space is *generally* a complex number. (Section Q5.6)

constructive interference: when two waves in phase superpose to form a wave with a greater amplitude. (Section Q2.1)

Coulomb repulsion coefficient a_C: the empirical constant for a term in the binding energy formula that expresses the reduction of the binding energy due to the electrostatic repulsion of the protons. (Section Q13.4)

critical: an adjective describing a sample of a fissile material if it is of an appropriate size and density so that the **multiplication factor** for fission reactions in the sample is 1,

meaning that a chain reaction will be (barely) self-sustaining. (Section Q15.5)

critical mass: the minimum mass of a given fissile material that when assembled into a sphere is critical. (Section Q15.5)

curie: a unit of radioactivity: $1 \, \text{Ci} = 3.7 \times 10^{10}$ Bq. (Section Q14.5)

cutoff frequency: the lowest frequency of light that is able to eject electrons from a given metal. (Section Q3.4)

Davisson-Germer experiment: the experiment that first demonstrated the wavelike character of electrons. The experiment involved directing an electron beam against a nickel crystal. The electrons reflect from rows of atoms in the crystal surface, creating an interference pattern that strongly enhances electron scattering at certain angles. (Section Q4.4)

de Broglie relation: the equation $p = h/\lambda$ that links the wavelength of a beam of particles that all have the same relativistic momentum p with the value of that momentum. (Section Q4.2)

de Broglie wavelength (of a particle): the wavelength that the de Broglie relation would assign to a beam of like particles. (Section Q4.2)

decay constant λ: the ratio of the probability that a nucleus will decay in an interval dt, divided by that dt. The definition is meaningful only if interval dt is so short that the probability of decay during that dt is very much less than 1. (Section Q14.5)

degenerate (energy level): an adjective describing an energy level if more than one distinct energy eigenfunction has that energy. (Section Q9.1)

destructive interference: the result when two waves out of phase superpose to form a wave with smaller amplitude (we use this term most often to describe situations where the two waves in fact cancel each other entirely). (Section Q2.1)

differential equation: an equation in which a function and one or more of its derivatives appear together. (The Schrödinger equation is a simple example.) (Section Q10.4)

diffraction: the word used to describe the way that a wave going through a slit fans out into circular waves beyond the slit. By historical convention, we use the word *interference* to describe what happens when circular waves from a countable number of discrete sources combine and interfere with each other, and *diffraction* to describe what happens when circular waves from an infinite and continuous array of sources combine and interfere. Thus we speak of two-slit *interference*, but single-slit *diffraction*, since in the latter case we consider the wave after it moves through the slit to be the sum of circular waves from an infinite number of sources. (Section Q2.1)

diffraction-limited: an adjective describing an optical instrument whose resolution is only limited by diffraction through its aperture. (Section Q2.4)

dynode: a plate inside the photomultiplier that ejects many electrons when struck by a single high-energy electron. The photomultiplier relies on electrons cascading from dynode to dynode to amplify the single electron liberated by the photon into a detectable pulse of electrons. (Section Q3.6)

eigenfunction (corresponding to a specific value of an observable): the wavefunction that a quanton would have if the probability of obtaining that observable value in an experiment were certain. An eigenfunction corresponding to a specific position is a spike-centered on that position. (*Eigen* is a German word meaning "characteristic.") An eigenfunction is a function that describes an eigenvector with an infinite number of components. (Section Q6.3)

eigenvector: the complex vector associated with a given value of an observable. (Section Q6.2)

Eigenvector rule: a rule in quantum mechanics stating that for every value that a quanton's observable might have, there is a corresponding complex vector which we call that value's **eigenvector.** (Section Q6.2)

electron capture (process): the weak-interaction process $e^- + p^+ \rightarrow n + v$. See **beta decay.** (Section Q13.1)

electron gun: a device for creating a beam of electrons all having the same (adjustable) kinetic energy. (Section Q4.3)

emission line: a spectrum line that results from excited electrons decaying to their ground state. (Section Q8.3)

energy eigenfunction $\psi_E(x)$: the wavefunction that a quanton would have if we knew with *certainty* that it had a specific, well-defined energy (as opposed to probabilities of various possible energies). If we do an experiment to evaluate the energy of a quanton, the quanton's original wavefunction will collapse to one of these eigenfunctions. (Section Q7.2)

energy eigenstate: the state of a quanton when its wavefunction is an energy eigenfunction. (Section Q8.2)

energy-level diagram: a way of pictorially representing the energy levels of a system, in which we use a horizontal bar to represent each energy level and place it in the appropriate vertical position to the right of a vertical scaled energy axis. (Section Q8.1)

energy levels: the values of a bound system's possible energies (imagined figuratively as being like levels on a staircase or parking garage). (Section Q8.1)

excited state: any energy eigenstate whose energy is above that of the ground state. (Section Q8.2)

exponential: the function e^x. (Section Q14.4)

fermion: a quanton whose spin quantum number s is a half-integer value (such as 1/2, 3/2, 5/2, . . .). Most subatomic particles (including protons, neutrons, electrons, neutrinos, quarks, and so on) are spin-$\frac{1}{2}$ fermions. (Section Q8.6)

fissile (material): a material whose nuclei can undergo fission and support a chain reaction. (Section Q15.5)

fission: the process of taking a very massive nucleus and splitting it into two parts, each of which has a greater binding energy per nucleon than the original nucleus. Since the net binding energy per nucleon involved in the process increases in this case, the process releases energy. (Section Q15.4)

Fourier theorem: a theorem stating that we can consider any periodic piecewise continuous wave function to be a sum of (possibly an infinite number of) sinusoidal waves with appropriately chosen frequencies and amplitudes. In particular, any arbitrary oscillation of a medium between reflecting boundaries can be constructed as a sum of sinusoidal oscillations having the frequencies and wavelengths of the medium's normal modes. (Section Q1.6)

fundamental frequency: the lowest possible frequency for a standing wave in a medium between two reflecting boundaries. This fundamental frequency is $f = v/2L$ if both boundaries are analogous to a fixed end of a string (or both are analogous to a free end) and $f = v/4L$ if one boundary is of each type, where v is the velocity of a wave through the medium and L is the distance between boundaries. (Section Q1.5)

fundamental mode: the **normal mode** with the lowest frequency. (Section Q1.5)

fusion: the process of combining two light nuclei to create a more massive nucleus with greater binding energy per nucleon. Since the net binding energy per nucleon involved in the process increases in this case, the process releases energy. (Section Q15.4)

gamma decay: a decay process by which a nucleon in an excited nuclear energy level undergoes a transition to a lower level without being transformed into another kind of nucleon. The transition radiates an energetic photon. (This decay process is analogous to an electron dropping to a lower energy level in an atom.) (Section Q14.3)

gamma particle: an early name for a high-energy photon emitted by an unstable nucleus. (Section Q12.6)

gravitational confinement: the method that stars use to initiate nuclear fusion. The weight of the star's outer layers compress the plasma at the star's core to the high temperature and density required for fusion. (Section Q15.6)

gray: the SI unit of radiation, defined to be the amount of radiation that deposits a total of 1 J/kg in a medium. Note that this means that 1 gray = 100 rad. (Section Q15.2)

ground state: the energy eigenstate corresponding to the lowest possible energy for a quanton in a given circumstance. (Section Q8.2)

half-life $t_{1/2}$: the time required for the number of undecayed nuclei (and thus the activity) of a radioactive sample to decay by one-half. Note that $t_{1/2} = (\ln 2)/\lambda$. (Section Q12.6)

harmonics: **normal modes** of a waving system that have a greater frequency than the fundamental mode, which has the lowest possible frequency. (Section Q1.5)

h-bar $\hbar$: a shorthand for $h/2\pi$, a common combination.

Huygens's principle: Given a wave front at time t, we can construct the position of the wave front at time $t + \Delta t$ by assuming that each point on the wave front emits circular waves that move outward from that point at the same speed v and have the strongest amplitude in the direction of motion of the original wave front. The new wave front is then a curve drawn tangent to these circular waves. (Section Q2.3)

hydrogenlike (eigenfunction): an energy eigenfunction for an atom that has essentially the same shape as a hydrogen energy eigenfunction but may be scaled down in size (due to a more strongly attracting nucleus). (Section Q9.2)

imaginary number: a quantity whose square is negative. (Section Q5.6)

imaginary part (of c): the quantity b in a complex number $c = a + ib$. (Section Q5.6)

inertial confinement: the name for the approach to fusion in which a small pellet of fuel is suddenly illuminated by a large number of laser beams that compress the pellet until it fuses. (Section Q15.6)

interior bonding coefficient a_I: the empirical constant in front of a term in the binding energy formula that expresses the binding energy in the bonds formed by a nucleon deep inside a nucleus. The coefficient is an empirical constant appearing in this term. (Section Q13.4)

intrinsic characteristics (of a quanton): quantities such as the mass and charge of a quanton that define the quanton and will not change as long as the quanton exists. (Section Q6.2)

ionize: the act of stripping an electron from an atom by a collision or other means.

isotopes: nuclei that have the same atomic number Z (and thus the same chemical properties) but different N. (Section Q12.1)

laser: a device that takes advantage of the process of stimulated emission to create a tightly directed beam of monochromatic and coherent light. The name is an acronym from *light amplification by stimulated emission of radiation*. (Section Q9.4)

linear and **nonlinear** (waves): waves that obey and do not obey the superposition principle, respectively. Real waves are generally linear to an excellent degree of approximation unless they are extremely strong. (Section Q1.3)

line wave: a two-dimensional wave whose crests at any instant of time mark out parallel straight lines. (Section Q2.1)

local wavelength (of a function): the quantity $\lambda(x)$ given such that $[\lambda(x)]^{-2} = -(d^2 f/dx^2)/4\pi^2 f$ for a function $f(x)$. This quantity specifies an amplitude-independent measure of the function's curvature that reduces to the wavelength when $f(x)$ is a sinusoidal wave. (Section Q10.2)

magnetic confinement: the name for the approach to fusion that uses magnetic fields to confine a plasma of hydrogen long enough and tightly enough for fusion reactions to take place. (Section Q15.6)

mass deficit Δm: the difference between the total mass of a system's parts when disassembled and its mass as a bound system: $\Delta m = m_{parts} - m_{sys}$. According to the theory of relativity, $E_b = \Delta mc^2$. (Section Q12.4)

mass number A: the total number of nucleons in a given nucleus, equal to the sum of Z and N. (Section Q12.1)

meltdown: a catastrophic accident involving a water-moderated fission reactor whose cooling water has ceased to circulate. If the heat produced by the reactor (which is produced from radioactivity in the reactor core even when the reaction is shut down) cannot be carried away by the coolant, the reactor interior could melt through the bottom of the containment vessel or could cause a steam explosion that breaches the containment vessel. (Section Q15.5)

metastable state: an energy eigenstate whose decay is "forbidden" by a selection rule. Such energy eigenstates will eventually decay, but since the forbidden transition has a much smaller probability per unit time than a normal transition, a quanton in a metastable state will remain in that state much longer than normal. (Section Q9.3)

moderator: a material that makes neutrons more effective at causing fission reactions by slowing them down. (Section Q15.5)

monochromatic (light): light whose waves have a single well-defined frequency and wavelength. Since the color of light is linked to its wavelength, this implies that the light will have a single, well-defined color (*monochromatic* means "one color"). (Section Q2.2)

monoenergetic (beam of particles): an adjective describing a beam of particles that all have the same kinetic energy K. If the particles are essentially free (not involved in any significant external interactions), this will mean that they also have the same momentum magnitude p, and the beam will thus have a well-defined wavelength $\lambda = h/p$. (Section Q4.3)

multiplication factor: the average number of neutrons from one fission reaction that successfully induces a new fission reaction. This must be greater than or equal to 1 for a chain reaction to be self-sustaining. (Section Q15.5)

natural logarithm $\ln x$: the inverse of the exponential function e^x. (Section Q14.4)

neutrino ν: a neutral, zero-mass lepton that interacts with other quarks and leptons *only* through weak interactions. (Section Q13.1)

neutron decay (process): the weak-interaction process $n \to p^+ + e^- + \bar{\nu}$. (Section Q13.1)

neutron number N: the number of neutrons in a nucleus. (Section Q12.1)

node: a position on a standing wave where the disturbance of the medium always remains zero as time passes. (Section Q1.5)

normalized (vector): a vector whose magnitude is 1. (Section Q5.6)

normal mode (of an oscillation): any one of the characteristic self-sustaining standing wave patterns for the oscillation of a medium between two reflecting boundaries. In many situations, these normal modes have frequencies that are integer multiples of the fundamental frequency. The mode whose frequency is the fundamental frequency is called the **fundamental mode.** The modes with higher frequency are called **harmonics** of the fundamental mode. (Section Q1.5)

nucleon: either a proton or a neutron. (Section Q12.1)

observables: quantities associated with a particle or quanton that may vary with time, such as position, momentum, kinetic energy, and angular momentum. (Section Q6.2)

orthogonal: an adjective describing two complex vectors whose dot product is zero. (Section Q5.6)

Outcome Probability rule: the rule in quantum mechanics that states that the probability of obtaining a particular result when an observable is determined is the absolute square of the dot product of the quanton's state vector with the eigenvector corresponding to that value. (Section Q6.2)

Pauli exclusion principle: a hypothesis asserting that *no two fermions can have exactly the same quantum wavefunction.* This implies that only two electrons (with opposing spins) can have the same spatial wavefunction $\psi(x)$ corresponding to a given energy eigenfunction. (Section Q8.6)

photocathode: a thin metal plate in a photomultiplier that ejects an electron (by the photoelectric effect) when struck by a photon. (Section Q3.6)

photoelectric effect: the ability of light to eject electrons from the surfaces of metals. (Section Q3.1)

photomultiplier: a device that is able to detect a single photon. (Section Q3.6)

photons: the particles of light in the model that Einstein used to explain the photoelectric effect. (Section Q3.5)

Planck's constant h: a universal physical constant that (among other things) appears as the constant of proportionality between the energy of a photon and the frequency of the light that it represents. Its measured value is 6.63×10^{-34} J $\cdot$ s $= 4.15 \times 10^{-15}$ eV $\cdot$ s. The value of hc (where c is the speed of light) is 1240 eV $\cdot$ nm. (Section Q3.5)

plasma: a gas in which essentially all atoms are ionized. This means that the components of the plasma (ionized atoms) and electrons are electrically charged. (Section Q15.6)

population inversion: any situation in which a greater number of atoms in a certain sample (population) are in the uppermost energy level of a pair than in the lower. Under normal circumstances, the *lower* state would be more populated; but when the upper state of the pair is a metastable state and the lower one is not, it becomes possible that random excitations will lead to an unusual concentration of atoms in the metastable state. (Section Q9.4)

positron e^+: the antiparticle to an electron. (Section Q13.1)

proton decay (process): the weak-interaction process $p^+ \to n + e^+ + \nu$. (Section Q12.1)

proton-proton chain: A linked sequence of fusion reactions that produces energy in the sun and many other medium- and low-mass stars. (Section Q15.6)

quantized (variable): an adjective describing a variable quantity (such as energy or radius) if it can only take on specific, discrete values. (Section Q7.3)

quanton: any physical entity (photon, electron, proton, neutron, atom) that in a given situation is small enough and/or slow enough to exhibit quantum behavior. (Section Q5.1)

quantum jump: see **transition.**

quantum mechanics: the theory that supersedes newtonian mechanics in describing the behavior of quantons. (Sections Q5.4, Q6.1)

quantum numbers: a set of integers that completely and uniquely index the energy eigenfunctions for a given bound system. For example, the energy eigenstate for an electron in a hydrogen atom (or multielectron atom) is uniquely specified by four quantum numbers n, ℓ, m, and m_s, which specify the shell number, angular momentum magnitude, angular momentum orientation, and spin orientation of the electron, respectively. (Section Q9.1)

rad: a unit of radiation, defined to be the amount of radiation that deposits a total ionization energy of 10^{-2} J/kg in a medium. (This is approximately equal to the roentgen.) It is an acronym of *r*adiation *a*bsorbed *d*ose. (Section Q15.2)

radioactive (nucleus): an unstable nucleus that sheds (radiates) quantons of one or more types as it seeks to get itself into a stable configuration. (Section Q12.6)

Rayleigh criterion: the criterion that states that the angular separation θ between two point light sources must be such that $\theta > \sin^{-1}(1.2\lambda/a)$ if an optical instrument with a circular aperture of radius a is to be able to resolve the sources as being separate. (Section Q2.4)

real number: an ordinary scalar quantity, in contrast to a **complex number** or an **imaginary number** (a number whose square is negative). (Section Q5.6)

real part (of c): a in a complex number written $c = a + ib$. (Section Q5.6)

refraction: the bending of light (or any other kind of wave) as it goes through the boundary between two media. (Problem Q2R.3.)

relative biological effectiveness (RBE) factor: a numerical factor that expresses how strongly ionizing radiation affects a certain kind of biological tissue. (Section Q15.2)

rem: a unit of radiation, defined to be the radiation dose in rads multiplied by a factor **RBE** that quantifies the relative biological effect of different types of radiation. *Rem* is an acronym for *r*oentgen *e*quivalent in *m*an. (Section Q15.2)

resonance: a characteristic of most oscillating systems describing the tendency to respond strongly to disturbances having frequencies close to one of the system's normal mode frequencies and weakly to disturbances having different frequencies. (Section Q1.7)

roentgen: a unit of radiation, defined to be the amount of radiation that produces a charge of 3×10^{-10} C in 1 cm^2 of air by ionization. (Section Q15.2)

scanning tunneling microscope: a device that takes advantage of electrons tunneling across the vacuum between a metal surface and a fine needle to create extremely high-resolution pictures of the surface (even to the point of displaying individual atoms on the surface). (Section Q11.3)

(time-independent) Schrödinger equation: a generalization of the de Broglie relation that applies to situations where a classical particle's momentum would not be constant. The Schrödinger equation is a simple differential equation whose solution for a given E is the energy eigenfunction $\psi_E(x)$ corresponding to that energy E. (Section Q10.3)

selection rule: a rule (such as $\Delta\ell = \pm 1$ for atomic transitions) that distinguishes probable transitions from very

improbable transitions. Transitions that violate a selection rule are usually so improbable that spectral lines corresponding to these transitions are not easily seen. (Section Q9.3)

semiempirical binding energy formula: an equation that allows us to compute (at least approximately) the binding energy E_b of a nucleus with $A > 20$ given only its values of A and Z. (Section Q13.6)

slit: a tall but very narrow opening in an opaque barrier. (Section Q2.1)

sound wave: a classical wave moving through air where the disturbance is a local increase or decrease in the density (and thus the pressure) of the air. (Section Q1.2)

spatial subset: the set of observables consisting of the position observable, the momentum observable, and any observable that can be calculated from these. (Section Q6.2)

spectral lines: the bright vertical lines that appear when the emitted light from an ensemble of excited quantons is sent through a vertical slit and then is spread horizontally according to wavelength by a prism or the like. The patterns of spectral lines are different for different kinds of systems. (Section Q8.3)

spectroscopic notation: a way of describing a particular energy level in a hydrogen (or multielectron) atom. The notation uses a number to represent the shell number n, a *letter* to represent the value of ℓ (where s, p, d, f correspond to $\ell = 0, 1, 2, 3$, respectively), and a numerical superscript to represent the number of electrons occupying that level. We can use a string of such units to represent the electron configuration for any atom. For example, the electron configuration of silicon is $1s^2 2s^2 2p^6 3s^2 3p^1$. (Section Q9.2)

spin: (1) an intrinsic property of many quantons (such as mass or charge) that makes the quanton behave under a wide range of circumstances as if it were spinning around an axis; (2) the quantity s $(= 0, \frac{1}{2}, 1, \frac{3}{2}, \ldots)$ that quantitatively specifies the magnitude of a quanton's spin angular momentum. (Section Q5.5)

spin subset: the set of observables corresponding to a quanton's three components of spin angular momentum and observable that we can calculate from these. (Section Q6.2)

spontaneous emission: the process by which a quanton in an energy eigenstate decays to a lower-energy state by emitting a photon that carries away the difference in energy between the two states. (Section Q8.2)

spontaneous fission: a process similar to alpha decay whereby a nucleus sheds a fragment that is *larger* than a helium nucleus. (Section Q14.2)

square wave: a wave whose value is a positive constant for one-half of its cycle and a negative constant for the other half. (Section Q1.6)

square well: a potential energy function consisting of a region of constant low potential energy flanked by two regions of constant high potential energy. The energy changes discontinuously at the boundaries between the regions. This potential energy function is a useful simplified model for a variety of situations. (Section Q11.2)

standing wave: a wave formed by the superposition of two sinusoidal waves that have the same amplitude and frequency but are traveling in opposite directions. A standing wave has the same sinusoidal shape at all times, but remains in a fixed position, simply oscillating in amplitude. (Section Q1.5)

State Vector rule: the rule in quantum mechanics that states that a quanton's physical state at any given instant is described by a complex vector $|\psi\rangle$ with a certain number of components. (Section Q6.2)

stationary state: loosely speaking, an adjective describing a wavefunction whose absolute square is a time-independent function. Energy eigenfunctions are stationary in this sense: all other wavefunctions are not. (Section Q7.2)

Stern-Gerlach device: a device that can determine a quanton's spin projection on any given axis. (Section Q5.5)

stimulated emission: a term describing the process in which a photon passing by an atom "shakes loose" another photon. To effectively stimulate emission, the passing photon must have an energy equal to the difference between the energy levels involved in the transition that creates the new photon. The new photon will thus have the same wavelength as the passing photon, and it turns out that its wavefunction will be exactly in step (in phase) with the passing photon as well. (Stimulated emission was predicted and discussed by Einstein in 1925.) (Section Q9.4)

strong (nuclear) interaction: the one of the four fundamental interactions that is responsible for binding the nucleons in a nucleus. (Section Q12.3)

subcritical: an adjective describing a sample of a fissile material if it has a size and density such that the **multiplication factor** for fission reactions in the sample is less than 1, meaning that a chain reaction will not sustain itself. (Section Q15.5)

supercritical: an adjective describing a sample of a fissile material if it has a size and density such that the **multiplication factor** for fission reactions in the sample is greater than 1, meaning that a chain reaction will grow rapidly. (Section Q15.5)

superposition principle: a principle stating that when two traveling waves overlap, the total wave function describing the combined wave is the simple algebraic sum of the two individual wave functions. (Section Q1.3)

Superposition rule: the rule in quantum mechanics that states that if we combine quantons having a state vector $|a\rangle$ with those having state $|b\rangle$ so that there is no way to tell which quanton had which state vector, then the combined quantons have a state which is some linear combination of these state vectors. (Section Q6.2)

surface correction coefficient a_S: the empirical coefficient in front of the term in the binding energy formula that expresses the correction to the interior bonding terms due to the fact that some nucleons in a nucleus are on the surface. (Section Q13.4)

tension wave: a classical wave moving along a stretched string, where the disturbance is a transverse displacement of the string away from its equilibrium position. (Section Q1.2)

thermonuclear bomb: an especially powerful nuclear weapon consisting of a normal fission bomb surrounded by a mixture of $^{2}_{1}$H and $^{3}_{1}$H. When the fission bomb explodes, it heats the hydrogen mixture, causing its nuclei to fuse. This greatly increases the energy that the bomb releases. (Section Q15.6)

Time-Evolution rule: the rule in quantum mechanics that states that to find out how a quantum state vector evolves in time, (1) write the state vector as a linear combination of energy eigenvectors and (2) attach the factor $e^{-iE_n t/\hbar}$ to each term, where E_n is the energy corresponding to that eigenvector. (Section Q6.2)

transition and quantum jump: two terms that describe a change in a quanton's energy eigenstate (the latter is more colloquial than the first). (Section Q8.2)

tunneling: a word describing the phenomenon whereby a quanton travels from one classically allowed region to another through a classically forbidden region. (Section Q11.3)

turning points: the points at which a classical particle with energy E moving in response to a potential energy $V(x)$ has zero kinetic energy: $E = V(x)$. Each such point is the boundary between a classically allowed and a classically forbidden region: a particle moving toward such a point comes instantaneously to rest at the point and then turns around and returns from whence it came. (Section Q7.1)

unified mass unit u: a convenient unit of atomic and nuclear mass, defined to be one-twelfth of the mass of $^{12}_{6}$C, meaning that $1\,u = 1.660559 \times 10^{-27}$ kg $= 931.48$ MeV$/c^2$. (Section Q12.1)

wave front: a curve connecting all points on a wave corresponding to a given crest of the wave. (Section Q2.3)

wavefunction (of a quanton at a given time)**:** a function $\psi(x)$ that expresses everything that we can know about the quanton beyond its intrinsic characteristics and that allows us to make (statistical) predictions about the quanton's associated variable quantities at present and into the future. (Section Q6.3)

wavelet: the circular wave emitted by a point on a wave front in Huygens's model. (Section Q2.3)

weak (nuclear) interaction: an interaction between particles with an extremely short range ($\approx 10^{-18}$ m) that acts between quarks and/or leptons independent of their charge or mass. The weak interaction also often has the effect of transforming the nature of the particles involved in the interaction. The weak interaction is one of the four fundamental interactions (along with the strong, electromagnetic, and gravitational interactions). (Section Q13.1)

work function W: a quantity expressing the minimum energy required to lift an electron out of a given metal. (Section Q3.5)

Index

Note: Page references followed by *f* and *t* refer to figures and tables, respectively.

Periodic Table of the Elements

Legend:
Atomic number — 1
Symbol — H
Atomic mass — 1.008

1 1A	2 2A	3 3B	4 4B	5 5B	6 6B	7 7B	8	9 8B	10	11 1B	12 2B	13 3A	14 4A	15 5A	16 6A	17 7A	18 8A
1 H 1.008																	2 He 4.003
3 Li 6.941	4 Be 9.012											5 B 10.81	6 C 12.01	7 N 14.01	8 O 16.00	9 F 19.00	10 Ne 20.18
11 Na 22.99	12 Mg 24.31											13 Al 26.98	14 Si 28.09	15 P 30.97	16 S 32.07	17 Cl 35.45	18 Ar 39.95
19 K 39.10	20 Ca 40.08	21 Sc 44.96	22 Ti 47.88	23 V 50.94	24 Cr 52.00	25 Mn 54.94	26 Fe 55.85	27 Co 58.93	28 Ni 58.69	29 Cu 63.55	30 Zn 65.39	31 Ga 69.72	32 Ge 72.59	33 As 74.92	34 Se 78.96	35 Br 79.90	36 Kr 83.80
37 Rb 85.47	38 Sr 87.62	39 Y 88.91	40 Zr 91.22	41 Nb 92.91	42 Mo 95.94	43 Tc (98)	44 Ru 101.1	45 Rh 102.9	46 Pd 106.4	47 Ag 107.9	48 Cd 112.4	49 In 114.8	50 Sn 118.7	51 Sb 121.8	52 Te 127.6	53 I 126.9	54 Xe 131.3
55 Cs 132.9	56 Ba 137.3	57 La 138.9	72 Hf 178.5	73 Ta 180.9	74 W 183.9	75 Re 186.2	76 Os 190.2	77 Ir 192.2	78 Pt 195.1	79 Au 197.0	80 Hg 200.6	81 Tl 204.4	82 Pb 207.2	83 Bi 209.0	84 Po (210)	85 At (210)	86 Rn (222)
87 Fr (223)	88 Ra (226)	89 Ac (227)	104 Rf (257)	105 Db (260)	106 Sg (263)	107 Bh (262)	108 Hs (265)	109 Mt (266)	110	111	112	(113)	114	(115)	116	(117)	

58 Ce 140.1	59 Pr 140.9	60 Nd 144.2	61 Pm (147)	62 Sm 150.4	63 Eu 152.0	64 Gd 157.3	65 Tb 158.9	66 Dy 162.5	67 Ho 164.9	68 Er 167.3	69 Tm 168.9	70 Yb 173.0	71 Lu 175.0
90 Th 232.0	91 Pa (231)	92 U 238.0	93 Np (237)	94 Pu (242)	95 Am (243)	96 Cm (247)	97 Bk (247)	98 Cf (249)	99 Es (254)	100 Fm (253)	101 Md (256)	102 No (254)	103 Lr (257)